AF865309

Generative AI for Connected and Autonomous Vehicles

IEEE Press

445 Hoes Lane
Piscataway, NJ 08854

Generative AI for Connected and Autonomous Vehicles

Kyungtae Han

IEEE Press
WILEY

Published by John Wiley & Sons, Inc., Hoboken, New Jersey.

Library of Congress Cataloging-in-Publication Data has been applied for:

Hardback ISBN: 9781394426683
ePDF ISBN: 9781394426706
ePub ISBN: 9781394426690
oBook ISBN: 9781394426713

Cover Design: Wiley
Cover Images: DKosig/Getty Images, metamorworks/Getty Images

To my parents

Contents

About the Author

Kyungtae Han, PhD, is a Senior Principal Researcher at Toyota Motor North America and formerly served as a Senior Research Scientist at Intel Labs. His research focuses on connected and automated vehicles, cyber-physical systems, edge computing, and generative artificial intelligence. He has authored over 90 publications and holds over 100 patents. His work has been recognized with several awards, including the SAE International Vincent Bendix Award, the IEEE Best Application Award, and the Intel Achievement Award.

Preface

Generative artificial intelligence (GenAI) has rapidly advanced the capabilities of autonomous systems, enabling new forms of perception, reasoning, interaction, and decision-making. This book presents a comprehensive treatment of GenAI for connected and autonomous vehicles (CAVs), integrating system architectures and practical methodologies that bridge research advances with real-world engineering needs.

While the term "generative AI" encompasses a broad family of models, this book places particular emphasis on large language models (LLMs) and their emerging role as the central reasoning and coordination engines within future mobility systems. At the same time, it provides foundational coverage of other important generative paradigms (diffusion models, generative adversarial networks, variational autoencoders, and multimodal architectures) that serve distinct roles in CAV applications. The book reflects a dual commitment to depth and clarity, ensuring that the presentation remains accessible to readers who are new to the field while also offering the rigor and completeness expected by researchers and practitioners.

Origins and Motivation

This book originated from a graduate-level course on GenAI for CAVs that I developed and taught across two offerings, in the spring of 2025 and the spring of 2026. The 15-session course attracted students with diverse engineering and computing backgrounds, and highlighted recurring themes in how learners engage with GenAI. In particular, students consistently expressed the need for a balanced integration of intuition, mathematical structure, and implementation detail. Their questions, insights, and challenges directly shaped the pedagogical philosophy of this text.

The same experience also revealed a gap in the literature: the individual generative models were well covered in isolation, but this book connects them and places LLM-centered systems alongside the diffusion, multimodal, and forecasting methods that handle perception and simulation.

Book Structure and Organization

The book is organized into seven parts spanning 22 chapters, structured across four conceptual layers: Generative and Multimodal Models, Model Interface, Agentic Systems and Protocols, and Adaptation and Validation. The "How to Read This Book" section provides a visual overview of this structure and offers guidance on selecting appropriate reading pathways organized by reader role (Academic, Professional, and Research Pathways). A detailed description of each part is provided in Chapter 1 (see Section 1.6).

Pedagogical Approach

The book adopts a three-strand pedagogical design intended to accommodate learners with varied strengths and interests. Conceptual Foundations begin each topic with motivating scenarios and high-level intuition that explain why a particular method matters in CAV applications. Mathematical Rigor formalizes key ideas through precise mathematical descriptions, enabling technically oriented readers to understand the assumptions, structures, and limitations underlying generative models. Practical Implementation provides hands-on code and example pipelines that accompany major concepts, using open-source models where appropriate to enable experimentation without relying on commercial application programming interfaces.

Each chapter is designed to be relatively self-contained, with key concepts restated when necessary, enabling flexible navigation based on specific needs and backgrounds.

Supplementary Resources

The book is accompanied by hands-on labs, and additional teaching materials are available to verified instructors. Details are provided in the Companion Resources section.

Intended Audience

This book is written for graduate students in computer science, electrical engineering, robotics, and automotive engineering; researchers working on generative models, autonomous systems, or human–AI interaction; industry practitioners developing AI-based perception, prediction, planning, or simulation systems; and engineers seeking to incorporate LLMs or other generative models into vehicle platforms, simulation loops, or connected mobility ecosystems.

While familiarity with machine learning is assumed, the book provides sufficient background to support readers transitioning from adjacent fields. The early chapters (Part I) offer accessible GenAI concepts, and mathematical details are presented with context to support readers with varying levels of mathematical background.

A Note on the Field's Evolution

GenAI is advancing at a rapid pace, and many of the systems discussed in this book, particularly those based on large language models and multimodal architectures, remain active areas of research. The purpose of this book is therefore not to catalog specific model versions or implementations, but to present the conceptual and architectural principles, design patterns, and system-level reasoning that remain applicable as the field evolves.

It is my hope that this book serves not only as a technical reference but also as a resource for future research and system development in the pursuit of safe, intelligent, and human-centered mobility.

Palo Alto, California, USA
July 2026

Kyungtae Han

Acknowledgments

I would like to express my sincere gratitude to all those who contributed, directly or indirectly, to the development of this book.

Toyota Motor North America: This book evolved from an internal course that I developed and taught at Toyota Motor North America. The course experience played a central role in shaping both the content and structure of this book. I extend my sincere appreciation to Onur Altintas for his leadership and encouragement throughout the development of this book.

Teaching Assistants: I am deeply grateful to the teaching assistants whose dedication and technical support were essential to the success of the course. Their contributions included leading hands-on sessions, conducting interactive discussions, and providing detailed code reviews that strengthened the practical components of this book.

Course Participants: The questions, feedback, and engagement of course participants were instrumental in refining the pedagogical approach of this book. Their curiosity and insights helped clarify how complex mathematical and technical concepts could be presented more effectively.

Reviewers and Colleagues: I thank the reviewers for their insightful comments and constructive suggestions, which improved the clarity and accuracy of the manuscript. I am especially grateful to John Kenney for his careful technical review of the vehicular communications material, whose detailed feedback sharpened the book's treatment of vehicle-to-everything technologies. I am also indebted to my colleagues for their thoughtful discussions and shared expertise that enriched the book's scope.

Family: I am deeply grateful to my family for their unwavering support, patience, and understanding throughout the extended writing process.

While I am solely responsible for the content and presentation of this book, the encouragement, feedback, and support from these individuals and organizations were invaluable in shaping and refining the material.

This book was prepared by the author in a personal capacity. The views expressed in this book are solely those of the author and do not necessarily reflect the views of the author's employer.

How to Read This Book

This book serves as both a foundational reference and a practical guide for generative AI in connected and autonomous vehicle (CAV) systems. The material progresses from the fundamentals to deployment and validation across multiple parts. Because the field spans multiple disciplines, including machine learning, multimodal modeling, and language systems, the chapters are structured to support readers with diverse backgrounds and goals.

Book Structure Overview

This book is organized into seven parts spanning 22 chapters, structured across four conceptual layers as shown in Figure 1. The **Generative and Multimodal Models** layer (Parts I and II) establishes the theoretical foundations: Part I introduces core concepts and large language model architectures, while Part II covers generative model families, including diffusion models, generative adversarial networks (GANs), variational autoencoders (VAEs), and vision-language models for multimodal representation. The **Model Interface** layer (Part III) examines language-based interaction through prompt engineering, retrieval-augmented generation, and function calling. The **Agentic Systems and Protocols** layer (Parts IV and V) demonstrates how generative AI enables autonomous agents, multi-agent systems, orchestration frameworks, and communication protocols, including model context protocol (MCP) and agent-to-agent (A2A) protocols for collective intelligence. The **Adaptation and Validation** layer (Parts VI and VII) addresses adaptation methods, including supervised fine-tuning, reinforcement learning fine-tuning, and knowledge distillation, as well as validation, safety, and ethical considerations through simulation, evaluation, and ethics frameworks.

Choosing Your Reading Pathway

Choose a reading pathway based on your background and objectives. While chapters can be read independently, some build naturally on earlier foundations. The pathways below are organized by reader role to help you quickly identify the most relevant guidance.

Academic Pathways

For graduate students seeking complete academic coverage, we recommend proceeding sequentially through all seven parts. This pathway provides comprehensive theoretical foundations and practical applications, suitable for coursework and research.

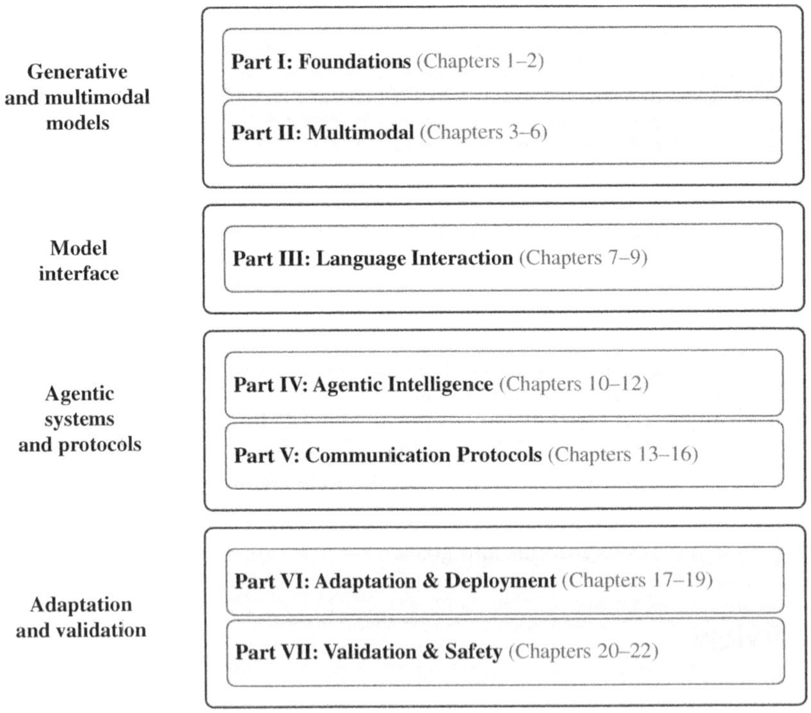

Figure 1 High-level architecture of the book: seven parts grouped into four conceptual layers, progressing from generative and multimodal models through the model interface and agentic systems to adaptation and validation.

For upper-division undergraduate students and readers focused on practical large language model (LLM) applications, begin with Part I, then proceed directly to Part III. Continue with Parts IV and V, and conclude with Part VII. This track centers on applying generative models: language interaction, agents, and the protocols that connect them. Part II can be deferred initially, as it is not a prerequisite for LLM-based systems, and Part VI, which covers model fine-tuning and adaptation, is more advanced and can be treated as optional for readers focused on applying rather than training generative models.

Professional Pathways

For industry practitioners developing generative AI systems for CAVs, start with Parts I and III, then proceed to Parts IV–VII. Part II can be consulted when working on specific applications requiring diffusion models, GANs, or VAEs.

For product and system architects, prioritize Parts I–V to master interface design, agentic system architecture, and communication protocols. Supplement with Parts VI and VII for adaptation strategies and validation considerations. This reading sequence emphasizes architectural decisions, protocol selection, and system integration patterns essential for building scalable CAV platforms.

Research Pathways

For AI researchers seeking theoretical depth, begin with Parts I and II, then explore Parts III through V, and review Parts VI and VII. Pay particular attention to advanced mathematical formulations marked in Advanced Note boxes throughout the book.

Readers with general computer science or engineering backgrounds will find Parts I, III, V, and VII accessible without extensive machine learning prerequisites. Parts II, IV, and VI assume familiarity with neural networks, optimization, and probability theory, though sufficient context is provided to support readers transitioning from adjacent fields.

The book can be used as a sequential textbook for graduate and advanced undergraduate courses, as a reference for specific topics, or as a technical guide for practitioners. By following the pathway most aligned with your background and goals, you can approach this book in the manner that best serves your needs.

Acronyms

A2A	Agent-to-Agent
ACC	Adaptive Cruise Control
ADAS	Advanced Driver Assistance System
AI	Artificial Intelligence
API	Application Programming Interface
AR	Autoregressive
ASIL	Automotive Safety Integrity Level
AV	Autonomous Vehicle
BERT	Bidirectional Encoder Representations from Transformers
BEV	Bird's Eye View
BLEU	Bilingual Evaluation Understudy
BLIP	Bootstrapping Language-Image Pre-training
BPE	Byte Pair Encoding
BSM	Basic Safety Messages
CAV	Connected and Autonomous Vehicle
CE	Cross-Entropy
CLIP	Contrastive Language-Image Pre-training
CNN	Convolutional Neural Network
CoT	Chain of Thought
DDIM	Denoising Diffusion Implicit Model
DDPM	Denoising Diffusion Probabilistic Model
DPO	Direct Preference Optimization
DSRC	Dedicated Short-Range Communications
ECU	Electronic Control Unit
ELBO	Evidence Lower Bound
EOS	End of Sequence
EV	Electric Vehicle
FFN	Feed-Forward Network
FIPA-ACL	Foundation for Intelligent Physical Agents—Agent Communication Language
FM	Flow Matching
GAN	Generative Adversarial Network
GELU	Gaussian Error Linear Unit
GenAI	Generative Artificial Intelligence

GPT	Generative Pre-trained Transformer
gRPC	Google Remote Procedure Call
GRPO	Group Relative Policy Optimization
GRU	Gated Recurrent Unit
HTTP	Hypertext Transfer Protocol
HTTPS	Hypertext Transfer Protocol Secure
JSON	JavaScript Object Notation
KD	Knowledge Distillation
KL	Kullback–Leibler
LCS	Longest Common Subsequence
LLM	Large Language Model
LM	Language Model
LN	Layer Normalization
LoRA	Low-Rank Adaptation
LSTM	Long Short-Term Memory
MAE	Mean Absolute Error
MAS	Multi-Agent System
MCP	Model Context Protocol
MLP	Multilayer Perceptron
MQTT	Message Queuing Telemetry Transport
MSE	Mean Squared Error
ODE	Ordinary Differential Equation
PEFT	Parameter-Efficient Fine-Tuning
PPO	Proximal Policy Optimization
RAG	Retrieval-Augmented Generation
RAGAS	Retrieval-Augmented Generation Assessment
REST	Representational State Transfer
RL	Reinforcement Learning
RLHF	Reinforcement Learning from Human Feedback
RNN	Recurrent Neural Network
RoPE	Rotary Positional Encoding
ROUGE	Recall-Oriented Understudy for Gisting Evaluation
RPC	Remote Procedure Call
RSA	Road Side Alert
SDE	Stochastic Differential Equation
SFT	Supervised Fine-Tuning
SPaT	Signal Phase and Timing
SSE	Server-Sent Events
TIM	Traveler Information Messages
TLS	Transport Layer Security
TRPO	Trust Region Policy Optimization
URI	Uniform Resource Identifier
VAE	Variational Autoencoder
V2I	Vehicle-to-Infrastructure
V2N	Vehicle-to-Network
V2V	Vehicle-to-Vehicle

V2X	Vehicle-to-Everything
VidLLM	Video Large Language Model
ViT	Vision Transformer
VLM	Vision-Language Model
VQ	Vector Quantization
VQA	Visual Question Answering

About the Companion Website

This book is accompanied by a companion website.

http://www.wiley.com/go/han/generativeai1

To access the website materials, click on the link under Related Resources on the book's page on wiley.com. You can also find this by searching www.wiley.com for the ISBN 9781394426683.

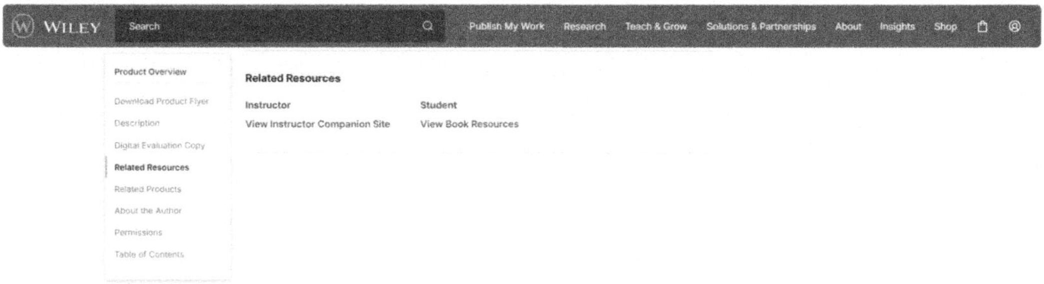

This website includes:

The companion website provides materials that support instructional use and self-directed study alongside the printed chapters.

For All Readers. Each chapter is paired with a hands-on lab: a self-contained, runnable exercise with a pinned software environment. The lab sample code is maintained in a publicly available repository that is updated independently of the printed book, so corrections and improvements can reach readers between editions. The companion website links to this repository.

For Instructors. Lecture slides (with editable source) and instructor notes are available to verified instructors through the publisher's password-protected site.

Part I

Foundations of Generative Artificial Intelligence

This part establishes the foundational concepts of generative artificial intelligence (AI) as applied to connected and autonomous vehicles. It introduces generative AI not as a collection of isolated models, but as a class of systems capable of semantic reasoning, structured generation, and context-aware decision support under real-world constraints. The discussion begins with a high-level overview of generative AI in the automotive domain and proceeds to a detailed examination of large language models, emphasizing their architecture, information flow, and operational characteristics. Together, these chapters provide the conceptual substrate upon which multimodal perception, agentic reasoning, and adaptive control mechanisms are built throughout the book.

1

Introduction to Generative AI for CAV Systems

1.1 Introduction: The Promise of Generative AI in CAVs

Imagine a vehicle that does more than drive itself. As it slows down for a sudden road obstruction, it informs the passenger: "There is construction ahead. I'm taking a safer detour recommended by real-time traffic updates." When a child unexpectedly crosses the street, the vehicle not only brakes, but explains: "A child just ran out from between parked cars. I reduced speed preemptively due to the movement detected on the sidewalk." When a passenger asks, "Can we make it to the airport in 30 minutes?" the vehicle analyzes real-time traffic, weather conditions, and route options, then responds: *"Yes, but we'll need to take the highway. Current traffic suggests a 25-minute journey, with a 5-minute buffer for unexpected delays."*

How can vehicles achieve such sophisticated understanding and communication? What makes these interactions possible, and how can systems be built that are capable of such nuanced reasoning?

By the end of this chapter, readers will understand how generative AI differs from traditional discriminative approaches, recognize the five major families of generative models and their roles in connected and autonomous vehicle (CAV) systems, appreciate the full scope of this book's coverage across 22 chapters, identify which model families are best suited for different CAV applications, and understand the trade-offs between proprietary and open-source models for CAV deployment.

1.2 The Paradigm Shift: Generative vs. Discriminative AI

The scenarios described at the opening of this chapter, a vehicle explaining its decisions, engaging in dialogue, and reasoning about complex situations, illustrate fundamental limitations of traditional autonomous vehicle systems. Current CAV architectures excel at perception and control: they can detect objects, recognize traffic signs, and execute maneuvers with high precision. However, they operate within fixed representational frameworks that cannot explain decisions, engage in dialogue, or adapt to novel situations. When a vehicle encounters an unexpected scenario, such as a construction zone with ambiguous lane markings, a pedestrian emerging from an occluded area, or a passenger requesting route information, traditional systems lack the representational capacity to reason about context, synthesize explanations, or communicate uncertainty. Unlike static applications, CAVs operate in dynamic, safety-critical environments where decisions must be explained, uncertainty must be communicated, and human interaction must feel natural.

These scenarios point to a form of machine intelligence that extends beyond perception and control. They require an ability to explain, to reason with uncertainty, and to communicate with humans in natural, contextual language. Such capabilities are becoming feasible today through the emergence of generative artificial intelligence.

Generative artificial intelligence (GenAI) refers to a class of models designed to synthesize novel content across a variety of data modalities (e.g., text, images, audio, video, and code). Unlike traditional AI approaches, which focus on classification, prediction, or control within predefined categories, GenAI systems learn to model the structure of complex data and produce outputs that are statistically consistent with the training distribution. These outputs are not retrieved from a finite repository but are constructed dynamically, based on learned representations and contextual inputs.

These limitations stem from the fundamental nature of how traditional AI systems work. In CAV systems, traditional AI models excel at classification tasks: identifying traffic signs, detecting pedestrians, or recognizing lane markings. These *discriminative models* learn to distinguish between predefined categories based on observed inputs. However, when a passenger asks, "Why did we slow down?" or when the system needs to explain its reasoning, CAV systems require a different capability: the ability to generate explanations, synthesize responses, and reason with context. This is where *generative models* excel. Understanding how these two paradigms differ is essential for choosing the right approach for different CAV applications. While discriminative models remain essential for perception tasks, the subsequent chapters focus on how generative models enable reasoning, communication, and other transformative capabilities that complement traditional perception systems.

To establish this foundation, we define several key concepts that will recur throughout this book. A *foundation model* is a large-scale pretrained model that serves as a base architecture for diverse downstream tasks. Foundation models are typically trained on broad, general-purpose datasets and exhibit emergent capabilities that enable adaptation to specific domains through fine-tuning, prompt engineering, or other enhancement techniques. *Semantic representation* refers to the encoding of meaning and relationships within data, enabling models to understand context, reason about relationships, and generate coherent outputs that reflect underlying structure rather than merely statistical patterns. These concepts are essential for understanding how generative models achieve capabilities beyond traditional discriminative systems: foundation models provide the architectural basis, while semantic representation enables the reasoning and understanding that distinguish generative from discriminative approaches.

The distinction between traditional and generative AI systems can be formalized by examining their modeling objectives and inference mechanisms. Traditional systems are typically trained as *discriminative models*, which learn to distinguish between predefined categories based on observed inputs. These models estimate the conditional probability distribution $p_\theta(y \mid x)$, where $x \in \mathcal{X}$ is the input, $y \in \mathcal{Y}$ is the output label, and θ denotes the model parameters. At inference time, a discriminative model outputs the most probable label:

$$y^* = \arg\max_{y \in \mathcal{Y}} p_\theta(y \mid x), \tag{1.1}$$

where y^* denotes the predicted class label chosen by the model.

In contrast, *generative models* are designed to produce new outputs that resemble the training distribution. Rather than selecting from predefined labels, these models synthesize structured content based on context. Generation is typically modeled by sampling from the conditional distribution:

$$\hat{x} \sim p_\theta(x \mid c), \tag{1.2}$$

where $c \in C$ represents a conditioning variable (context, e.g., prompt, sensor history, image, or label), and $\hat{x}$ is the generated output.

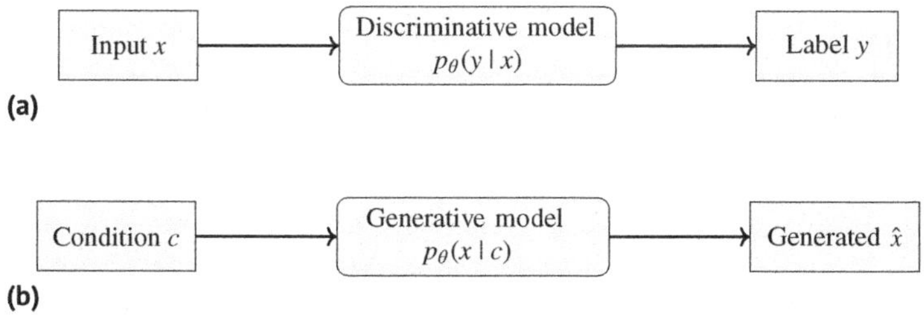

Figure 1.1 Schematic comparison of discriminative and generative modeling. (a) Discriminative models learn a direct mapping from input to label. (b) Generative models synthesize data instances from a learned conditional distribution.

Table 1.1 Discriminative versus generative AI models across modeling objective, typical tasks, output form, and representative architectures.

Aspect	**Discriminative AI**	**Generative AI**
Model objective	Learn $p(y \mid x)$: label prediction	Learn $p(x)$ or $p(x \mid c)$: data synthesis
Typical tasks	Classification, detection, regression	Text generation, image synthesis, audio modeling
Output type	Discrete class label or continuous value	Structured data instance (e.g., text, image, code)
Input-output relationship	Fixed mapping from input to label	Many-to-many mapping conditioned on input context
Training objective	Cross-entropy, mean squared error	Log-likelihood or adversarial/divergence-based objective
Representative models	Logistic regression, SVM, ResNet, BERT (for classification)	GPT, DALL·E, StyleGAN, VQ-VAE, Sora

The two modeling paradigms differ not only in formulation but also in learning objective, task scope, and inference behavior. Figure 1.1 illustrates the inference pipelines of both model types. Table 1.1 contrasts these dimensions.

Advanced Note 1.1 Training Objectives

Discriminative models learn to predict labels y from inputs x by modeling the conditional probability $p_\theta(y \mid x)$. They are trained to increase the accuracy of their predictions by minimizing the following objective over a dataset $\mathcal{D} = \{(x_i, y_i)\}_{i=1}^{N}$:

$$\mathcal{L}_{\text{disc}}(\theta) = -\mathbb{E}_{(x,y)\sim\mathcal{D}}[\log p_\theta(y \mid x)],$$

where $\mathcal{D} = \{(x_i, y_i)\}_{i=1}^{N}$ denotes the training dataset.

Generative models, by contrast, learn to synthesize new data instances that resemble the training distribution. They model either $p(x)$ or $p(x \mid c)$, where $c \in \mathcal{C}$ is a conditioning variable (context, e.g., prompt, sensor history, or label). The model is trained to increase the likelihood of generating correct samples:

$$\mathcal{J}_{\text{gen}}(\theta) = \mathbb{E}_{x\sim\mathcal{D}}[\log p_\theta(x \mid c)].$$

(Continued)

Advanced Note 1.1 (Continued)

While $\mathcal{L}_{\text{disc}}$ is a conventional loss function minimized during training, $\mathcal{J}_{\text{gen}}$ is a training objective that is maximized to improve sample quality. The two reflect fundamentally different learning goals.

Remarks:

- Both objectives use the finite dataset $\mathcal{D}$ as an approximation of the true data distribution P_{data}. In practice, P_{data} is never directly accessible.
- The discriminative objective focuses on learning decision boundaries between classes, while the generative objective aims to capture the full data distribution.

1.3 Transformative Capabilities in CAVs

This section examines how generative models address the limitations of traditional CAV systems, connecting the theoretical foundation to practical capabilities through concrete applications.

Generative AI overcomes the representational rigidity of traditional systems by introducing five transformative capabilities that directly address the limitations of perception-only architectures. Each capability is realized through specific GenAI applications that demonstrate the practical impact of this technology:

Semantic reasoning enables generative models to understand context, relationships, and meaning rather than merely recognizing patterns. Unlike discriminative models that classify pixels into labels, generative models process the *meaning* of a scene. This capability manifests in several concrete applications: *generative text responses* that explain vehicle behavior in natural language (e.g., "I'm slowing down because I detected construction ahead and merging traffic from the right"), *visual question answering (VQA)* that allows passengers to query scenes (e.g., "Is the crosswalk clear?" or "How many vehicles are in front?"), and *grounded scene explanations* that link textual descriptions to specific visual regions. When a CAV system encounters an unexpected scenario, such as a construction zone with ambiguous lane markings, generative models can synthesize explanations, reason about relationships between objects, and generate contextually appropriate responses. This capability is essential for building trust and enabling human-vehicle interaction, as explored in Chapters 7–9 (prompt engineering, RAG, function calling) and Chapters 5–6 in Part II (vision-language models and video-language models for scene understanding and temporal reasoning).

World modeling refers to the ability to construct internal representations of the environment that capture not just current observations but also potential future states, alternative scenarios, and causal relationships. Generative models learn to model the joint distribution of observations, enabling them to reason about what might happen next, what could have happened differently, and how actions affect outcomes. This capability is realized through *generative scene synthesis* for simulation and testing, where diffusion and flow matching generate synthetic sensor sequences representing rare but critical scenarios (e.g., sudden hailstorms with low visibility and malfunctioning traffic lights). This enables a CAV to ask, "What happens if that car runs the red light?" This is a critical step for defensive driving. *Trajectory forecasting* models predict future vehicle paths, while *scenario generation* creates diverse driving situations for robust system evaluation.

This capability is crucial for CAV systems that must plan routes, anticipate hazards, and make decisions under uncertainty, as detailed in Chapters 3–4 (diffusion, transport, and specialized models).

Multimodal fusion enables generative models to integrate information across diverse data types, including text, images, audio, and sensor streams, into unified representations. Traditional CAV systems process each modality separately, requiring explicit fusion mechanisms that may miss subtle cross-modal relationships. Generative models, particularly vision-language models (VLMs) and multimodal LLMs, learn to represent and reason about relationships across modalities, enabling applications such as *V2X message alignment* (matching real-time camera views with Vehicle-to-Everything (V2X) alerts), *occlusion-aware reasoning* (handling alerts about objects not visible to the ego camera, such as "pedestrian behind truck"), and *scene-to-V2X consistency checking* (validating sensor perception against received messages). These capabilities enable more natural interaction and richer understanding of complex driving scenarios, as explored in Chapters 5–6 (vision-language models and video-language models).

Uncertainty expression is a critical capability for safety-critical systems. Instead of a confidence score (e.g., "0.85"), generative models can articulate ambiguity through *generative text explanations* that communicate reasoning: "I am unsure if that is a pedestrian or a shadow because the lighting is poor." Generative models can synthesize explanations that communicate uncertainty, confidence, and reasoning processes, enabling passengers and operators to understand system behavior and make informed decisions. This capability extends to *uncertainty-aware responses* in dialogue systems and *probabilistic scene descriptions* that express multiple plausible interpretations of ambiguous situations. These applications are essential for safety-critical CAV deployment, as discussed in Chapter 9 (function calling with uncertainty) and Chapter 22 (ethics and safety).

Policy adaptation refers to the ability to adjust behavior based on context, user preferences, and changing conditions. Traditional CAV systems rely on fixed policies that may not adapt well to individual preferences, regional driving styles, or novel situations. Generative models can learn from demonstrations, incorporate feedback, and generate adaptive policies that balance safety, efficiency, and user preferences. This capability is realized through *generative policy synthesis* that creates behavior plans from natural language instructions, *preference learning* that adapts driving style based on user feedback, and *context-aware decision generation* that adjusts behavior for regional norms or weather conditions. These applications enable personalized and adaptive CAV systems, as explored in Chapters 17–18 (fine-tuning and reinforcement learning for preference alignment and behavioral adaptation) and Chapters 10–12 (agentic systems with adaptive reasoning).

These capabilities position generative models as a transformative technology for CAV systems, enabling not just perception and control but also explanation, interaction, and adaptive reasoning. The concrete applications, from generative text responses and VQA to scene synthesis and multimodal fusion, demonstrate the practical impact of GenAI in real-world CAV deployment. The next section examines how these capabilities are realized through specific model architectures.

1.4 Generative Model Families

This section examines the major families of generative models and their roles in CAV systems. Understanding these families and their trade-offs is essential for making informed design decisions based on specific CAV application requirements, including speed, quality, and computational constraints.

Table 1.2 Taxonomy of generative model families with representative CAV applications.

Model family	Key characteristics	CAV applications
Autoregressive models	LLMs, sequential predictors; generate sequences step by step	Text generation, visual question answering (VQA), trajectory forecasting, natural language interaction
Diffusion models	Iterative denoising process; high-fidelity generation	Scene synthesis, rare-event simulation, dataset augmentation, high-fidelity sensor data generation
Flow matching	Deterministic flows; learns velocity fields for direct transport	Fast scene generation, real-time synthesis, efficient simulation pipelines
Variational autoencoders (VAEs)	Encoder-decoder architecture; probabilistic latent space	Data compression, sensor fusion, anomaly detection, V2X communication
Generative adversarial networks (GANs)	Generator-discriminator architecture; adversarial training	Sensor simulation, domain adaptation, real-time image enhancement

Generative models can be organized into five primary families, each reflecting a distinct strategy for approximating complex data distributions. Table 1.2 presents a taxonomy of these families with their key characteristics and CAV applications. The following subsections examine each family in detail.

1.4.1 Autoregressive Models

Autoregressive models form a broad class of generative architectures that generate sequences step by step, where each element depends on all previous elements. This sequential generation approach underlies two distinct subgroups of models that serve different roles in CAV systems.

The first subgroup comprises *autoregressive transformer models* [1], which constitute the foundation of modern large language models (LLMs) [2]. These models generate structured sequences one token at a time, conditioning each prediction on all previous outputs. Their ability to model long-range dependencies, reason with context, and interface naturally with multimodal inputs makes them central to the language-centric systems developed in later chapters. In CAV systems, autoregressive Transformer models enable natural language interaction, explaining decisions, and reasoning about complex scenarios. For example, when a vehicle encounters an unexpected road condition, an LLM can generate a clear explanation: "I'm slowing down because I detected construction ahead and merging traffic from the right." Chapter 2 provides comprehensive coverage of these Transformer-based architectures, while Chapters 7–9 examine how they can be enhanced through prompt engineering, retrieval-augmented generation, and function calling to meet specific CAV requirements.

The second subgroup includes *domain-specific autoregressive models for sequential prediction*, such as models used for trajectory forecasting, sensor time-series dynamics, or audio-based diagnostics in CAV systems. Although these models share the same autoregressive factorization $p(x_t|x_{<t})$, they serve operational purposes distinct from those of large-scale language models. They typically operate within real-time perception and prediction pipelines where low latency is critical, enabling tasks such as predicting vehicle trajectories, forecasting sensor patterns, and detecting anomalies in real-time. These models are examined in Chapter 4, where they appear alongside VAEs and GANs as domain-specific models tailored to CAV applications.

While both subgroups share the mathematical formulation of autoregressive modeling, their *inference latency* and *application scope* in a CAV are fundamentally different: Transformer-based LLMs provide high-latency reasoning and natural language interaction, while domain-specific sequential models provide low-latency control and real-time prediction. This latency trade-off is a critical consideration for CAV deployment, as explored in Part VI (Chapters 17–19), which addresses practical deployment considerations including fine-tuning, distillation, and edge optimization, and Part VII (Chapters 20–22), which addresses validation, evaluation, and safety considerations.

By distinguishing these two subgroups, this chapter clarifies how autoregression serves both language modeling and sequential prediction roles in CAV systems. Throughout this book, the term "autoregressive model" may refer to either Transformer-based LLMs (when discussing language and reasoning) or domain-specific sequential models (when discussing perception and prediction), with context making the distinction clear.

1.4.2 Diffusion Models

Diffusion-based models approach generation as a process of iterative refinement, beginning with random noise and gradually denoising it through learned transformations to produce coherent outputs. Conceptually, diffusion models work by *slowly removing fog from a sensor image*: the forward process gradually corrupts clean data by adding noise (like fog obscuring a camera view), and the reverse process learns to remove that noise step-by-step to recover the original scene. This iterative denoising approach has gained prominence in image synthesis, particularly with models like stable diffusion [3].

The diffusion process operates through two complementary stages: a *forward diffusion process* that incrementally adds noise to data until it becomes pure noise, and a *reverse denoising process* that learns to iteratively remove noise to reconstruct realistic samples. This iterative approach requires many steps (typically 50–1000) but achieves high sample quality through gradual refinement.

In CAV systems, diffusion models are primarily used for high-fidelity offline simulation, dataset augmentation, and rare-event synthesis, where computational cost is less restrictive, but high fidelity is paramount. Consider a development team testing a new perception system. They need to evaluate performance under rare but critical conditions: a sudden hailstorm combined with low visibility and a malfunctioning traffic light. Capturing such scenarios in real-world data collection is nearly impossible, as they occur too infrequently and are too dangerous to stage. Diffusion models solve this problem by generating synthetic sensor sequences that faithfully reproduce these complex, multimodal conditions, enabling systematic testing without risk. Chapter 3 ("Diffusion and Flow Matching for CAV Systems") provides comprehensive coverage of diffusion models, their mathematical foundations, and their applications in CAV systems.

1.4.3 Flow Matching

Flow matching represents the *evolution of diffusion models*, addressing their primary limitation: slow sampling speed. While diffusion models are the current standard for high-fidelity simulation, flow matching represents the future of real-time generation due to its efficiency.

A growing class of generative techniques reframes data synthesis as a *deterministic transport problem*. Rather than denoising a noisy sample through a stochastic reverse diffusion process, these models learn a time-dependent vector field that transports a simple base distribution (typically Gaussian noise) into the target data distribution. This approach achieves high sample quality with substantially fewer sampling steps than classical diffusion models.

Think of generating a data point from noise like navigating a ship through fog. **Diffusion models** drift randomly while correcting course, eventually finding the destination but taking a zigzag path. **Flow matching**, however, calculates the direct, shortest-path current (the vector field) that flows straight from the starting noise to the final data point. This 'straight-line' approach is mathematically more complex to derive but much faster to execute.

The key advantage of flow matching is its *straight-line trajectories* from noise to data, in contrast to the stochastic paths taken by diffusion models. This deterministic nature enables faster sampling: while diffusion models require many iterative denoising steps (often 50–1000 steps), flow matching can generate samples through a single computation or just a few steps, dramatically reducing computational cost. The most widely used formulation is *flow matching* [4], which learns a velocity field that transports samples along linear interpolation paths from noise to data. *Rectified flow* is a closely related variant that explicitly enforces straight paths, improving numerical stability and enabling efficient sampling with fewer steps (typically 10–50 instead of 50–1000). Recent variants that collapse the generative trajectory into fewer steps enable even faster sampling, though they typically trade some sample fidelity for speed.

Flow matching is rapidly influencing the generative AI landscape. Modern variants of stable diffusion, including stable diffusion 3, employ flow matching or hybrid diffusion-flow training. Large-scale video models (e.g., sora-like architectures) rely on flow matching mechanisms to maintain temporal coherence while reducing sampling cost. Within CAV systems, flow matching offers a principled foundation for high-fidelity, scalable scene synthesis and rare-event generation. Its deterministic nature and reduced sampling overhead make it increasingly attractive for generative simulation pipelines, complementing the stochastic methods discussed in Chapter 3. As edge hardware becomes more powerful, flow matching may be deployed in real-time CAV applications, though it currently excels in offline simulation scenarios.

1.4.4 Variational Autoencoders

Variational autoencoders (VAEs) [5] are probabilistic generative models that learn to encode input data into a latent space and decode it back to the data space. VAEs explicitly model latent variables through encoder-decoder architectures: the encoder maps input data into a probabilistic latent space (typically parameterized as a Gaussian distribution), and the decoder reconstructs or generates data from latent samples. This probabilistic formulation enables VAEs to learn structured latent representations that support compression, fusion, and anomaly detection. The key advantage of VAEs is their ability to learn meaningful latent spaces that capture underlying data structure while enabling diverse generation through sampling from the latent distribution.

In CAV systems, VAEs serve diverse roles, including sensor data compression for efficient V2X communication, multi-sensor fusion through shared latent representations, and anomaly detection by identifying samples that deviate from the learned latent distribution. Their stable training properties and structured latent spaces make them valuable for applications requiring probabilistic modeling and representation learning. Chapter 4 examines VAEs in detail, including their mathematical foundations and CAV applications.

1.4.5 Generative Adversarial Networks

Generative adversarial networks (GANs) [6] employ a two-player minimax game between a generator network and a discriminator network. The generator learns to produce realistic samples by mapping random noise (latent variables) directly to data space, while the discriminator learns to distinguish real from generated samples. This adversarial training process drives the generator to improve until it produces samples that the discriminator cannot reliably classify. Unlike VAEs, which explicitly model latent distributions, GANs use latent variables implicitly and learn implicit data distributions through adversarial optimization.

While GANs are less common for text generation, they remain vital for *sensor simulation* and *domain adaptation* in CAV systems, which are their primary use cases. For example, GANs can generate synthetic sensor data to augment training datasets, adapt models across different sensor configurations or environmental conditions, and enhance low-quality sensor images in real-time. Their single-pass generation capability and ability to produce sharp, high-frequency details make them well-suited for real-time perception enhancement tasks. In CAV systems, GANs serve diverse roles, from lightweight, real-time capable components in perception pipelines to components in offline data generation and analysis systems. Chapter 4 examines GANs in detail, including their mathematical foundations, training challenges, and CAV applications.

Each family represents a different trade-off between training stability, sample quality, and computational efficiency, but all address the same fundamental challenge of learning complex data distributions from finite samples. The choice of which generative approach to use depends on the specific requirements of the CAV application, as explored in detail throughout this book.

1.5 The Generative AI Ecosystem

1.5.1 Evolution of Generative AI Paradigms

The evolution of AI has progressed from rule-based systems to machine learning, deep learning, and ultimately generative modeling. Each paradigm shift addressed limitations of previous approaches: machine learning enabled probabilistic modeling but required feature engineering; deep learning automated feature extraction but remained primarily discriminative; generative AI introduced synthesis capabilities, and, in the case of large autoregressive models, enabled explanation, dialogue, and adaptive reasoning essential for CAV systems.

The evolution of generative AI has been driven by a series of foundational breakthroughs, each contributing to the emergence of increasingly capable and generalizable models. These milestones reflect progress in both architectural innovation and application breadth, leading to the five model families introduced in this chapter:

- **2013–2014:** Variational autoencoders (VAEs) [5] and generative adversarial networks (GANs) [6] established two fundamental approaches to generative modeling. VAEs introduced probabilistic latent representations, while GANs demonstrated adversarial training for high-quality image synthesis.
- **2017:** The Transformer architecture [1] redefined sequence modeling and laid the foundation for large-scale autoregressive language models. Its self-attention mechanism enabled parallelization and improved long-range dependency modeling, transforming natural language processing.
- **2018–2020:** This period saw the release of influential models such as BERT [7] (2019, bidirectional language representation), GPT-2 [8] (2019, unidirectional text generation), GPT-3 [2] (2020, large-scale language generation), and StyleGAN [9] (2020, photorealistic image synthesis). These models expanded the applicability of representation learning and generative modeling across language and vision domains.
- **2021–2022:** Diffusion models emerged as a powerful framework for high-fidelity generation, with models like DALL·E [10] (2021, text-to-image generation), DALL·E 2 [11] (2022), and stable diffusion [3] (2022) demonstrating exceptional image synthesis capabilities. The introduction of CLIP [12] (2021, contrastive vision-language pretraining), Codex [13] (2021,

code generation), and ChatGPT [14] (2022, conversational AI) marked the transition toward instruction-tuned and publicly accessible generative systems.

- **2023–2024:** Recent developments focused on multimodality and efficiency. Models such as GPT-4 [15], Claude 2 [16], and Gemini 1.5 [17] integrated capabilities across text, image, and audio. Diffusion-based video generation (e.g., Sora [18]) and flow matching [4] addressed sampling efficiency, while open-source models from DeepSeek and Mixtral broadened access to generative technologies.
- **2025–2026:** Developments emphasized enhanced multimodal integration, improved reasoning performance, and continued momentum in open and open-weight model ecosystems. Major releases came from the GPT-5.5 [21], Gemini 3.5 [20], and Claude 5 [22] families (e.g., Claude Sonnet 5), alongside the DeepSeek-R1 [19] reasoning model, which featured mixture-of-experts architectures, stronger reasoning capabilities, and substantially expanded context windows reaching hundreds-of-thousands to low-million token ranges. During this period, generative systems also began evolving from conversational assistants into persistent, tool-using agents that plan across multiple steps and invoke external tools.

Figure 1.2 presents a visual timeline of these milestones, illustrating the accelerating pace of innovation in generative AI, from foundational architectures to today's multimodal systems. Each milestone contributed to the development of the five model families (autoregressive, diffusion, flow matching, VAEs, and GANs) that form the foundation of modern generative AI for CAV applications.

1.5.2 Development Paradigms and Access Modalities

The contemporary generative AI landscape can be understood through two complementary dimensions: *development paradigms* (proprietary vs. open-source) and *access modalities* (remote API vs. local deployment). For CAVs, the choice between these options has profound implications for system design, balancing performance, safety, latency, privacy, and cost. These dimensions differ significantly in terms of development scale, accessibility, deployment constraints, and user control, as examined below.

1.5.2.1 Development Paradigms: Proprietary vs. Open-source

The generative AI ecosystem is shaped by two complementary development paradigms: proprietary models and open-source models.

Proprietary generative AI models are characterized by restricted access to model weights, architecture details, and training data. These models are typically developed by major industrial research

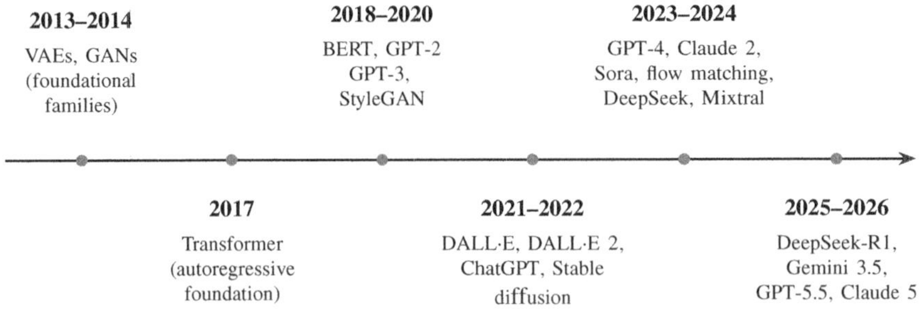

Figure 1.2 Timeline of key milestones in the development of generative AI, highlighting major architectural breakthroughs and the emergence of the five model families (autoregressive, diffusion, flow matching, VAEs, and GANs) that form the foundation of modern generative AI for CAV applications.

laboratories, such as OpenAI, Google DeepMind, and Anthropic, which leverage extensive computational resources and large-scale proprietary datasets. Proprietary models are usually offered exclusively via cloud-based APIs, providing robust performance, commercial-grade reliability, and strong alignment with safety standards. Notable examples include GPT (OpenAI), Gemini (Google DeepMind), and Claude (Anthropic). While these models offer high accuracy and user-friendly interfaces, they limit access to internal mechanisms and customizability, requiring users to rely on API access rather than local deployment.

Open-source generative AI models are characterized by public access to architectural details, pretrained weights, and training configurations, enabling local deployment and full customization. These models are developed by a diverse ecosystem, including major technology companies (e.g., Meta with its Llama series), specialized AI companies (e.g., Mistral AI and DeepSeek), academic institutions, and independent research communities. They support a wide range of local deployment options and customization techniques, including parameter-efficient fine-tuning (e.g., LoRA, or low-rank adaptation), quantization, and domain-specific retraining. Representative open-source models include LLaMA (Meta), Mistral (Mistral AI), DeepSeek, and Mixtral. These efforts have played a critical role in democratizing generative AI research and expanding its use in privacy-sensitive and resource-constrained environments. Notably, some organizations (e.g., Meta) develop both proprietary and open-source models, reflecting different strategic approaches to different model capabilities and safety considerations.

Table 1.3 summarizes the key differences between proprietary and open-source generative AI models. Safety-critical systems often require local deployment to ensure privacy, reduce latency, and maintain operation without internet connectivity. However, proprietary models accessed via remote APIs may offer superior performance for non-critical applications such as route planning or entertainment. Chapter 19 examines knowledge distillation techniques that can help bridge this gap by creating efficient models that combine the performance of large proprietary models with the deployment flexibility of open-source alternatives.

Career Context: CAV engineers typically use **proprietary APIs** for prototyping and non-critical features (e.g., infotainment) but rely on **open-source/local** models for safety-critical components to ensure low latency and privacy. This practical distinction reflects the fundamental trade-offs between ease of integration and system-level control that shape real-world CAV development.

Table 1.3 Proprietary versus open-source generative AI models.

Attribute	Proprietary models	Open-source models
Performance	Optimized using extensive compute and curated data	Competitive performance via efficient open training methods
Access	Available via cloud APIs with usage restrictions	Full local access to model weights and configurations
Customization	Limited to prompt tuning and API-level control	Supports full fine-tuning, LoRA, quantization, and retraining
Cost	Requires paid API usage or subscription	Free to use; infrastructure-dependent costs only
Privacy	Data is transmitted to external servers (data privacy concern)	Data remains local, ensuring privacy compliance
Innovation model	Centralized, proprietary innovation cycle	Community-driven experimentation and rapid iteration

1.5.2.2 Access Modalities: Remote API vs. Local Deployment

In addition to their development origins, generative AI models can be categorized by where they execute and how they are accessed in practical applications. The key distinction is whether the model runs on remote cloud infrastructure or on local hardware, regardless of whether access is via API or direct integration.

Remote API access refers to models that execute on external cloud servers, accessed over the internet through web-based APIs. Proprietary models are most commonly accessed this way, which abstracts away the complexity of model deployment and maintenance. This modality enables rapid prototyping and scalable integration into production systems. However, it also imposes constraints on customization, incurs usage costs, and requires continuous internet connectivity for inference. Data must be transmitted to external servers, raising privacy concerns. Such models are well-suited for scenarios where reliability and speed are prioritized over transparency, local control, or privacy.

Local deployment refers to models that execute on local infrastructure (desktop, server, or edge device), whether accessed via direct integration or through local APIs (e.g., Ollama, LM Studio). Open-source models support full local deployment, allowing researchers and developers to host models on private infrastructure. Toolkits such as Hugging Face Transformers, Ollama, and LM Studio provide local API interfaces that abstract model interaction while keeping execution local. This access mode enables privacy-preserving applications, domain-specific adaptation, and offline functionality. These capabilities are essential in safety-critical domains such as autonomous driving. Note that local deployment can still use API interfaces for convenience, but the model execution and data remain entirely local.

These two access modalities reflect contrasting design choices between ease of integration and system-level control. The operational characteristics of these approaches are compared in Table 1.4.

Table 1.4 Remote API access versus local deployment of generative AI models.

Attribute	Remote API access	Local deployment
Execution environment	Model runs in external cloud server	Model runs on local infrastructure or edge device
Access interface	Web-based API (HTTP/HTTPS)	Direct integration or local API (e.g., Ollama)
Customization	Limited to prompt engineering	Full customization (e.g., fine-tuning, quantization)
Privacy	Data is transmitted externally	Data remains entirely local
Connectivity requirement	Requires continuous internet access	Operates offline once deployed
Deployment complexity	Minimal (via API call)	Requires infrastructure and setup
Latency	Depends on network and server availability	Can be optimized for low-latency, real-time response
Use case suitability	Ideal for cloud-integrated and commercial SaaS applications	Preferred in privacy-sensitive or real-time systems (e.g., autonomous vehicles)
Cost model	Pay-per-use or subscription	One-time compute cost, no recurring API fees

1.6 Summary and Roadmap

1.6.1 Chapter Summary

This chapter established the conceptual foundation for understanding generative AI in CAV systems. The chapter began by contrasting generative and discriminative AI paradigms, demonstrating how generative models extend beyond classification and prediction to enable explanation, dialogue, and contextual reasoning. Unlike discriminative models that select from predefined labels based on input patterns, generative models synthesize novel, contextually appropriate outputs conditioned on context.

The chapter identified five transformative capabilities that generative AI brings to CAV systems: **semantic reasoning** (understanding context and meaning for natural language interaction), **world modeling** (constructing internal representations of current and future states for planning and simulation), **multimodal fusion** (integrating information across text, images, audio, and sensor streams), **uncertainty expression** (articulating ambiguity and confidence through generative explanations), and **policy adaptation** (adjusting behavior based on context, preferences, and changing conditions). These capabilities address fundamental limitations of traditional perception-only CAV architectures, enabling systems that can explain decisions, communicate with passengers, and adapt to novel situations.

The chapter examined five primary families of generative models, each serving distinct roles in CAV applications: **autoregressive models** (including Transformer-based LLMs for natural language interaction and sequential prediction), **diffusion models** (iterative denoising for high-fidelity scene synthesis and rare-event simulation), **flow matching** (deterministic flows for fast, real-time generation), **variational autoencoders (VAEs)** (probabilistic latent representations for compression and sensor fusion), and **generative adversarial networks (GANs)** (adversarial training for sensor simulation and domain adaptation). Understanding these families and their trade-offs is essential for making informed design decisions based on specific CAV requirements, including speed, quality, and computational constraints.

The chapter also explored the practical realities of the contemporary generative AI landscape, examining the trade-offs between **proprietary and open-source development paradigms** and between **remote API access and local deployment**. Proprietary models accessed via cloud APIs offer robust performance and commercial reliability but limit customization and raise privacy concerns. Open-source models enable full local deployment, supporting privacy-preserving applications, domain-specific adaptation, and offline functionality essential for safety-critical CAV systems. The choice between these options has profound implications for system design, balancing performance, safety, latency, privacy, and cost.

These foundations establish why generative AI matters for CAVs and provide the conceptual framework for understanding how different model families and capabilities serve distinct roles in intelligent vehicle systems. The subsequent chapters in Part I and Part II will examine the architectural and mathematical foundations that enable these capabilities, while later parts explore how these models are integrated into agentic systems, communication protocols, and deployment strategies.

1.6.2 The Roadmap Ahead

This book provides comprehensive coverage of generative AI for CAV systems through a structured progression spanning 22 chapters organized into seven major parts, structured across four conceptual layers: generative and multimodal models, model interface, agentic systems and protocols, and adaptation and validation. Understanding this organization helps readers navigate the material effectively and choose appropriate pathways based on their background and goals.

Prerequisites: This book assumes familiarity with basic machine learning concepts (e.g., neural networks, training and inference, probability distributions). However, the foundational chapters (Part I) provide sufficient background for readers new to generative AI. Each chapter is designed to be relatively self-contained, with key concepts restated when necessary, enabling flexible navigation based on specific needs.

Reading Pathways: For guidance on selecting appropriate reading pathways based on your role and objectives, see the "How to Read This Book" section, which provides pathway recommendations organized by reader role (academic, professional, and research pathways).

Part I: Foundations of Generative Intelligence for CAVs (Chapters 1–2) establishes the theoretical and practical foundations of generative AI. This part introduces core concepts (Chapter 1) and examines large language model architectures (Chapter 2). These chapters provide the building blocks for understanding how generative models achieve semantic reasoning, structured generation, and context-aware decision support.

Part II: Multimodal Representation and World Grounding (Chapters 3–4, 5–6) focuses on generative models that construct coherent internal representations of the physical world from raw sensory data. This part covers diffusion and flow matching (Chapter 3), GANs, VAEs, and autoregressive models for sensor and temporal modeling (Chapter 4), vision-language models for scene understanding (Chapter 5), and video large language models for temporal reasoning (Chapter 6). These chapters establish the representational foundation that supports reasoning, planning, and interaction in intelligent CAV systems.

Part III: Language-based Interaction, Memory, and Tool Use (Chapters 7–9) examines how language models are used to shape interaction, retrieve contextual information, and invoke structured system actions within CAV systems. This part covers prompt engineering for CAVs (Chapter 7), retrieval-augmented generation for context grounding and memory (Chapter 8), and function calling and tool use in GenAI-driven CAV systems (Chapter 9). These techniques enable LLMs to interface with CAV systems, access real-time data, and execute actions.

Part IV: Agentic Intelligence and Coordination (Chapters 10–12) demonstrates how generative AI enables autonomous agents and multi-agent systems. This part covers GenAI agent architectures (Chapter 10), GenAI multiagent systems (Chapter 11), and orchestration frameworks for agentic AI (Chapter 12). This part shows how LLMs can be extended into agentic systems that reason, plan, and coordinate effectively in CAV environments.

Part V: Communication Protocols and Collective Intelligence (Chapters 13–16) focuses on the communication protocols and architectural frameworks that support collective intelligence in CAV ecosystems. This part covers communication protocols for agentic AI in connected vehicles (Chapter 13), model context protocol (MCP) for agentic systems (Chapter 14), agent-to-agent protocols for collaborative intelligence (Chapter 15), and integrated CAV agent architectures with MCP and A2A (Chapter 16). These chapters frame protocols as semantic interfaces between autonomous agents.

Part VI: Adaptation and Deployment (Chapters 17–19) addresses adaptation mechanisms that refine and optimize system behavior after foundational models and architectures have been established. This part covers supervised fine-tuning and parameter-efficient adaptation (Chapter 17), reinforcement learning fine-tuning of generative models (Chapter 18), and knowledge distillation for edge deployment in CAV systems (Chapter 19). These methods are presented as complementary strategies for behavioral adaptation.

Part VII: Validation and Safety (Chapters 20–22) examines the methodologies and frameworks necessary to ensure that GenAI-driven CAV systems are reliable, interpretable, and safe. This part covers simulation platforms for GenAI-driven CAVs (Chapter 20), evaluation of generative AI systems in connected and autonomous vehicles (Chapter 21), and ethical and safety considerations in GenAI-CAV systems (Chapter 22). This part positions validation and safety as integral components of system design.

The journey ahead progresses from understanding individual model components to leveraging their capabilities, building complex agentic systems, and deploying them safely in real vehicles. Each chapter builds on the foundations established here, progressively moving from theory to practice, from individual components to integrated systems. The choice of model family, deployment strategy, and integration approach depends on specific CAV requirements, balancing performance, safety, latency, and cost. This book provides the knowledge and tools needed to make these critical decisions and build the next generation of intelligent, communicative, and trustworthy autonomous vehicles.

References

1 Ashish Vaswani, Noam Shazeer, Niki Parmar, Jakob Uszkoreit, Llion Jones, Aidan N. Gomez, Łukasz Kaiser, and Illia Polosukhin. Attention is all you need. In *Advances in Neural Information Processing Systems*, 2017. https://papers.nips.cc/paper_files/paper/2017/file/3f5ee243547dee91fbd053c1c4a845aa-Paper.pdf

2 Tom Brown, Benjamin Mann, Nick Ryder, Melanie Subbiah, Jared D. Kaplan, Prafulla Dhariwal, Arvind Neelakantan, Pranav Shyam, Girish Sastry, Amanda Askell, et al. Language models are few-shot learners. *Advances in Neural Information Processing Systems*, 33:1877–1901, 2020.

3 Robin Rombach, Andreas Blattmann, Dominik Lorenz, Patrick Esser, and Björn Ommer. High-resolution image synthesis with latent diffusion models. In *Proceedings of the IEEE/CVF Conference on Computer Vision and Pattern Recognition*, pages 10684–10695, 2022.

4 Yaron Lipman, Ricky T. Q. Chen, Heli Ben-Hamu, Maximilian Nickel, and Matt Le. Flow matching for generative modeling. *arXiv preprint arXiv:2210.02747*, 2023.

5 Diederik P. Kingma and Max Welling. Auto-encoding variational Bayes. *arXiv preprint arXiv:1312.6114*, 2013.

6 Ian J. Goodfellow, Jean Pouget-Abadie, Mehdi Mirza, Bing Xu, David Warde-Farley, Sherjil Ozair, Aaron Courville, and Yoshua Bengio. Generative adversarial nets. In *Advances in Neural Information Processing Systems*, 2014.

7 Jacob Devlin, Ming-Wei Chang, Kenton Lee, and Kristina Toutanova. BERT: pre-training of deep bidirectional transformers for language understanding. In *Proceedings of the 2019 Conference of the North American Chapter of the Association for Computational Linguistics: Human Language Technologies, Volume 1 (Long and Short Papers)*, pages 4171–4186, 2019.

8 Alec Radford, Jeffrey Wu, Rewon Child, David Luan, Dario Amodei, and Ilya Sutskever. Language models are unsupervised multitask learners, 2019.

9 Tero Karras, Samuli Laine, Miika Aittala, Janne Hellsten, Jaakko Lehtinen, and Timo Aila. Analyzing and improving the image quality of StyleGAN. In *Proceedings of the IEEE/CVF Conference on Computer Vision and Pattern Recognition*, pages 8110–8119, 2020.

10 Aditya Ramesh, Mikhail Pavlov, Gabriel Goh, Scott Gray, Chelsea Voss, Alec Radford, Mark Chen, and Ilya Sutskever. Zero-shot text-to-image generation. In *International Conference on Machine Learning*, pages 8821–8831, 2021.

11 Aditya Ramesh, Prafulla Dhariwal, Alex Nichol, Casey Chu, and Mark Chen. Hierarchical text-conditional image generation with clip latents. *arXiv preprint arXiv:2204.06125,* 2022.

12 Alec Radford, Jong Wook Kim, Chris Hallacy, Aditya Ramesh, Gabriel Goh, Sandhini Agarwal, Girish Sastry, Amanda Askell, Pamela Mishkin, Jack Clark, et al. Learning transferable visual models from natural language supervision. In *International Conference on Machine Learning*, pages 8748–8763, 2021.

13 Mark Chen, Jerry Tworek, Heewoo Jun, Qiming Yuan, Henrique Ponde de Oliveira Pinto, Jared Kaplan, Harri Edwards, Yuri Burda, Nicholas Joseph, Greg Brockman, et al. Evaluating large language models trained on code. *arXiv:2107.03374,* 2021.

14 OpenAI. ChatGPT: optimizing language models for dialogue, 2022. https://openai.com/blog/chatgpt/.

15 OpenAI. GPT-4 Technical Report, 2023. https://openai.com/research/gpt-4.

16 Anthropic. Claude 2, 2023. https://www.anthropic.com/news/claude-2.

17 Google DeepMind. Gemini 1.5: unlocking multimodal understanding across millions of tokens of context, 2024. https://deepmind.google/technologies/gemini/.

18 OpenAI. Video generation models as world simulators, 2024. https://openai.com/research/video-generation-models-as-world-simulators.

19 DeepSeek-AI. Deepseek-r1: incentivizing reasoning in LLMs through reinforcement learning, 2025. https://www.nature.com/articles/s41586-025-09422-z, Accessed: 2025-12-17.

20 Google DeepMind. Gemini 3.5: Frontier intelligence with action, 2026. https://blog.google/innovation-and-ai/models-andresearch/gemini-models/gemini-3-5.

21 OpenAI. Introducing GPT-5.5, 2026. https://openai.com/index/introducing-gpt-5-5.

22 Anthropic. Introducing Claude Sonnet 5, 2026. https://www.anthropic.com/news/claude-sonnet-5.

2

Large Language Models: Architecture and Operation

2.1 Introduction

2.1.1 Contextual Ambiguity in Language

Imagine a driver approaching an intersection where a traffic light has malfunctioned. The driver asks their vehicle's voice assistant: "What should I do here?" The assistant must interpret this ambiguous request by considering multiple contextual factors: the vehicle's current speed, surrounding traffic conditions, local traffic laws, and the driver's previous commands. This scenario exemplifies the fundamental challenge that LLMs must address: understanding meaning not from isolated words, but from the complex network of contextual relationships in which language occurs.

Human language is inherently ambiguous. Many words exhibit multiple meanings, with their intended interpretation determined not by the word alone but by the surrounding context in which they occur. This property, known as *polysemy*, poses a central challenge in computational language understanding and is particularly critical in safety-sensitive applications such as CAVs.

Consider the following sentence:

> *"The EV was left to charge overnight at the station."*

Here, the word *charge* refers to the act of storing electrical energy. However, in other contexts, *charge* can signify an entirely different concept. Table 2.1 presents a variety of example sentences that illustrate how the interpretation of a single word varies with its usage.

While human readers effortlessly resolve such ambiguities by integrating surrounding clues (e.g., "EV," "station," "overnight"), traditional language models often fail to make such distinctions. Early approaches to natural language processing represented words as fixed vectors, independent of context. Techniques such as Word2Vec [1] and GloVe [2] captured semantic similarity but assigned a single vector to each word, thereby conflating distinct meanings into a single, context-independent vector.

This shortcoming motivated the development of contextual language models that generate dynamic word representations conditioned on surrounding text. Such models must not only detect polysemy but also resolve it on the fly using local linguistic context. To accomplish this, modern language models rely on a mechanism known as *self-attention*.

Table 2.1 Polysemy of the word *charge* in different contexts.

Sentence	Interpretation of *charge*
The EV was left to **charge** overnight at the station.	Store electrical energy
The suspect was formally **charged** in court.	Legal accusation
The waiter will **charge** you for extra toppings.	Request for payment
The soldiers began to **charge** the enemy.	Physical attack

2.1.2 The Self-attention Revolution

Self-attention enables a model to weigh the relevance of every token (a discrete unit of text, such as a word or word fragment) in the input sequence when computing the representation of a particular token. For example, in the sentence "The EV was left to charge overnight at the station," the model must determine that the word *charge* is semantically aligned with energy usage, not legal or military contexts. This determination is made by assigning higher attention weights to tokens like *EV*, *station*, and *overnight*, and lower weights to semantically irrelevant tokens.

This ability to model contextual relevance lies at the core of the transformer architecture [3]. Rather than relying on token-by-token sequential processing via recurrence, the transformer employs self-attention to capture dependencies between any pair of tokens, regardless of their distance in the sequence. This design achieves both computational efficiency and scalability, enabling the training of models on large corpora with high fidelity to linguistic structure.

In the context of CAVs, the ability to model contextual relationships is essential. Verbal commands issued to an in-vehicle assistant often involve ambiguous terms whose interpretation depends on environmental cues or prior dialogue. A failure to correctly resolve such ambiguity could lead to inappropriate or unsafe vehicle behavior. Consequently, robust context modeling is not merely a desirable feature but a fundamental prerequisite for safe and reliable operation in language-enabled automotive systems.

2.2 Large Language Models

2.2.1 Definition and Core Principles

A large language model (LLM) is a deep neural network trained on massive text corpora with the objective of predicting the next token in a sequence. Unlike rule-based systems that rely on handcrafted grammar rules or explicit ontologies, LLMs are data-driven. They learn linguistic, semantic, and factual patterns from the statistical structure of language data.

Modern LLMs are typically based on the transformer architecture and contain hundreds of millions to trillions of parameters. Their scale enables them to capture nuanced properties of human language, including syntactic structure, discourse coherence, and contextual disambiguation. While LLMs are often described as text generators, they are more fundamentally *reasoning engines* that model relationships, infer patterns, and make decisions based on context. This reasoning capability makes them valuable not only for conversational AI but also for planning, decision support, and complex reasoning tasks in CAV applications. These models are pretrained on broad datasets such as books, articles, code repositories, and web content, and are often adapted through additional tuning for specific domains or applications.

2.2.2 Representative Model Architectures

To provide concrete context for the concepts discussed in this chapter, we briefly introduce several prominent LLM architectures that exemplify these core principles.

The current landscape of large language models includes several prominent examples that demonstrate different architectural approaches and capabilities. Representative models include generative pretrained transformer (GPT) [4, 5], an autoregressive decoder-only model designed for conversational AI and general-purpose language generation; Claude [6], an alignment-focused model emphasizing safe, interpretable, and human-aligned outputs through specialized training methodologies; Gemini [7], a multimodal architecture extending beyond text to handle vision and other modalities within a unified framework; large language model meta AI (LLaMA) [8], an open-weight model family enabling research and fine-tuning by providing access to model weights and training procedures; and DeepSeek [9], a mixture-of-experts architecture demonstrating efficient scaling by activating subsets of expert networks per forward pass, achieving strong performance in reasoning, mathematics, and programming tasks.

While these models differ in architecture, training methods, and licensing terms, they share a common foundation in token-based language modeling, relying on probabilistic next-token prediction to generate fluent and coherent text.

2.2.3 Learning to Predict the Next Token

During training, a large language model learns to estimate the probability of the next token given all preceding tokens. A token is a discrete unit of text that the model processes, which may be a word, part of a word, or punctuation. For example, the word *unbelievable* might be segmented into tokens such as `"un"`, `"bel"`, and `"ievable"`, while a simple word like *car* may be represented as a single token. Tokenization strategies vary across model families, but all convert raw text into sequences of discrete elements suitable for neural computation.

Formally, given a sequence of tokens $\{x_1, x_2, \dots, x_T\}$, the model is trained to approximate the joint probability distribution as a product of conditional probabilities:

$$P(x_1, x_2, \dots, x_T) = \prod_{t=1}^{T} P(x_t \mid x_1, x_2, \dots, x_{t-1}). \tag{2.1}$$

At each time step t, the model predicts the conditional probability distribution $P(x_t \mid x_1, \dots, x_{t-1})$ over the vocabulary of possible next tokens. Training involves minimizing the discrepancy between the predicted distribution and the true next token, typically using cross-entropy loss (a measure that penalizes the model when it assigns low probability to the correct token).

Example 2.1:

Input: *"The car stopped at the"*
Target: *Predict the probability of tokens such as "light," "intersection," "end," and others.*
Training Objective: *Assign high probability to the correct next token.*

While the outputs of LLMs are ultimately human-readable words and sentences, the internal prediction process operates entirely at the token level (predicting one token at a time rather than entire words). A single word may correspond to one token or to multiple subword tokens, depending on the tokenization method.

Modeling language at the token level enhances flexibility and generalization. It allows LLMs to handle rare words, morphologically complex forms, and multilingual text within a unified probabilistic framework.

2.2.4 Generating Sequences by Autoregressive Decoding

During inference, a large language model generates output by sampling tokens sequentially, one at a time, from the learned conditional distributions. This process, known as *autoregressive decoding*, constructs coherent sequences by progressively expanding the context with each new token.

The generation process begins with an initial context $C_0 = \{x_1, x_2, \dots, x_n\}$, typically consisting of the input prompt. At each step $t \geq n + 1$, the model processes all tokens in the context $C_{t-1} = \{x_1, \dots, x_{t-1}\}$ through all layers, resulting in refined hidden states $\mathbf{h}_1, \dots, \mathbf{h}_{t-1}$. The hidden state $\mathbf{h}_{t-1}$ at the last position (which aggregates information from all tokens in the context) is used to predict a probability distribution over the vocabulary, from which the next token x_t is sampled:

$$x_t \sim P(\cdot \mid C_{t-1}) = P(\cdot \mid x_1, \dots, x_{t-1}), \tag{2.2}$$

where C_{t-1} represents the sequence of tokens generated thus far. After sampling x_t, the context is updated:

$$C_t = C_{t-1} \cup \{x_t\}. \tag{2.3}$$

The newly generated token x_t is then embedded and processed through all layers, resulting in its own refined hidden state $\mathbf{h}_t$ that will be used to predict the subsequent token x_{t+1}.

This iterative procedure continues until a termination condition is met, such as generating a special end-of-sequence token or reaching a maximum output length.

To illustrate this process, consider the following step-by-step example:

1. **Initial Context:** `["The", "car", "slowed"]`
2. **Model Predicts:** `"at"`
3. **Updated Context:** `["The", "car", "slowed", "at"]`
4. **Model Predicts:** `"the"`
5. **Updated Context:** `["The", "car", "slowed", "at", "the"]`
6. **Model Predicts:** `"intersection"`, and so on

It is important to note that although the generated sequence grows token-by-token, the transformer architecture does not maintain persistent internal state across generation steps. Conceptually, at each step, the model processes the entire context C_{t-1} through a full forward pass. However, practical implementations employ key-value caching (discussed in detail below) to avoid recomputing intermediate representations for previously processed tokens, making this process computationally efficient. This architectural statelessness distinguishes transformer-based decoding from earlier architectures like recurrent neural networks (RNNs) [10], which carried hidden state across time steps.

This stateless design simplifies architectural implementation and training. While the full forward pass through all layers is executed at each step, caching strategies mitigate the computational cost of processing long sequences.

2.3 The Rise of Transformers in Language Modeling

2.3.1 From N-grams to RNNs to Transformers

N-gram models [11], the predecessors to neural language models, predict the next word based on the frequency of word sequences in training data. For example, a bigram model predicts the next word given only the previous word, while a trigram model uses the previous two words. These models suffer from the curse of dimensionality and cannot capture long-range dependencies effectively. Prior to the introduction of transformers, most language models were built using RNNs [10] or their gated variants such as long short-term memory (LSTM) networks [12]. These models process input sequences sequentially, updating an internal hidden state token-by-token as they move through the input. Although RNNs marked a significant advancement over earlier n-gram models, they suffer from fundamental limitations.

One of the most severe limitations is their inability to capture long-range dependencies. This stems from the *vanishing gradient problem*, where gradients become exponentially smaller as they are backpropagated through many time steps, making it difficult for the model to learn long-range relationships. Although gated mechanisms like LSTMs and gated recurrent units (GRUs) [13] attempt to mitigate this problem by using specialized gates to control information flow, they still struggle to retain information across very long sequences. Furthermore, because RNNs are inherently sequential, they cannot fully leverage the parallel computation capabilities of modern hardware accelerators such as GPUs.

The transformer architecture, introduced by Vaswani et al. in the seminal paper *"Attention is All You Need"* (2017), addressed these limitations by replacing recurrence with self-attention [3]. This architectural shift led to dramatic improvements in performance and scalability across a wide range of sequence modeling tasks, including machine translation, text generation, and conversational AI.

2.3.2 Key Innovations in the Transformer Design

The core innovation of the transformer is its self-attention mechanism, which enables each token in a sequence to attend to every other token. Rather than compressing the entire input history into a single fixed-size vector, the transformer allows each token to compute its representation as a weighted combination of all other tokens. This allows for more flexible and contextually aware modeling, particularly for long and complex sequences.

Additionally, because the transformer processes all tokens simultaneously, it enables full parallelization during training. This characteristic not only reduces computation time but also facilitates the use of larger batch sizes and more expressive model architectures.

The original transformer architecture employs an encoder-decoder structure. The encoder uses unmasked self-attention, allowing each token to attend freely to all others in the input sequence. The decoder, by contrast, uses masked self-attention, ensuring that each position can only attend to previous positions and itself. This design enforces causality during output generation and prevents information leakage from future tokens.

Models focused purely on language generation, such as GPT, adopt only the decoder component of the transformer. They retain the masked self-attention mechanism originally used in the transformer decoder to ensure causal generation, while discarding the encoder. This design simplifies the architecture for tasks where generation proceeds strictly from left to right, such as text continuation, dialogue, and instruction following.

2.3.3 Decoder-only Transformers for Language Generation

The decoder-only architecture warrants further examination, as it forms the foundation of most modern large language models.

Modern autoregressive language models such as GPT are based on the decoder component of the original transformer architecture. These models discard the encoder entirely and utilize a stack of masked self-attention layers to perform both encoding of the input prompt and generation of new tokens.

It is important to clarify that *"decoder-only"* does not imply the absence of input processing. Rather, the same stack of masked transformer layers is responsible for both contextualizing the prompt and generating output tokens. This unified architecture builds context and produces subsequent tokens in an autoregressive manner, using left-to-right processing enforced by causal attention masking. In this setup, each token attends only to itself and preceding tokens (*unidirectional context*), future positions are masked during attention (*causal masking*), and no separate encoder is used: the prompt and generated sequence pass through the same shared layers.

The defining characteristics of GPT-style transformers include:

- **Generative:** Capable of free-form text generation
- **Pretrained:** Trained on large corpora using next-token prediction as the objective
- **Transformer-based:** Built on stacked decoder blocks with masked self-attention

This GPT-style transformer structure supports efficient, scalable language generation and underpins many of the most widely deployed large language models.

2.4 The LLM Generation Pipeline

Having established the historical context and architectural foundations of transformers, we now turn to a detailed examination of how these components work together during the generation process. From the model's perspective, generation is not a monolithic action but a pipeline involving transformation, refinement, and decision-making. The components form an integrated system where the output of one stage becomes the input to the next, progressively transforming raw text into contextualized representations and ultimately into probability distributions over the vocabulary. This pipeline perspective reveals how seemingly independent mechanisms (such as tokenization and attention) are actually interdependent, with each stage building upon the work of previous stages.

2.4.1 End-to-end Flow of GPT-style Transformers

Figure 2.1 illustrates the high-level flow of the generation pipeline in decoder-only transformers such as GPT. The process begins with tokenization and ends with probabilistic sampling. Each stage in this pipeline corresponds to a distinct architectural component, which we explore in detail in the following sections.

The stages depicted in Figure 2.1 are:

- **Tokenization and Embeddings:** Converting input text into numerical vectors that capture semantic meaning
- **Positional Encoding:** Injecting sequence order into token representations to enable position-aware processing
- **Self-attention Mechanism:** Modeling contextual relationships between tokens to create contextually aware representations

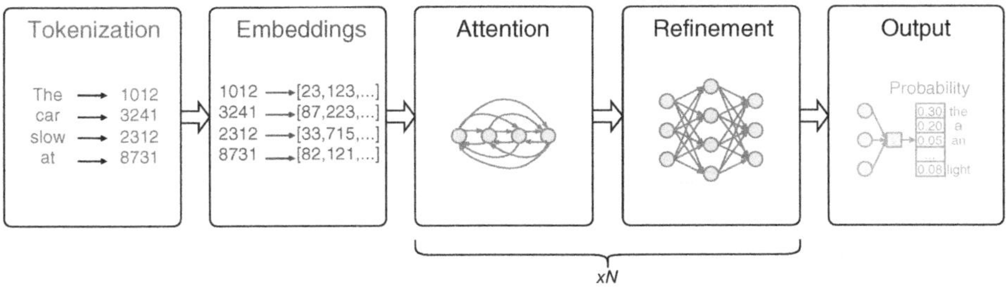

Figure 2.1 High-level view of the GPT generation pipeline.

- **Feedforward Refinement and Layer Stacking:** Deepening token representations through nonlinear transformations and multi-layer composition
- **Output Generation:** Predicting and sampling the next token in sequence based on the refined contextualized representations

2.4.2 Information Flow Through the Pipeline

To understand how these stages work together, consider the transformation of a simple input through the pipeline. Suppose we provide the prompt "The car slowed" to the model, which must predict the next token.

Stage 1: Tokenization and embeddings transforms the text string into a sequence of token indices (e.g., `["The", "car", "slowed"]`), then maps each token to a learned embedding vector. At this point, each token has a static semantic representation that captures its general meaning but lacks context or position information.

Stage 2: Positional encoding adds position information to each embedding. The token "car" now carries both its semantic meaning (from the embedding) and its position in the sequence (from positional encoding). This enables the model to distinguish between "The car slowed" and "Slowed the car," even though they contain the same tokens.

Stage 3: Self-attention processes all tokens simultaneously, allowing each token to attend to all others. The token "slowed" can now attend to "car" to understand that it refers to the vehicle's action, not some other entity. This creates contextually aware representations where each token's meaning is refined based on its relationships with other tokens in the sequence.

Stage 4: Feedforward refinement and layer stacking applies nonlinear transformations and processes the sequence through multiple layers. Each layer further refines the representations, with early layers capturing local patterns and deeper layers integrating information across the entire sequence. In the final layer, each token carries a rich representation that incorporates both its semantic content and its role within the full context.

Stage 5: Output generation takes the final representation of the last token ("slowed") and projects it into the vocabulary space, producing a probability distribution over all possible next tokens. The model might assign high probability to tokens like "at," "down," or "suddenly," reflecting its understanding that these are likely continuations given the context "The car slowed."

This pipeline operates not just during initial prompt processing, but also during autoregressive generation. When the model generates a new token (e.g., "at"), that token is fed back through the entire pipeline, allowing the model to maintain context and coherence across the generated sequence.

The following sections examine each stage in detail, explaining not just what each component does but how it contributes to the overall goal of generating fluent, contextually appropriate language.

2.5 Stage 1: Tokenization and Embeddings

2.5.1 From Text to Tokens

LLMs process input as sequences of discrete units known as *tokens*, rather than raw character streams or full words. Tokenization is the process of converting text into these fundamental units, which serve as the model's basic computational elements.

A token can be a complete word (e.g., `store`), a subword fragment (e.g., `un`, `bel`, `ievable`), or a punctuation mark or token with leading whitespace (e.g., `!`, `" the"`).

Subword tokenization strategies, such as Byte Pair Encoding (BPE) [14] and Unigram Language Models [15], are commonly used in modern LLMs. BPE iteratively merges the most frequent adjacent character pairs to create subword units, while unigram models learn a probabilistic subword vocabulary. These approaches provide a balance between vocabulary compactness and expressiveness, allowing the model to construct rare or novel words from more frequent subword components.

Formally, tokenization can be expressed as a function that maps a text string to a sequence of token indices:

$$\text{Tokenize}(s) = [t_1, t_2, \dots, t_n] \quad \text{where } t_i \in \{1, 2, \dots, |V|\}, \tag{2.4}$$

where s is the input string, V is the vocabulary of all possible tokens, and $|V|$ is the vocabulary size. Each token t_i is represented as an integer index corresponding to its position in the vocabulary. Table 2.2 provides illustrative examples of how input text is segmented into tokens under common subword tokenization schemes.

The choice of tokenization affects both generalization and efficiency. Since a single word may correspond to multiple subword tokens, the model learns to compose meaning from parts. In contrast, other words such as `"car"` may be represented as a single token. The segmentation behavior is governed by the tokenizer's vocabulary, frequency statistics, and training procedure.

2.5.2 Token Embeddings

Once tokenization is complete, each token is mapped to a fixed-length, high-dimensional vector through a process called *embedding*. This vector captures learned semantic properties of the token, acquired during pretraining across billions of words.

Let V denote the vocabulary of all tokens and d_{model} the embedding dimension. The embedding layer can be represented as a matrix $E \in \mathbb{R}^{|V| \times d_{\text{model}}}$, where each row corresponds to the vector

Table 2.2 Examples of tokenized text.

Input text	Token sequence
`"Hello world"`	`[Hello, _world]`
`"unbelievable"`	`[un, bel, ievable]`

Note: "_" represents a space character.

Table 2.3 Illustrative token embeddings (simplified).

Token	Token ID	Embedding vector
`"car"`	3201	[0.12, −0.45, ... , 0.78]
`"truck"`	8672	[0.15, −0.42, ... , 0.80]

representation of a token. The embedding function maps each token index to its corresponding vector:

$$\text{Embed}(t_i) = E[t_i] = \mathbf{e}_i \in \mathbb{R}^{d_{\text{model}}}. \tag{2.5}$$

Table 2.3 shows simplified token embedding vectors, where semantically related words occupy nearby positions in the learned vector space.

For a sequence of tokens $[t_1, t_2, \dots, t_n]$, the embedding layer produces a sequence of vectors:

$$[\mathbf{e}_1, \mathbf{e}_2, \dots, \mathbf{e}_n] \quad \text{where } \mathbf{e}_i = \text{Embed}(t_i) \in \mathbb{R}^{d_{\text{model}}}. \tag{2.6}$$

Semantically similar tokens are typically close to each other in this vector space. For example, embeddings for `"car"` and `"truck"` may differ only slightly, while `"car"` and `"banana"` are positioned far apart. These relationships are not explicitly programmed but emerge from the model's training objective of predicting tokens in context.

However, embeddings alone provide only static semantic information. To function as a reasoning engine, the model must understand how tokens relate to each other within a specific context. For instance, the word *charge* means something different in "charge the battery" versus "charge the suspect." The model needs a mechanism to dynamically adjust token representations based on surrounding context, enabling it to reason about relationships, resolve ambiguities, and make contextually appropriate decisions. This transition from static embeddings to dynamic contextual reasoning is achieved through positional encoding and attention mechanisms.

Having represented tokens semantically, the model must now encode their order in the sequence. While embeddings capture token meaning, they contain no information about sequence position. This limitation is addressed by positional encoding.

2.6 Stage 2: Positional Encoding

2.6.1 Why Position Must Be Explicit

The transformer architecture processes all tokens in parallel, unlike RNNs, which process tokens sequentially and encode position implicitly. This parallel design accelerates training and inference but creates a fundamental challenge: without additional signals, the model has no way to know the position of each token in the sequence.

To resolve this, transformers incorporate an additional input called the *positional encoding*, which is combined with each token embedding. This encoding allows the model to distinguish between, for instance, "The car passed the bus" and "The bus passed the car," even though the same tokens are used in both sentences.

2.6.2 Sinusoidal Position Encoding

The original transformer, as introduced by Vaswani et al., used deterministic sinusoidal functions to encode position. Unlike learnable positional embeddings, which are limited to positions encountered during training, sinusoidal functions can mathematically encode any position. For example, if a model is trained on sequences of maximum length 512 tokens, the sinusoidal functions can compute positional encodings for positions 1000 or 2000, even though these positions were never seen during training. However, performance typically degrades when processing sequences significantly longer than the training length.

For each position p and each dimension i of the positional encoding vector, the values are computed as:

$$\mathrm{PE}_{(p,2i)} = \sin\left(\frac{p}{10000^{2i/d_{\mathrm{model}}}}\right), \tag{2.7}$$

$$\mathrm{PE}_{(p,2i+1)} = \cos\left(\frac{p}{10000^{2i/d_{\mathrm{model}}}}\right), \tag{2.8}$$

where d_{model} is the dimensionality of the model's embeddings. This creates a set of sinusoidal waveforms of varying frequencies across dimensions, such that each position in the sequence has a unique and smooth encoding pattern. Furthermore, the use of sine and cosine makes it possible for the model to infer relative positions based on linear operations.

These encodings offer two key properties: they are **deterministic** and do not require learning, which enables extrapolation to positions beyond those seen during training; and they enable the model to reason about **relative position** because the difference between two positions can be represented as a linear function of their encodings. This capability to model long-range dependencies is essential for CAV applications, where multi-turn driver dialogues and connected-vehicle communication require understanding relationships between tokens separated by many intervening words.

2.6.3 Combined Input Representation

For sinusoidal positional encoding, the positional encodings are added elementwise to the token embeddings to produce the model's final input:

$$\text{Input Vector}_{(p)} = \text{Embedding}_{(p)} + \mathrm{PE}_{(p)}. \tag{2.9}$$

This combined vector, now containing both semantic content and positional information, serves as the input to the attention mechanism. With tokens represented as position-aware embeddings, the model can now contextualize each token by attending to others in the sequence.

While some later models have adopted learnable positional embeddings, the sinusoidal scheme remains widely studied due to its elegance, interpretability, and historical significance in early transformer models. However, sinusoidal positional encoding has limitations when processing sequences far beyond the training length. The model's performance often degrades as attention patterns learned during training may not generalize well to these unseen positions. This limitation led to the development of more robust alternatives like rotary position embedding (RoPE) [16], which offers better extrapolation capabilities and more stable attention patterns across long distances.

Advanced Note 2.1 Rotary Position Embedding

Modern large language models have largely adopted rotary position embedding (RoPE) [16] as the preferred positional encoding method. Unlike sinusoidal encoding, which adds position information to embeddings, RoPE applies rotation transformations to token embeddings based on their position in the sequence. For a token embedding $\mathbf{x} \in \mathbb{R}^{d_{\text{model}}}$ at position p, RoPE applies a rotation matrix $R_{\Theta,p}$ that transforms the embedding through dimension-pair rotations. The key advantage is that relative positions between tokens are encoded through relative rotations, making attention scores naturally depend on relative positions rather than absolute positions. This enables better extrapolation to sequences longer than training length and more stable attention patterns across long distances, allowing modern models to effectively utilize context windows of 32k+ tokens.

2.7 Stage 3: Attention and Contextualization

2.7.1 Why Attention Matters for Language and Reasoning

The attention mechanism is the central innovation that distinguishes the transformer architecture from earlier sequence models and enables it to function as a reasoning engine. Rather than relying on a fixed-size hidden state, attention enables each token to construct its own contextualized representation by attending to other tokens in the sequence. This dynamic allocation of attention allows the model to capture both local and long-range dependencies, making it capable of reasoning about relationships across the entire input context.

In CAV applications, this reasoning capability enables the model to connect disparate pieces of information: a driver's command, sensor readings, traffic conditions, and previous dialogue turns. The model can reason about which information is relevant to the current context, just as attention determines which tokens are relevant to understanding each word's meaning.

In natural language, the meaning of a word often depends on other words that may be arbitrarily distant. For example, in the sentence:

The driver who passed the bus slowed down near the intersection.

Here, the correct semantic linkage is between *driver* and *slowed*, not *bus* and *slowed*. Attention makes such distinctions possible by assigning higher relevance weights to contextually important tokens.

This capability is particularly important in LLMs, where preserving semantic coherence across long sequences is essential.

2.7.2 Self-attention

The self-attention mechanism computes a new representation for each token by weighting and aggregating information from all tokens in the sequence. This is achieved through three learned projections for each token:

- **Query (Q)**—captures what the token is looking for in the context.
- **Key (K)**—encodes the content each token makes available to others.
- **Value (V)**—provides the information that is actually aggregated if selected.

For CAV applications, a relevant analogy is *sensor fusion*, where the attention mechanism filters and combines information from multiple sources:

- **Query (Q)** represents the question: "What is the status of the traffic light?"
- **Key (K)** represents labels for all objects in the scene: car, light, pedestrian, tree, etc.
- **Value (V)** represents the actual data associated with those objects: the light's color and state, the car's position, etc.

The attention mechanism works by comparing each query to all keys to find the most relevant matches, then retrieving the corresponding values. Just as sensor fusion filters a scene to focus only on relevant objects (e.g., the traffic light), attention filters the input sequence to focus on contextually relevant tokens. The attention score between two tokens is computed as the similarity between the query of one token and the key of another. These scores are used to form a weighted sum of the value vectors, which produces a new, context-aware representation for the querying token. In CAV applications, this enables the model to disambiguate driver intent (e.g., distinguishing "turn on the AC" from "turn on the AC adapter") or reference map features by attending to relevant context across the input sequence.

Mathematically, this process begins by transforming the input embeddings into the three representations. Given the combined input vectors from the previous stage (embeddings plus positional encoding), we represent them as a matrix $X \in \mathbb{R}^{n \times d_{\text{model}}}$ where n is the sequence length and d_{model} is the embedding dimension. Each row $\mathbf{X}_p$ of X contains the embedding vector for the token at sequence position p. From this, we define:

$$Q = XW^Q, \tag{2.10}$$

$$K = XW^K, \tag{2.11}$$

$$V = XW^V, \tag{2.12}$$

where W^Q, W^K, and W^V are trainable weight matrices with dimensions $d_{\text{model}} \times d_k$, and d_k is typically chosen to ensure that the dot product scales appropriately.

The attention output for each token is then computed by comparing its query to all keys using scaled dot products:

$$\text{Attention}(Q, K, V) = \text{softmax}\left(\frac{QK^\top}{\sqrt{d_k}}\right)V. \tag{2.13}$$

This operation produces a weighted average over the value vectors, where the weights are determined by the similarity between queries and keys. The scaling factor $\sqrt{d_k}$ helps stabilize training by controlling the variance of the dot products.

Advanced Note 2.2 Key-value Caching for Efficient Inference

During autoregressive generation, the model processes an expanding sequence one token at a time. Since the context grows incrementally (only one new token is added per step), most key and value computations are redundant. Modern transformer implementations employ *key-value caching* to store computed key and value vectors for all previously processed tokens, avoiding recomputation when the context expands. At step t, the model computes only the new query $Q_t = X_t W^Q$ for the latest token and attends to all cached keys and values: $\text{Attention}(Q_t, K_{1:t}, V_{1:t}) = \text{softmax}(Q_t K_{1:t}^\top / \sqrt{d_k}) V_{1:t}$. Only K and V are cached because queries are used exactly once per step, while keys and values are reused in every future step. This optimization reduces attention computation complexity from $O(n^2)$ to $O(n)$ per generation step, making long-context generation computationally feasible.

2.7.3 Masked Self-attention for Autoregressive Generation

In decoder-only architectures, the model must generate output one token at a time without accessing future tokens. This is enforced during training and inference using *masked self-attention*, which prevents each token from attending to any subsequent tokens in the sequence.

Formally, before applying the softmax in the attention computation, a mask is applied to the matrix $QK^\top$ such that all elements corresponding to future positions are set to $-\infty$. After softmax, these positions receive zero probability, ensuring the model's predictions remain causally consistent.

As illustrated in Figure 2.2, the model enforces causality by masking attention to future tokens during both training and inference. Masked attention is what makes autoregressive language generation feasible. It guarantees that each output token is conditioned only on past context, mimicking the constraints faced during actual generation.

2.7.4 Multi-head Attention

A single attention mechanism might be sufficient to capture certain relationships, such as subject-verb agreement. However, natural language is multifaceted, and different types of dependencies, such as syntactic, semantic, and temporal, may coexist in a single sentence.

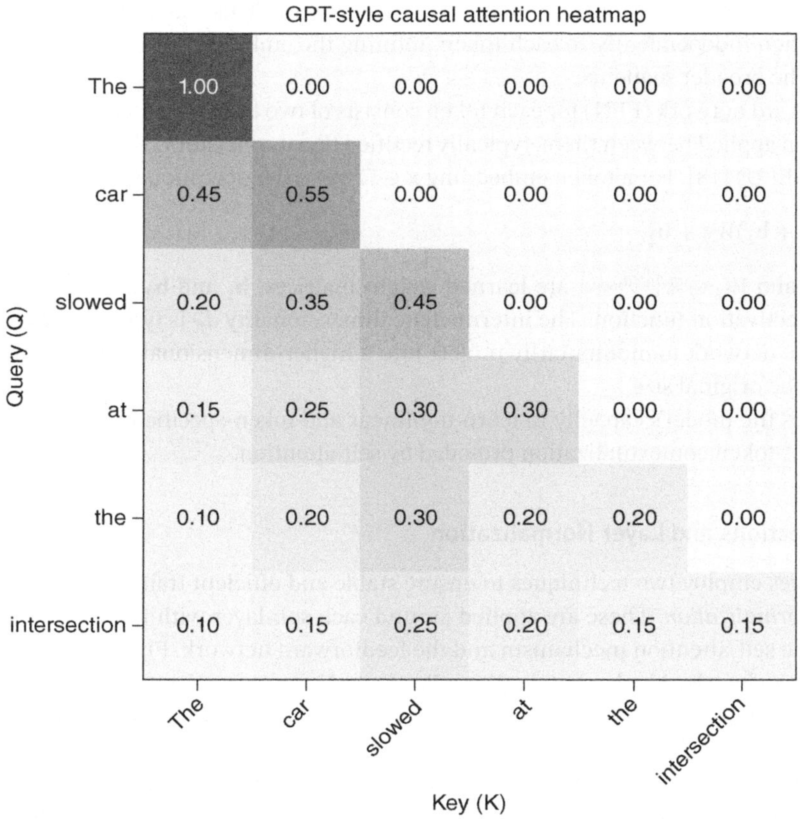

Figure 2.2 Masked self-attention ensures that each token attends only to itself and previous tokens.

To enhance representational capacity, the transformer uses *multi-head attention*. Instead of computing a single attention distribution, it computes multiple attention heads in parallel. Each head learns to focus on different aspects of the input. The outputs of these heads are then concatenated and passed through a linear projection to form the final representation.

For h heads, the outputs are computed as:

$$\text{head}_i = \text{Attention}(Q_i, K_i, V_i), \quad i = 1, \dots, h, \tag{2.14}$$

$$\text{MultiHead}(Q, K, V) = \text{Concat}(\text{head}_1, \dots, \text{head}_h)W^O, \tag{2.15}$$

where $W^O \in \mathbb{R}^{hd_k \times d_{\text{model}}}$ is a trainable projection matrix. In standard implementations, $d_k = d_{\text{model}}/h$ so that $hd_k = d_{\text{model}}$, ensuring the output dimension matches the input dimension. This allows the outputs from multiple heads to be merged into a unified vector of the original dimensionality d_{model}. By combining multiple attention patterns in this manner, the model develops a richer, more nuanced understanding of each token's contextual role.

2.8 Stage 4: Refinement and Stacking

2.8.1 Feedforward Transformation

Having contextualized each token through attention, the model now refines these representations through a token-wise transformation known as the feedforward network. This component applies a nonlinear transformation independently to each token, refining the embedding based on information extracted from the broader sequence.

Formally, the feedforward network (FFN) for each token consists of two linear layers with a nonlinear activation function applied between them, typically rectified linear unit (ReLU) [17] or Gaussian error linear unit (GELU) [18]. For a token embedding $\mathbf{x} \in \mathbb{R}^{d_{\text{model}}}$, the operation is defined as:

$$\text{FFN}(\mathbf{x}) = \sigma(\mathbf{x}W_1 + \mathbf{b}_1)W_2 + \mathbf{b}_2, \tag{2.16}$$

where $W_1 \in \mathbb{R}^{d_{\text{model}} \times d_{\text{ff}}}$ and $W_2 \in \mathbb{R}^{d_{\text{ff}} \times d_{\text{model}}}$ are learned weight matrices, $\mathbf{b}_1$ and $\mathbf{b}_2$ are bias vectors, and σ denotes the activation function. The intermediate dimensionality d_{ff} is typically larger than d_{model}, enabling the network to momentarily project into a higher-dimensional space before compressing back into the original size.

This module enhances the model's capacity to learn nonlinear and token-specific refinements, complementing the inter-token contextualization provided by self-attention.

2.8.2 Residual Connections and Layer Normalization

Transformer architectures employ two techniques to ensure stable and efficient training: *residual connections* and *layer normalization*. These are applied around each sub-layer within the decoder block, including both the self-attention mechanism and the feedforward network. Figure 2.3 illustrates the structure of this decoder block, showing how Pre-LayerNorm normalization is applied before each sub-layer, with residual connections preserving the input signal throughout the transformation.

In GPT-style transformers, which follow a decoder-only structure, the dominant design pattern uses **Pre-LayerNorm** (Pre-LN), where normalization is applied before each sub-layer. This is in contrast to the original transformer formulation, which used post-layer normalization. The Pre-LN configuration improves gradient flow and stabilizes training in very deep networks.

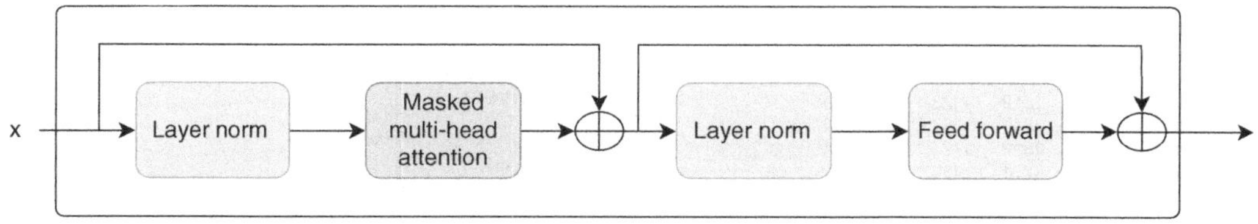

Figure 2.3 Decoder block in GPT-style transformer with Pre-LayerNorm: each sub-layer receives a normalized input.

In the Pre-LN variant, each sub-layer follows the structure:

$$\mathbf{x}_{\text{out}} = \mathbf{x} + \text{SubLayer}(\text{LayerNorm}(\mathbf{x})), \tag{2.17}$$

where `SubLayer` may refer to either masked multi-head attention or the feedforward network. The residual connection adds the original input back to the sub-layer output, preserving signal and aiding optimization.

Layer normalization itself is defined as:

$$\text{LayerNorm}(\mathbf{x}) = \frac{\mathbf{x} - \mu}{\sqrt{\sigma^2 + \epsilon}} \cdot \gamma + \beta, \tag{2.18}$$

where μ and σ^2 denote the mean and variance across the feature dimension, and γ, β are learned scale and shift parameter vectors.

This architecture ensures that each layer receives well-scaled inputs, mitigates covariate shift, and enables the training of models with hundreds of layers.

2.8.3 Why Stacking Improves Expressiveness

The true expressive capacity of the transformer architecture arises not from any individual layer but from the compositional effect of stacking multiple identical decoder blocks. Each block contains self-attention, feedforward transformation, residual connections, and layer normalization. In GPT-style transformers, the number of such layers ranges from several dozen to more than one hundred.

To understand how stacking works, it is helpful to trace how token representations evolve through the layers. Each token begins as a fixed embedding from the lookup table—a static semantic representation that encodes the token's general meaning independent of context. However, as the token passes through successive layers, its representation is progressively refined and contextualized. At each layer, the token's embedding is updated based on its interactions with other tokens in the sequence through attention mechanisms, and then refined through feedforward transformations. This process continues layer-by-layer: early layers capture local syntactic patterns and morphological relationships, while deeper layers integrate information from across the entire sequence, building rich contextual representations that reflect the token's role and meaning within the full context.

Each successive layer builds upon the output of the previous one, incrementally refining the token representations. This hierarchical composition enables the model to move from surface-level patterns to deeper semantic and syntactic abstractions. Empirical studies have shown that early layers tend to capture morphological and local syntactic dependencies, while deeper layers are responsible for modeling long-range context, discourse structure, and task-specific reasoning.

Figure 2.4 illustrates this multi-layer refinement process, emphasizing the repeated application of attention and feedforward layers across multiple stacked transformer blocks.

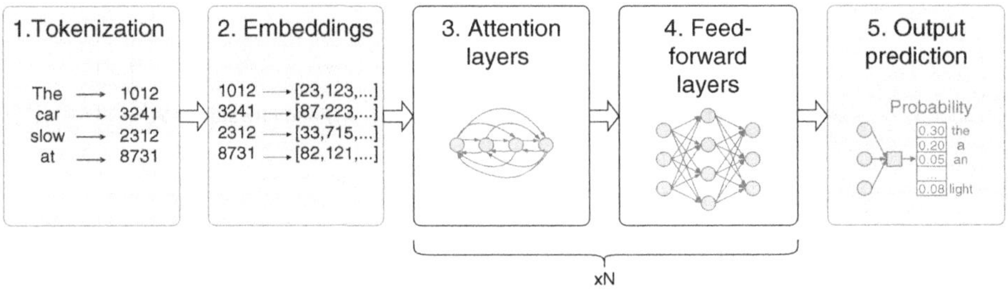

Figure 2.4 Layer stacking in transformer-based language models. Each token passes through multiple rounds of attention and feedforward processing (*xN*) to progressively refine its representation. Early layers capture syntax, while deeper layers encode semantics and logic.

This progression can be conceptually visualized as a transformation pipeline:

$$\mathbf{h}^{(0)} \rightarrow \texttt{[Layer 1]} \rightarrow \texttt{[Layer 2]} \rightarrow \cdots \rightarrow \texttt{[Layer N]} \rightarrow \mathbf{h}^{(N)},$$

where $\mathbf{h}^{(0)}$ represents the initial token embeddings and $\mathbf{h}^{(N)}$ represents the final contextualized hidden states after processing through all N layers.

At each stage, contextual interactions (via attention) and intra-token refinement (via feedforward networks) are applied, enabling the model to build ever richer representations. By the time a token reaches the final layer, it carries a deeply contextualized encoding that reflects its meaning and role across the entire sequence. This final representation $\mathbf{h}^{(N)}$ contains information aggregated from all previous layers and all tokens in the context, making it a rich, context-aware encoding suitable for downstream tasks such as next-token prediction.

2.9 Stage 5: Output and Sampling

Having processed tokens through multiple layers of attention and refinement, the model must now convert these contextualized representations into the next token prediction. This final stage transforms hidden states into probability distributions over the vocabulary.

2.9.1 From Hidden State to Token Probability

At the final stage of a transformer decoder, each token position t is associated with a refined hidden state vector $\mathbf{h}_t \in \mathbb{R}^{d_{\text{model}}}$. This vector represents the final layer's output $\mathbf{h}_t^{(N)}$ for position t, which has been progressively refined through all N layers. It encodes contextual information about the sequence prefix $x_{\leq t}$, accumulated through masked self-attention across all previous positions. The enhanced embedding $\mathbf{h}_t$ implicitly contains useful information aggregated from all tokens in the context, making it a rich representation for prediction.

However, the model's objective is not to generate hidden states, but to produce the next token x_{t+1} by computing a probability distribution over the vocabulary. To accomplish this, the model projects the final hidden state $\mathbf{h}_t$ into the vocabulary space using a learned output matrix $W_{\text{out}} \in \mathbb{R}^{d_{\text{model}} \times |V|}$ and bias vector $\mathbf{b} \in \mathbb{R}^{|V|}$. The resulting logits $\mathbf{z}_t \in \mathbb{R}^{|V|}$ with the vocabulary V are computed as:

$$\mathbf{z}_t = \mathbf{h}_t W_{\text{out}} + \mathbf{b}, \tag{2.19}$$

where each element $z_{t,i}$ represents an unnormalized score for vocabulary token v_i. These logits are converted into probabilities using the softmax function:

$$P(x_{t+1} = v_i \mid x_{\leq t}) = \frac{\exp(z_{t,i})}{\sum_{j=1}^{|V|} \exp(z_{t,j})}. \tag{2.20}$$

The resulting probability distribution expresses the model's estimate of which token is most likely to follow the current sequence prefix. This distribution reflects the contextualized knowledge accumulated in $\mathbf{h}_t$ through all layers: the initial embedding from the lookup table, the contextual relationships learned via attention, and the semantic refinements from feedforward networks. During inference, we sample from this probability distribution (not from $\mathbf{h}_t$ directly) to select the next token x_{t+1}. Once generated, x_{t+1} will itself be embedded and processed through all layers to obtain its own refined representation $\mathbf{h}_{t+1}$, which will be used to predict the subsequent token.

2.9.2 Sampling Methods

Once the model has computed a probability distribution over the vocabulary, it must select the next token to generate. This process, known as *sampling*, determines the character and diversity of the model's output. Several sampling strategies are available, each with distinct trade-offs between determinism, coherence, and creativity.

2.9.2.1 Greedy Decoding

The most straightforward strategy is greedy decoding, in which the model selects the token with the highest predicted probability:

$$x_{t+1} = \arg\max_{v \in V} P(x_{t+1} = v \mid x_{\leq t}). \tag{2.21}$$

This decision is made based on the probability distribution generated from the final hidden state $\mathbf{h}_t$, which encodes all preceding tokens up to position t. Although this approach produces deterministic outputs, it often yields repetitive or overly generic completions.

2.9.2.2 Top-k Sampling

Top-k sampling restricts the candidate pool to the k most probable tokens and renormalizes the distribution within this subset. This limits low-quality options while preserving diversity. The parameter k governs the size of the sampling space.

2.9.2.3 Top-p (Nucleus) Sampling

Top-p sampling selects the smallest set of tokens whose cumulative probability mass exceeds a specified threshold p (typically between 0.8 and 0.95). This dynamic approach adapts to the model's confidence and is particularly effective in open-ended generation.

2.9.2.4 Temperature Scaling

Temperature scaling adjusts the sharpness of the probability distribution prior to sampling by applying a temperature parameter $T > 0$ to the logits: $P_i = \exp(z_i/T)/\sum_j \exp(z_j/T)$. Lower temperatures ($T < 1$) sharpen the distribution for more focused outputs, while higher temperatures ($T > 1$) flatten it for increased diversity. The choice of sampling strategy should reflect the requirements of the application. Deterministic decoding is preferred in tasks requiring consistency and accuracy, such as formal report generation or code synthesis. In contrast, stochastic sampling techniques enhance fluency and originality in open-domain tasks such as storytelling or dialogue systems.

2.10 Model Limitations and Comparative Landscape

While the transformer architecture and its components enable powerful language generation capabilities, it is important to understand their inherent limitations. This section examines these constraints, explains their architectural roots, and provides a framework for evaluating and selecting models based on application requirements. Understanding these limitations is essential for designing effective LLM-based systems, as discussed in the context of connected and autonomous vehicles.

2.10.1 Limitations of Autoregressive LLMs

Despite their impressive generative capabilities, large language models based on autoregressive transformers exhibit several foundational limitations: the tendency to *hallucinate* (generate plausible but factually incorrect or unsupported content), the finite *context window*, and the absence of *persistent state* during inference. These constraints arise from fundamental architectural choices made in the transformer design, as discussed in Sections 2.2.4 and 2.6. Understanding why these limitations exist and how they can be mitigated is crucial for effective system design.

2.10.1.1 Hallucination

LLMs can generate plausible-sounding but factually incorrect or unsupported content, a phenomenon known as *hallucination*. This occurs because models are trained to predict the next token based on statistical patterns in training data, without direct access to ground truth or real-time information. In safety-critical applications such as CAVs, hallucinations can lead to incorrect explanations of vehicle behavior, false information about traffic conditions, or misleading responses to driver queries. The model may confidently generate information that appears correct but lacks factual grounding, particularly when asked about topics outside its training data or when required to reason about real-time sensor data. Hallucination and related safety concerns, including detection methods and mitigation strategies, are examined in detail in Chapter 22.

2.10.1.2 Fixed Context Window

Each LLM has a maximum number of tokens it can process in a single input sequence. This limit is fundamentally constrained by the architecture's positional embedding configuration and training setup, as introduced in Section 2.6. The context window limitation stems from several architectural factors. First, positional encoding constraints require that encodings be computed for every position in the sequence. As discussed in Section 2.6, for sinusoidal encodings, while the mathematical functions can theoretically encode any position, the model's learned attention patterns are optimized for the training sequence length. For learnable positional embeddings, the embedding matrix has a fixed size determined during training. Second, computational complexity presents a fundamental barrier: self-attention mechanisms, as described in Section 2.7, have quadratic complexity with respect to sequence length, meaning that processing longer sequences requires significantly more computational resources, both in terms of memory and computation time. Third, training data distribution shapes model behavior: models are trained on sequences of specific lengths, and their learned representations are optimized for these lengths. Extending beyond training length often leads to performance degradation, even when mathematically possible.

For instance, large proprietary models may support context windows ranging from 128k to over 1M tokens, while open-source models typically support 8k–128k tokens depending on their architecture and training configuration. Recent architectural innovations have enabled some models to process sequences exceeding 1M tokens, though computational costs increase significantly with longer contexts. When input sequences exceed this limit, the model must truncate or omit earlier tokens, leading to potential loss of relevant context. This limitation is particularly consequential in tasks such as legal document analysis, technical report summarization, or extended human-machine dialogue, where consistency across long spans of text is essential. In in-vehicle assistants, the fixed context window constrains how much conversation history can be maintained, requiring careful management of dialogue context to preserve important earlier exchanges.

2.10.1.3 Stateless Generation

As introduced in Section 2.2.4, autoregressive LLMs perform inference using a single forward pass and do not maintain memory across invocations. This stateless design was a deliberate architectural choice that offers significant advantages. The stateless architecture enables parallel training across sequences, as each forward pass is independent, which contrasts with RNNs that require sequential processing and cannot be easily parallelized. This training efficiency was crucial for scaling transformer models to the massive sizes we see today. Additionally, computational simplicity results from each generation step being a self-contained computation, making the architecture easier to implement, debug, and optimize. Finally, scalability is enhanced by the absence of persistent state, which allows models to scale to massive sizes without managing complex state transitions across time steps.

However, this design choice comes with trade-offs. Any conversational or sequential context must be explicitly re-included in each new prompt. The model does not carry internal state from one step to the next unless external memory mechanisms such as history buffers or retrieval-augmented generation (RAG, discussed in detail in Chapter 8) are employed. This statelessness creates real-world constraints for in-vehicle assistants: each driver command must include sufficient context, or the system must maintain external conversation history buffers to enable natural multi-turn interactions. For example, in a multi-turn in-vehicle dialogue, a user might say, "Set it to 70," referring to a previous command like "Set the temperature." If the earlier utterance is not included in the current prompt, the model has no means of resolving the reference to "it," potentially resulting in incorrect or ambiguous behavior.

2.10.1.4 Mitigation Strategies

These architectural constraints underscore the importance of complementary mechanisms to improve robustness and continuity in language-driven systems. Several mitigation strategies have emerged to address these limitations, falling into two broad categories: *external augmentation* and *architectural extensions*.

External augmentation strategies work around limitations by supplementing the model with additional systems. Retrieval-augmented generation (RAG, covered in detail in Chapter 8) enables systems to query external knowledge bases to retrieve relevant context that exceeds the model's context window, then incorporate this information into the prompt. External memory modules allow systems to maintain conversation history or task-specific context in external storage, selectively including relevant portions in each prompt. Prompt engineering (discussed in Chapter 7) can maximize the utility of the available context window by prioritizing the most relevant information through careful prompt design.

Architectural extensions modify the model itself to better handle long contexts. As discussed in Section 2.6, improved positional encoding schemes such as rotary position embedding (RoPE) enable better extrapolation to sequences longer than training length. Efficient attention mechanisms, including sparse attention patterns and hierarchical processing, can reduce the computational cost of processing long sequences. These architectural innovations can extend effective context windows, though often with trade-offs in computational cost or attention quality.

These mitigation strategies are essential for deploying LLMs in production systems, particularly in applications requiring long-term memory or multistep reasoning. Understanding these limitations and their mitigation strategies is crucial for making informed decisions when selecting models for specific applications.

2.10.2 Model Selection Framework

The landscape of large language models includes a wide variety of architectures differing in scale, training methodology, context length, and performance characteristics. Rather than simply listing model specifications, this section provides a framework for evaluating and selecting models based on application requirements, architectural trade-offs, and deployment constraints.

2.10.2.1 Architectural Patterns and Trade-offs

Modern LLMs employ several architectural patterns that affect their characteristics. Dense architectures use all parameters for every input token, offering predictable behavior and easier optimization, but requiring full computational resources for all inputs. Decoder-only autoregressive models represent a prominent example of this pattern. Mixture-of-experts (MoE) architectures [19] activate only a subset of expert networks per forward pass, reducing computational cost while maintaining large model capacity. Typical MoE configurations activate two of eight experts per forward pass, effectively using a fraction of total parameters during inference, enabling larger models with manageable inference costs. Training methodologies also vary: standard pretraining focuses on next-token prediction, while reinforcement learning from human feedback (RLHF) [20] and other alignment techniques improve safety and helpfulness.

Representative examples of modern LLMs include proprietary models (GPT, Claude, Gemini families) with large context windows and multimodal capabilities, and open-source models (LLaMA, Mixtral, DeepSeek families) that enable customization and fine-tuning.

2.10.2.2 Selection Criteria

When choosing an LLM for a specific application, practitioners should consider several key dimensions: context requirements (long documents need larger context windows), deployment constraints (edge vs. cloud), licensing and customization needs (open-source vs. proprietary), and cost considerations (model size affects inference cost). Model evaluation methodologies and selection frameworks are discussed in Chapter 21.

2.10.3 Model Selection for CAV Applications

These general principles apply across domains, but safety-critical applications such as connected and autonomous vehicles impose additional constraints that require specialized consideration.

In the context of connected and autonomous vehicles, model selection requires careful consideration of multiple competing constraints: real-time performance, accuracy, safety, cost, and deployment environment. This subsection provides a decision framework and concrete example to guide model selection for CAV applications.

Consider an in-vehicle voice assistant that must handle driver commands, answer questions about vehicle status, and provide navigation assistance. The system must operate reliably even when network connectivity is unavailable.

The requirements include latency under 500 ms for voice responses, multi-turn dialogue support, offline operation capability, correct handling of safety-critical commands, and cost-effective fleet deployment. The solution employs a hybrid approach: a primary on-device model (e.g., open-source models with approximately 8B parameters) for fast, reliable edge inference, and a cloud-based fallback (e.g., large proprietary models) for complex queries requiring large context or multimodal understanding. An external conversation history buffer maintains dialogue context beyond the model's context window, selectively including relevant history in each prompt.

This case study illustrates key trade-offs: edge deployment offers reliability and low latency but limits capability, while cloud deployment provides superior performance but introduces latency and connectivity dependencies. The hybrid approach balances these constraints, demonstrating how model selection criteria—capability, cost, latency, and reliability—apply to safety-critical CAV applications.

2.11 Summary

This chapter examined the internal structure and functional principles of large language models, focusing on the transformer architecture that underpins modern LLM systems. We began by establishing the fundamental challenge of contextual understanding in language processing and how the self-attention mechanism addresses this challenge. The chapter traced the historical evolution from n-gram models through RNNs to the transformer architecture, highlighting how architectural innovations enabled the scaling of language models to unprecedented sizes.

The chapter introduced the LLM generation pipeline as a framework for understanding how these models process and generate text. This pipeline consists of five stages: tokenization and embeddings, positional encoding, attention and contextualization, refinement and stacking, and output generation. Each stage transforms the representation progressively, from raw text to rich, contextually grounded token predictions. The detailed examination of each stage revealed how seemingly independent mechanisms work together to enable sophisticated language understanding and generation.

While LLMs are powerful tools for natural language processing, they are also constrained by fundamental limitations: the tendency to hallucinate, finite context windows (ranging from 8k to over 1M tokens depending on the model), and stateless inference. These limitations arise from architectural choices that enable scalability and efficiency but create challenges for applications requiring long-term memory or multistep reasoning. The chapter discussed mitigation strategies falling into two categories: external augmentation (RAG, external memory, prompt engineering) and architectural extensions (improved positional encodings, efficient attention mechanisms). A model selection framework was provided to guide practitioners in choosing appropriate models based on application requirements, deployment constraints, and cost considerations.

These limitations and mitigation strategies are especially relevant in high-stakes applications such as connected and autonomous vehicles, where contextual understanding, continuity, and reliability are essential for safe interaction. Understanding both the internals of LLM architecture and their inherent limitations is crucial for designing effective language-driven systems in CAV applications.

References

1 Tomas Mikolov, Kai Chen, Greg Corrado, and Jeffrey Dean. Efficient estimation of word representations in vector space. In *International Conference on Learning Representations*, 2013.

2 Jeffrey Pennington, Richard Socher, and Christopher D. Manning. Glove: global vectors for word representation. In *Proceedings of the 2014 Conference on Empirical Methods in Natural Language Processing (EMNLP)*, pages 1532–1543, 2014.

3 Ashish Vaswani, Noam Shazeer, Niki Parmar, Jakob Uszkoreit, Llion Jones, Aidan N. Gomez, Łukasz Kaiser, and Illia Polosukhin. Attention is all you need. In *Advances in Neural Information Processing Systems*, 2017. https://papers.nips.cc/paper_files/paper/2017/file/3f5ee243547dee91fbd053c1c4a845aa-Paper.pdf.

4 Alec Radford, Jeffrey Wu, Rewon Child, David Luan, Dario Amodei, and Ilya Sutskever. Language models are unsupervised multitask learners, 2019.

5 Tom Brown, Benjamin Mann, Nick Ryder, Melanie Subbiah, Jared D. Kaplan, Prafulla Dhariwal, Arvind Neelakantan, Pranav Shyam, Girish Sastry, Amanda Askell, et al. Language models are few-shot learners. *Advances in Neural Information Processing Systems*, 33:1877–1901, 2020.

6 Anthropic. Claude 2, 2023. https://www.anthropic.com/news/claude-2.

7 Google DeepMind. Gemini 1.5: unlocking multimodal understanding across millions of tokens of context, 2024. https://deepmind.google/technologies/gemini/.

8 Hugo Touvron, Thibaut Lavril, Gautier Izacard, Xavier Martinet, Marie-Anne Lachaux, Timothée Lacroix, Baptiste Rozière, Naman Goyal, Eric Hambro, Faisal Azhar, et al. LLaMA: open and efficient foundation language models. *arXiv preprint arXiv:2302.13971*, 2023.

9 Zhihong Shao, Damai Dai, Daya Guo, Bo Liu, and Zihan Wang. Deepseek-v2: a strong, economical, and efficient mixture-of-experts language model. *arXiv preprint arXiv:2405.04434*, 2024.

10 David E. Rumelhart, Geoffrey E. Hinton, and Ronald J. Williams. Learning representations by back-propagating errors. *Nature*, 323(6088):533–536, 1986.

11 Claude E. Shannon. A mathematical theory of communication. *The Bell System Technical Journal*, 27(3):379–423, 1948.

12 Sepp Hochreiter and Jürgen Schmidhuber. Long short-term memory. *Neural Computation*, 9(8):1735–1780, 1997.

13 Kyunghyun Cho, Bart Van Merriënboer, Caglar Gulcehre, Dzmitry Bahdanau, Fethi Bougares, Holger Schwenk, and Yoshua Bengio. Learning phrase representations using RNN encoder-decoder for statistical machine translation. *arXiv preprint arXiv:1406.1078*, 2014.

14 Rico Sennrich, Barry Haddow, and Alexandra Birch. Neural machine translation of rare words with subword units. In *Proceedings of the 54th Annual Meeting of the Association for Computational Linguistics (Volume 1: Long Papers)*, pages 1715–1725, 2016.

15 Taku Kudo. Subword regularization: improving neural network translation models with multiple subword candidates. In *Proceedings of the 56th Annual Meeting of the Association for Computational Linguistics (Volume 1: Long Papers)*, pages 66–75, 2018.

16 Jianlin Su, Murtadha Ahmed, Yu Lu, Shengfeng Pan, Wen Bo, and Yunfeng Liu. Roformer: enhanced transformer with rotary position embedding. Neurocomputing, 568, 2024.

17 Vinod Nair and Geoffrey E. Hinton. Rectified linear units improve restricted Boltzmann machines. In *Proceedings of the 27th International Conference on Machine Learning (ICML-10)*, pages 807–814, 2010.

18 Dan Hendrycks and Kevin Gimpel. Gaussian error linear units (GELUs). *arXiv preprint arXiv:1606.08415*, 2016.

19 Noam Shazeer, Azalia Mirhoseini, Krzysztof Maziarz, Andy Davis, Quoc Le, Geoffrey Hinton, and Jeff Dean. Outrageously large neural networks: the sparsely-gated mixture-of-experts layer. *arXiv preprint arXiv:1701.06538*, 2017.

20 Long Ouyang, Jeff Wu, Xu Jiang, Diogo Almeida, Carroll L. Wainwright, Pamela Mishkin, Chong Zhang, Sandhini Agarwal, Katarina Slama, Alex Ray, et al. Training language models to follow instructions with human feedback. In *Proceedings of the 36th International Conference on Neural Information Processing Systems*, NIPS '22, Red Hook, NY, USA, 2022.

Part II

Multimodal Representation and World Grounding

Autonomous driving requires a coherent internal representation of the physical world that integrates perception, semantics, and temporal context. This part focuses on generative models that construct such representations from raw sensory data, enabling vehicles to move beyond reactive perception toward semantically grounded understanding. The chapters cover diffusion and flow-matching methods, adversarial and variational generative methods, and vision–language and video–language models, with particular emphasis on contrastive representation learning as a mechanism for aligning heterogeneous modalities. Collectively, these models form the representational foundation that supports reasoning, planning, and interaction in intelligent connected and autonomous vehicle systems.

3

Diffusion and Flow Matching

3.1 Introduction: Why Diffusion and Flow Matching for CAV Systems

The operational landscape of connected and autonomous vehicles (CAVs) increasingly demands the ability to anticipate, interpret, and respond to a wide spectrum of environmental conditions. Traditional data collection pipelines, even when supported by large-scale fleets, remain constrained by the rarity of safety-critical events and the practical difficulty of capturing diverse combinations of weather, illumination, and traffic behaviors. These limitations have motivated generative approaches that can synthesize realistic, controllable driving data on demand rather than waiting for rare conditions to occur naturally.

Among these approaches, diffusion models and flow matching have emerged as a compelling response [1], providing a principled means to generate synthetic sensor data, construct diverse driving scenarios, and produce multimodal trajectory predictions for planning and prediction modules. Beyond image synthesis, these models have become foundational tools for high-fidelity generative modeling, enabling applications that require realistic sensor simulation, scenario generation, and predictive modeling under diverse operating conditions. Their ability to balance generation quality with controllability makes them especially well suited to the needs of CAV systems.

3.1.1 Motivating Scenario

Consider an autonomous vehicle approaching a curved segment of a rural road during early morning hours. Dense fog reduces visibility, moisture accumulates on the windshield, and intermittent sun glare appears as the vehicle rounds the bend. Although each factor is common in isolation, their joint occurrence is difficult to capture comprehensively in real-world datasets. Furthermore, such combinations often contribute to perception uncertainty and downstream decision-making errors. A development team may wish to construct a targeted set of sensor sequences that reproduce these interactions in a controlled manner to diagnose perception failures, improve robustness, or test new planning strategies.

These models provide a mechanism to generate synthetic sensor observations that faithfully reflect these multimodal conditions. By learning the structure of complex visual and geometric data, they can generate or modify scenes while maintaining high levels of realism. This capability supports scenario construction and safety evaluation under conditions that might otherwise be inaccessible or occur too infrequently in naturalistic driving to support systematic experimentation.

Conceptually, diffusion models work by *slowly removing fog from a sensor image*: the forward process gradually corrupts clean data by adding noise, analogous to fog obscuring a camera view, and the reverse process learns to remove that noise step-by-step to recover the original scene. This iterative denoising process makes them particularly well-suited for CAV tasks where high-fidelity generation is essential.

3.1.2 Why Diffusion Specifically

These models excel at high-fidelity synthesis of complex visual and geometric patterns, making them particularly effective for CAV sensor data generation. Their iterative refinement process enables fine-grained control over generation, supports conditional generation on structured inputs such as semantic maps and text descriptions, and captures multimodal uncertainty essential for diverse scenario construction. Traditional diffusion models require many denoising steps, typically 50–1000, making them ideal for applications where quality takes precedence over speed, such as sensor enhancement, trajectory generation, behavior modeling, scene completion, and simulation. Recent advances in flow matching, including rectified flows, learn deterministic paths that require fewer steps, typically 10–50, enabling faster generation for large-scale CAV applications.

While simulation remains a central use case, the mathematical tools developed in this chapter extend to many components of modern CAV pipelines, including perception recovery, multimodal prediction, planning under uncertainty, and controllable generative scene modeling.

3.2 Conceptual Foundations

3.2.1 Distribution Transformation

All generative models that transform noise into data share a fundamental unifying concept: *transforming one probability distribution into another*. This core principle, first formalized by Sohl-Dickstein et al. [2], underlies modern diffusion-based approaches, which learn to reverse a noise-adding process, and flow matching, which learns direct paths from noise to data. The starting distribution is typically Gaussian noise, which is simple, well-understood, and easy to sample, while the target distribution is a complex data distribution such as images, sensor data, or driving scenes. The transformation mechanism is a learned process that gradually moves samples from noise to data.

The techniques differ primarily in three dimensions: the transformation mechanism, which may use stochastic random steps or deterministic direct paths; the number of steps required, ranging from 1000+ steps to 1–4 steps; and the learning approach, which may involve noise prediction that learns to reverse a corruption process, velocity fields that learn direct transport paths, or direct mappings that learn one-step transformations. Figure 3.1 illustrates this unifying concept.

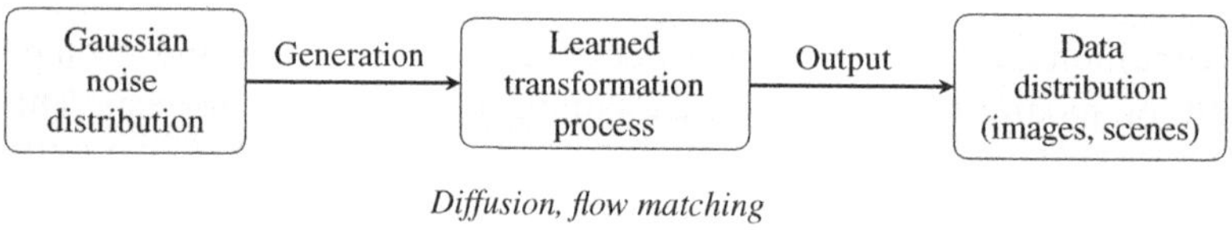

Figure 3.1 Unifying concept of generative models that transform noise into data through a learned transformation process.

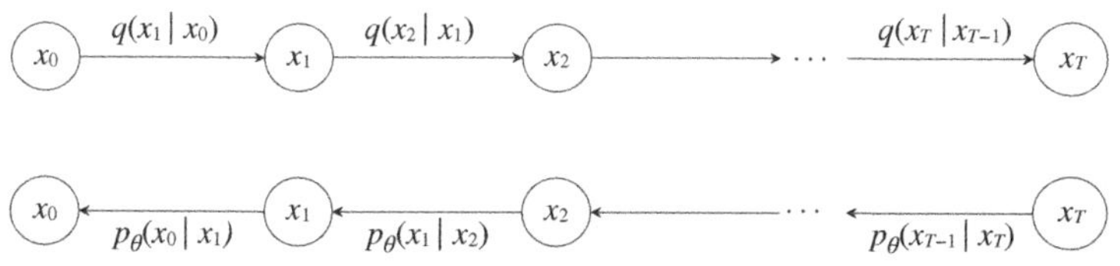

Figure 3.2 Forward diffusion process (top) and learned reverse denoising process (bottom).

3.2.2 Diffusion-denoising Paradigm

The fundamental idea behind diffusion models is to decompose the generative process into two complementary components: a *forward diffusion process* and a *reverse denoising process*. The forward process incrementally injects noise into a data sample until the distribution approaches simple, isotropic noise. In contrast, the reverse process aims to reconstruct the original data by iteratively removing noise. This paradigm was introduced by Sohl-Dickstein et al. [2], who established the theoretical foundation for learning data distributions through a gradual corruption and denoising process. The reverse steps are not analytically tractable and must be learned from data, typically through a neural network that predicts either the denoised sample or the noise component at each timestep.

The overall structure of the forward and reverse Markov chains is summarized in Figure 3.2. This visualization emphasizes that diffusion models do not learn a single global mapping from noise to data but instead learn local denoising transitions that, when composed, traverse the data manifold.

3.3 Mathematical Foundations

Throughout this chapter, we explicitly note when notation conventions differ to maintain clarity across frameworks. The following subsections present the mathematical foundations of diffusion models and flow matching separately, allowing each approach to be understood independently.

3.3.1 Diffusion Mathematics

Diffusion models implement stochastic reverse-time transport. The forward process gradually corrupts clean data by adding noise, and the reverse process learns to remove that noise step-by-step to recover the original scene.

Notation: In diffusion models, x_0 denotes clean data and x_T denotes noise. The reverse process (generation) goes from x_T (noise) to x_0 (data). In the unified framework: $x_T \leftrightarrow p_0$ (noise), $x_0 \leftrightarrow p_1$ (data). Note that this notation convention differs from flow matching (Section 3.3.2), where x_0 denotes noise and x_1 denotes data, reflecting the different mathematical formulations of these approaches.

3.3.1.1 Forward Process: Destroying Structure

The forward process gradually corrupts a clean data sample x_0 (e.g., an image or LiDAR frame) by adding Gaussian noise at each step [3]. At timestep $t = 0$, we have the clean image; at timestep $t = T$, we have pure noise. The key insight is that we can compute the noisy version at any timestep t directly from the original data, without simulating all intermediate steps:

$$x_t = \sqrt{\bar{\alpha}_t}\, x_0 + \sqrt{1 - \bar{\alpha}_t}\, \epsilon, \tag{3.1}$$

This equation mixes the original data x_0 with random noise $\epsilon \sim \mathcal{N}(0, I)$ to produce the noisy version x_t, where $\bar{\alpha}_t$ controls how much of the original signal remains versus how much noise has been added. The parameter $\bar{\alpha}_t = \prod_{s=1}^{t} \alpha_s$ is the cumulative product of step-wise coefficients $\alpha_s = 1 - \beta_s$, where β_s is the noise variance added at step s. As t increases, $\bar{\alpha}_t$ decreases (moving from 1 toward 0), resulting in more noise and less original signal. When $t = T$, $\bar{\alpha}_T \approx 0$, so x_T is essentially pure noise. This closed-form expression enables efficient training: we can randomly sample a timestep t and immediately compute the noisy version x_t using this single equation, without simulating all intermediate steps.

The sequence of parameters $\{\bar{\alpha}_t\}$ (or equivalently, $\{\beta_t\}$) is called the *noise schedule*. Different schedules (linear, cosine, learned) control how quickly noise is added. A linear schedule increases β_t linearly, adding noise at a constant rate. A cosine schedule uses a cosine function for $\bar{\alpha}_t$, which adds noise slowly at early timesteps and faster at later timesteps, often leading to better results.

3.3.1.2 Reverse Denoising Process: Removing Noise (Generation)

During generation, we reverse the forward process: starting with pure noise x_T, we iteratively remove noise to recover a clean sample. The key insight is that *the reverse process must be learned*, as it cannot be computed analytically. The model learns to predict the noise component ϵ that was added:

$$\epsilon_\theta(x_t, t) \approx \epsilon, \tag{3.2}$$

where $\epsilon_\theta(x_t, t)$ is a neural network that takes a noisy sample and timestep as inputs. If the model can accurately predict the noise, it can compute the predicted mean $\mu_\theta(x_t, t)$ of the less noisy sample x_{t-1}:

$$\mu_\theta(x_t, t) = \frac{1}{\sqrt{\alpha_t}}\left(x_t - \frac{\beta_t}{\sqrt{1 - \bar{\alpha}_t}}\,\epsilon_\theta(x_t, t)\right), \tag{3.3}$$

where $\mu_\theta(x_t, t)$ is the predicted mean of the reverse distribution $p_\theta(x_{t-1} \mid x_t)$. Equation (3.3) effectively undoes the forward process by removing the predicted noise component from the noisy sample, moving one step closer to the clean data.

3.3.1.3 Training: Predicting Noise

The training objective minimizes the error between predicted and actual noise. For each training example, we randomly sample a timestep t, compute the noisy version x_t using Equation (3.1), and train the network to predict the noise. By learning to predict noise accurately across all timesteps, the model implicitly learns the structure of the data distribution.

Advanced Note 3.1 Diffusion Training Objective

The diffusion model training objective minimizes the squared error between predicted and actual noise:

$$\mathcal{L}(\theta) = \mathbb{E}_{x_0, t, \epsilon}\left[\| \epsilon - \epsilon_\theta(x_t, t)\|^2\right], \tag{3.4}$$

where $x_0 \sim p_{\text{data}}$ is a clean data sample, $t \sim \mathcal{U}(1, T)$ is a randomly sampled timestep, $\epsilon \sim \mathcal{N}(0, I)$ is the noise added during the forward process, and x_t is computed via the marginal distribution in Equation (3.1). The neural network $\epsilon_\theta(x_t, t)$ learns to predict the noise component, enabling the reverse denoising process during generation.

3.3.1.4 Conceptual Understanding

The noise prediction objective is equivalent to learning the *score function* $\nabla_{x_t} \log p(x_t)$, which points in the direction of higher data density. This score function guides the denoising process: at each step, the model moves the sample toward regions where real data is more likely. For CAV applications, this means the model learns to generate realistic weather variations, lighting conditions, and scene structures by following the learned score function through the data manifold.

The generation process typically requires 50–1000 iterative denoising steps, making diffusion models computationally expensive but capable of generating highly detailed, realistic outputs. This makes them ideal for CAV applications like sensor enhancement, trajectory generation, scenario generation, and data augmentation, where quality takes precedence over speed.

In practice, two main sampling variants exist: denoising diffusion probabilistic models (DDPM) [3] use a fully stochastic reverse process with many steps (typically 1000), where each step adds randomness. Denoising diffusion implicit models (DDIM) [4] use the same trained model but with deterministic sampling, requiring fewer steps (typically 50–200) while maintaining similar quality. The same trained diffusion model can be used with different sampling schedules, demonstrating that deterministic sampling can achieve similar quality with far fewer steps.

3.3.2 Flow Matching Mathematics

Flow matching provides a deterministic alternative to diffusion. While diffusion models and flow matching differ in training objectives, both can use deterministic probability-flow ODEs that fit within the same continuous-time dynamical system framework (diffusion models through conversion from their native SDE formulation, flow matching natively). Instead of reversing a noisy corruption process, flow matching learns a velocity field that *directly pushes noise toward data*. Conceptually, think of this as *steering a vehicle along a smooth trajectory*: rather than removing fog step-by-step, flow matching learns to directly navigate from noise to realistic sensor data along a predictable path. Rectified flow is not a separate generative process, but a training refinement of flow matching that progressively straightens transport trajectories.

Notation: In flow matching, x_0 denotes noise and x_1 denotes data. This differs from diffusion notation (Section 3.3.1), where x_0 denotes data and x_T denotes noise, reflecting the different mathematical formulations. In the unified framework: $x_0 \leftrightarrow p_0$ (noise), $x_1 \leftrightarrow p_1$ (data).

A sample evolves according to an ODE:

$$\frac{dx}{dt} = v_\theta(x, t), \tag{3.5}$$

where $v_\theta(x, t)$ is a neural network that learns the velocity field. An ODE is a differential equation that relates a function and its derivatives with respect to a single variable, time t, describing how the sample $x(t)$ changes deterministically over time. Although flow matching and rectified flow share the same generative ODE at inference time, they differ in how the velocity field is learned: flow matching optimizes a general velocity-matching objective, whereas rectified flow explicitly constrains the target velocity to encourage near-linear transport trajectories. The difference lies in the training loss that defines v_θ, not in the sampling equation itself.

A common choice for the target trajectory between noise x_0 and data x_1 is linear interpolation:

$$x(t) = (1 - t)x_0 + tx_1,$$

for which the target velocity field is simply $v^*(x(t), t) = x_1 - x_0$. Flow matching allows flexible transport trajectories, which may be curved in practice. The model learns to follow paths that minimize the transport cost but does not enforce perfectly straight trajectories. Rectified flow modifies the training objective to explicitly bias the solution toward straight-line paths by constraining the target velocity, reducing curvature, and improving numerical stability. This enables efficient sampling with significantly fewer steps (10–50 instead of 50–1000).

Advanced Note 3.2 Training Objectives: Flow Matching vs. Rectified Flow

The flow matching training objective minimizes the squared error between predicted and target velocities:

$$\mathcal{L}_{\text{FM}} = \mathbb{E}_{x_0,x_1,t}[\|v_\theta(x(t),t) - v^*(x(t),t)\|^2],$$

where $x_0 \sim p_{\text{noise}}, x_1 \sim p_{\text{data}}, x(t) = (1-t)x_0 + tx_1$, and $v^*(x(t),t)$ is the ideal transport velocity along the interpolation path. This general formulation allows the learned trajectories to be curved.

Rectified flow modifies the training objective to explicitly encourage straight-line paths. A simplified rectified flow objective is:

$$\mathcal{L}_{\text{RF}} = \mathbb{E}_{x_0,x_1,t}[\|v_\theta(x(t),t) - (x_1 - x_0)\|^2],$$

where the target velocity is directly set to $x_1 - x_0$ (the constant velocity that produces a straight path). More generally, rectified flow iteratively updates the target velocity during training so that trajectories become straighter across training rounds. The key difference is that flow matching learns a general velocity field, while rectified flow explicitly constrains the target to encourage near-linear transport.

3.3.2.1 Conceptual Understanding

Flow matching reframes generative modeling as a direct transport problem rather than a reverse-time denoising process. The model learns a deterministic velocity field that specifies how samples evolve through data space, producing smooth generative trajectories from noise to data. This deterministic structure reduces sampling cost and makes flow matching an attractive candidate for large-scale CAV simulation tasks that require efficient generation of high-fidelity samples. The key distinction from diffusion models lies in both the governing dynamics and the training objective: flow matching natively uses deterministic ODEs and directly optimizes for velocity matching, while diffusion models natively use stochastic SDEs (though they can be converted to deterministic probability-flow ODEs for sampling) and optimize for noise prediction or score matching.

Table 3.1 summarizes the key differences between diffusion models and flow matching at the level of mathematical formalisms.

Table 3.1 Comparison of mathematical formalisms for generative transport.

Model	Mathematical object learned	Transport type	Steps	Interpretation
Diffusion	Score/noise function $\epsilon_\theta(x_t, t)$	Stochastic reverse SDE	50–1000	Gradually denoise (remove fog)
Flow matching	Velocity field $v_\theta(x, t)$	Deterministic ODE	10–50	Push noise to data (smooth steering)

Advanced Note 3.3 A Unified Mathematical Perspective

While diffusion models and flow matching are presented separately above, they can be understood within a unified mathematical framework. All generative models in this chapter learn a *trajectory* $x(t)$ that transforms noise distribution p_0 (Gaussian) at $t = 0$ to data distribution p_1 (realistic sensor data) at $t = 1$.

The mathematical framework that unifies these approaches is the differential equation given in Equation (3.5), where $v_\theta(x, t)$ is a neural network that learns how to move samples through data space. This equation provides a unified conceptual framework at the level of governing dynamics, though each approach implements it differently: diffusion models can be converted to deterministic probability-flow ODEs (derived from their SDE formulation), while flow matching directly learns deterministic ODEs. The key insight is that *how* v_θ is interpreted and learned distinguishes the approaches.

Diffusion models learn v_θ through score matching or noise prediction objectives, which can be converted to deterministic probability-flow ODEs for sampling. Flow matching learns v_θ directly through velocity matching objectives. Both can use deterministic ODEs for generation (diffusion models through conversion from their native SDE formulation, flow matching natively), but their training objectives differ: diffusion models optimize for denoising (removing fog step-by-step), while flow matching optimizes for direct transport (steering along a smooth trajectory).

This unified viewpoint reveals that generative modeling is fundamentally about learning *how to navigate from simple to complex distributions*. The approaches differ in their native formulations (diffusion uses stochastic SDEs, flow matching uses deterministic ODEs) and training objectives, leading to different trade-offs in speed, quality, and implementation complexity that directly impact CAV applications.

3.4 Architectural Choices in Diffusion and Transport Models

The mathematical framework of diffusion and flow matching specifies *what* must be learned (e.g., noise $\epsilon_\theta(x_t, t)$, scores $\nabla_{x_t} \log p(x_t)$, or velocity fields $v_\theta(x_t, t)$), but it does not dictate *how* these functions are parameterized. The choice of neural network architecture is driven by three main requirements: the ability to represent hierarchical structure in high-dimensional data (capturing both global road layout and local object details), efficient handling of conditioning signals (text, semantic maps, driving metadata), and computational scalability for large images, multi-sensor inputs, or video sequences. This section provides a brief, unified view of the architectural choices used in diffusion and flow matching, with a focus on how they relate to CAV sensor data and applications rather than on layer-level implementation details.

3.4.1 UNet Architecture

Early and widely used diffusion models adopt a UNet-style encoder-decoder architecture to parameterize the noise prediction function $\epsilon_\theta(x_t, t)$. UNet was originally introduced for medical image segmentation [5], but its characteristics map naturally to diffusion. The encoder progressively downsamples the input, capturing large-scale structure such as road topology, horizon lines, and

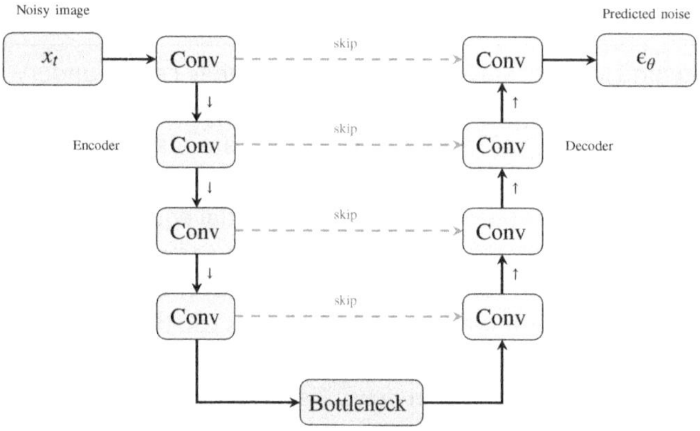

Figure 3.3 UNet architecture for diffusion models with encoder-decoder structure and skip connections.

background context. Skip connections concatenate feature maps from encoder layers with corresponding decoder layers, preserving fine details like lane markings, object boundaries, and small obstacles that might otherwise be lost during downsampling. Convolutional blocks exploit spatial locality and local correlations, which are abundant in camera images and bird's-eye-view (BEV) representations used in CAV perception.

Figure 3.3 illustrates the UNet architecture. In the diffusion setting, the UNet takes a noisy sample x_t (or its latent-space version z_t), a timestep encoding t, and optional conditioning (e.g., text embeddings, semantic maps), and outputs a prediction of the noise component. The same network is reused at every denoising step, with t injected via positional/time embeddings and conditioning typically injected via cross-attention or feature-wise affine modulation. From a CAV perspective, the key takeaway is conceptual: *any* architecture used in diffusion must be able to process sensor data at multiple scales while preserving local details critical for safety, such as the exact placement of vehicles, pedestrians, and lane boundaries. UNet satisfied this requirement for early image-based diffusion models and remains prevalent in many open-source and production systems.

3.4.2 Transformer-based Backbones

As generative models have scaled in both data and parameter count, there has been a trend toward transformer-based backbones that unify text, image, and other modalities within a single token-based representation. Modern diffusion and flow matching systems often employ vision transformers for image tokens in latent space, multimodal diffusion transformers that jointly process text tokens and image tokens, and video diffusion transformers that model space and time jointly for sensor streams. In these architectures, the diffusion or flow process operates on a sequence of latent tokens rather than on a 2D feature map. The same mathematical framework still applies (noise prediction, score estimation, or velocity fields), but the underlying network shifts from convolutions with skip connections to self-attention blocks with cross-attention for conditioning.

For CAV applications, transformer-based backbones offer several advantages. Text prompts, semantic scene graphs, BEV representations, and auxiliary metadata (weather, traffic density) can be embedded into a common token space, enabling stronger multimodal fusion. Large-scale training on diverse image, video, and driving datasets benefits from architectures that have already proven effective at web scale in language and vision, providing better scalability. Video or multi-frame sensor streams can be treated as spatio-temporal token grids, enabling temporally coherent scenario generation. In many modern systems, the traditional UNet is replaced or complemented by such transformer-based backbones, particularly when integrating diffusion or rectified flow with large language models and vision-language models.

3.5 Applications in CAV Systems

Diffusion models find practical application across multiple domains in CAV development [1]. This section provides a focused overview of key application areas, emphasizing practical usage patterns and integration strategies.

3.5.1 Scene, Sensor, and Temporal Data Generation

Diffusion models excel at generating realistic camera imagery, LiDAR point clouds, and multi-frame sensor sequences for data augmentation and robustness testing. For camera imagery, they generate road scenes with variations in lighting (day/night transitions), weather conditions (fog, rain, snow), and environmental effects (glare, reflections). Diffusion models can be conditioned on text descriptions or semantic maps to allow precise control over scene attributes. For LiDAR point clouds, they synthesize geometric structures using BEV projections or by encoding 3D point cloud data into latent representations, with applications including augmenting minority classes (vulnerable road users, rare vehicles) and creating synthetic geometric configurations.

Video diffusion models extend this capability to temporal sequences, producing multi-frame outputs where consecutive frames exhibit coherent motion and consistent scene evolution. Common approaches include modeling video as a spatio-temporal volume or generating frames sequentially with conditioning on previously generated frames. This temporal modeling is essential for synthesizing realistic driving sequences where temporal smoothness is critical for evaluating perception and planning systems.

Multi-sensor consistency is ensured through joint latent spaces or cross-modal conditioning, which is critical for fusion-based perception systems. Generated data must be evaluated for both visual fidelity and physical plausibility, with task-based evaluation (e.g., downstream perception performance) providing practical quality metrics.

3.5.2 Scenario Generation for Testing and Evaluation

Diffusion models enable scalable scenario generation for CAV testing and evaluation. They generate rare event synthesis by creating safety-critical scenarios (near-collisions, sudden lane intrusions) that are difficult to capture in naturalistic data. Conditional generation allows creation of plausible variations of rare events.

Diffusion models operate on structured representations such as semantic scene representations (lane graphs, occupancy grids, agent states) to generate scenarios that respect road geometry, traffic rules, and kinematic constraints. This enables rapid test case generation from high-level descriptions or sparse inputs without manual asset configuration. Diffusion outputs can serve as synthetic sensor streams, scene descriptors, or agent trajectories in CAV platforms, either replacing or augmenting physics-based rendering.

3.5.3 Trajectory and Planning

Diffusion models can model multimodal future behaviors and generate candidate trajectories. They perform trajectory prediction by generating diverse trajectory samples that capture uncertainty in agent intent and interactions, supporting risk evaluation and decision-making. Diffusion-based planners learn implicit driving priors from demonstrations to generate candidate action sequences, which are evaluated using safety constraints and vehicle dynamics. Multi-agent modeling represents joint trajectories and interaction patterns (yielding, lane changes) for safety evaluation and social prediction model training.

Note: Computational overhead makes diffusion-based planning impractical for real-time on-vehicle deployment, limiting these applications to offline research contexts. Generated trajectories require validation to ensure physical feasibility and safety compliance.

3.5.4 Integration with LLMs and VLMs

Diffusion models integrate with large language models and vision-language models to form multimodal generative pipelines. For text-conditioned generation, large language models produce natural language scene descriptions that diffusion models translate into sensor data via CLIP embeddings [6]. For example, a prompt like "Generate a multilane highway scene with moderate traffic and light fog" is converted into corresponding camera/LiDAR frames.

Integrated workflows involve large language models interpreting scenario descriptions and producing structured specifications, diffusion models synthesizing scenes, and vision-language models verifying consistency and extracting semantic labels, forming iterative pipelines for scalable scenario generation. CAV development tasks include scenario expansion, where large language model descriptions drive diffusion synthesis; data augmentation, where natural language instructions generate synthetic sensor data; interactive debugging, where large language model analysis enables targeted scene reproduction; and safety test design, where large language model edge case identification guides diffusion generation.

Key challenges include ensuring physical and semantic constraint compliance when conditioning on abstract language inputs and maintaining temporal and spatial consistency across modalities. These multimodal pipelines require careful validation and constraint enforcement mechanisms.

3.6 Limitations and Considerations

Although diffusion models provide powerful generative capabilities for CAV systems, several important limitations must be considered. Traditional diffusion models (DDPM, DDIM) typically require many iterative denoising steps (50–1000), making them computationally expensive. They are primarily used for CAV applications where generation speed is less critical than output quality, such as sensor enhancement, trajectory generation, behavior modeling, scene completion, dataset augmentation, and simulation. Fast variants that collapse the generative trajectory into fewer steps enable faster generation with some quality trade-offs.

Diffusion models do not inherently enforce physical or semantic constraints, so generated scenes may violate road geometry, object kinematics, or traffic rules, requiring additional validation and post-processing for safety-critical applications. Generated data may introduce biases, artifacts, or domain gaps that affect downstream model performance, necessitating careful curation, validation, and balanced mixing with real data.

Diffusion models lack formal guarantees about output behavior, and synthetic scenarios may not always be accepted for safety certification without rigorous validation frameworks, requiring transparent documentation and systematic evaluation. Models trained on limited datasets may fail to generalize to different environments or sensor configurations, necessitating training on diverse datasets and domain adaptation strategies.

Responsible deployment requires robust validation pipelines, physics-based constraints, and hybrid systems that combine generative models with traditional safety mechanisms. These measures ensure that diffusion-based methods contribute positively to CAV development while minimizing risks.

3.7 Summary

Diffusion models and flow matching provide complementary approaches to generative modeling that transform noise distributions into complex data distributions, offering high-fidelity generation capabilities for CAV systems. Both approaches share the fundamental objective of transporting samples from a simple distribution p_0 (Gaussian noise) to a complex distribution p_1 (realistic sensor data), but they differ in their native formulations and training objectives: diffusion models natively use stochastic SDEs and optimize for noise prediction or score matching (Section 3.3.1), while flow matching natively uses deterministic ODEs and optimizes for velocity matching (Section 3.3.2). An advanced unified mathematical perspective connecting these approaches is presented in Advanced Note 3.3.

These models excel at CAV applications, including sensor enhancement, trajectory generation, behavior modeling, scene completion, scenario generation, dataset augmentation, and rare-event synthesis, where high fidelity and controlled variation are essential. Architectural choices (Section 3.4) range from UNet for early diffusion models to transformer-based backbones for modern multimodal systems.

Practical deployment requires careful parameter selection and validation frameworks to ensure physical plausibility and safety compliance. When used responsibly, these models complement other generative models, including LLMs, generative adversarial networks (GANs), and variational autoencoders (VAEs), in modern CAV development pipelines, providing essential capabilities that would be infeasible through naturalistic data collection alone.

References

1 Katharina Winter, Abhishek Vivekanandan, Rupert Polley, Yinzhe Shen, Christian Schlauch, Mohamed-Khalil Bouzidi, Bojan Derajic, Natalie Grabowsky, Annajoyce Mariani, Dennis Rochau, et al. Generative AI for autonomous driving: a review. *arXiv preprint arXiv:2505.15863*, 2025.

2 Jascha Sohl-Dickstein, Eric Weiss, Niru Maheswaranathan, and Surya Ganguli. Deep unsupervised learning using nonequilibrium thermodynamics. In *International Conference on Machine Learning*, pages 2256–2265, 2015.

3 Jonathan Ho, Ajay Jain, and Pieter Abbeel. Denoising diffusion probabilistic models. *Advances in Neural Information Processing Systems*, 33:6840–6851, 2020.

4 Jiaming Song, Chenlin Meng, and Stefano Ermon. Denoising diffusion implicit models. *arXiv preprint arXiv:2010.02502*, 2020.

5 Olaf Ronneberger, Philipp Fischer, and Thomas Brox. U-Net: convolutional networks for biomedical image segmentation. In *Medical Image Computing and Computer-Assisted Intervention–MICCAI 2015*, pages 234–241, Springer, 2015.

6 Alec Radford, Jong Wook Kim, Chris Hallacy, Aditya Ramesh, Gabriel Goh, Sandhini Agarwal, Girish Sastry, Amanda Askell, Pamela Mishkin, Jack Clark, et al. Learning transferable visual models from natural language supervision. In *International Conference on Machine Learning*, pages 8748–8763, 2021.

4

GANs, VAEs, and Autoregressive Models

4.1 Introduction

Autonomous vehicles operating in real-world environments face a fundamental challenge: perception systems must operate reliably across diverse conditions while meeting strict latency requirements. A vehicle's perception pipeline must process sensor data in milliseconds, adapting to changing weather, lighting, and road conditions while maintaining accuracy. This requirement creates a tension between computational efficiency and generative quality that shapes the design of generative models for CAV systems.

Three generative model families address this challenge through distinct paradigms that enable fast inference: generative adversarial networks (GANs) [1], variational autoencoders (VAEs) [2], and autoregressive models. GANs introduced adversarial training, where competition between networks drives learning. VAEs pioneered probabilistic latent representations, enabling structured compression and generation. Autoregressive models established sequential factorization as a powerful framework for modeling dependencies. These approaches have shaped the field since their introduction and remain integral components of state-of-the-art systems.

These three components occupy distinct positions within the broader generative modeling landscape, reflecting a fundamental computational efficiency trade-off that distinguishes fast-inference models from high-fidelity alternatives. While newer approaches like diffusion models have gained prominence for high-fidelity synthesis, GANs, VAEs, and autoregressive models excel in applications where inference speed is critical. In CAV systems, they enable real-time perception enhancement, sensor data compression, trajectory forecasting, and domain adaptation. Beyond real-time applications, they also contribute to offline data generation, simulation, and analysis pipelines. Their principles and components are frequently integrated into larger systems, demonstrating their enduring influence on the field.

Figure 4.1 illustrates how these models complement each other within CAV systems. High-fidelity models (left) support offline simulation and scenario generation, while fast-inference models (right, stacked vertically) handle real-time sensor processing, domain adaptation, and sequential prediction tasks. The bidirectional arrow indicates the fundamental trade-off between inference speed and generative quality.

The following sections explore each model family in detail, examining their mathematical foundations, training procedures, and practical applications. By understanding these fundamental approaches, readers will gain insight into how modern generative AI systems are constructed and how different model families can be combined to address complex real-world challenges.

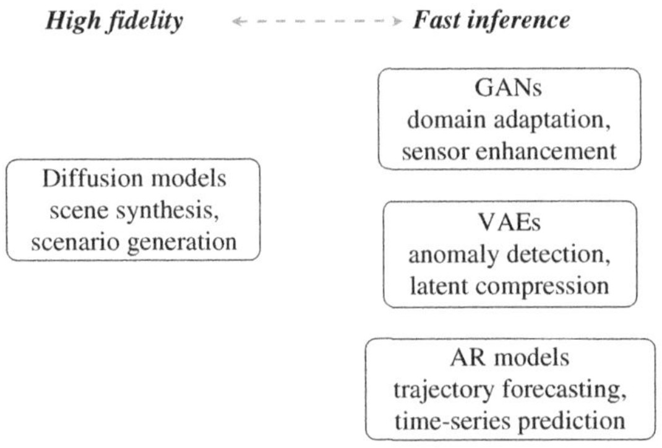

Figure 4.1 Functional roles of generative models within the CAV autonomy stack.

4.2 Generative Adversarial Networks

GANs represent one of the most influential developments in generative modeling over the past decade. While newer methods like diffusion models have become popular for creating high-quality images, GANs remain essential building blocks in many systems and applications, including CAV systems. GANs excel at tasks that require sharp, detailed outputs, such as adapting images between different conditions (e.g., changing weather or lighting), enhancing sensor data quality, and restoring degraded images. Their proficiency in structured transformations like style transfer, weather translation, and resolution enhancement makes them valuable tools for perception pipelines in autonomous vehicles. In what follows, we develop a conceptual understanding of how GANs work, establish their mathematical foundations, explore their practical applications in domain adaptation and sensor enhancement, and conclude with a discussion of their limitations.

4.2.1 Conceptual Understanding

GANs work by training two neural networks that compete with each other. The first network, called the *generator*, learns to create new data samples that appear realistic. The second network, called the *discriminator*, learns to distinguish between real data and the samples created by the generator.

The generator takes random noise as input and transforms it into synthetic samples, such as generating realistic camera images from scratch. The discriminator receives both real data from the training dataset and the generator's outputs, and it learns to classify which samples are real and which are generated. During training, the generator improves at creating samples that fool the discriminator, while the discriminator improves at detecting generated samples. This competition between the two networks drives both to improve, eventually resulting in a generator that can produce samples that are nearly indistinguishable from real data.

The key advantage of this approach is that the generator learns to create realistic data without needing to explicitly model the probability distribution of the training data. Instead, the discriminator acts as a learned quality checker that provides feedback to the generator, guiding it toward creating more realistic outputs. This makes GANs particularly effective at capturing fine details and sharp edges in images, which can be challenging for other generative models. However, training GANs requires careful attention because the balance between the generator and discriminator is delicate, requiring careful tuning of learning rates and network architectures.

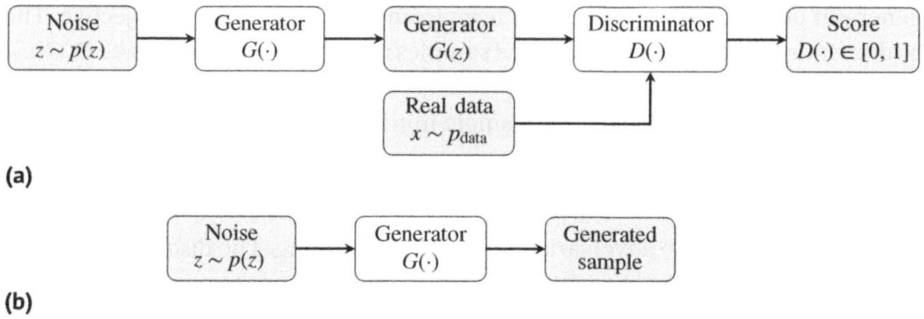

Figure 4.2 (a) Training: both generator and discriminator are active, (b) inference: only the generator is needed. The GAN architecture during training and inference.

It is important to note that while both networks are essential during training, only the generator is needed at inference time. Once training is complete, the discriminator has served its purpose of guiding the generator's learning, and new samples can be generated by simply passing random noise through the trained generator. This makes GANs computationally efficient for deployment, as only a single forward pass through the generator network is required to produce new samples.

Figure 4.2 illustrates the GAN architecture during both training and inference phases. During training (a), both the generator G and discriminator D are active. The generator transforms noise z into synthetic outputs, while the discriminator receives both generated and real data, learning to distinguish between them. During inference (b), only the generator is needed to produce new samples.

4.2.2 GAN Mathematical Formulation

To formalize the intuitive understanding we have developed, we now turn to the mathematical formulation of GANs. A GAN consists of two neural networks: a generator G and a discriminator D. The generator takes random noise z as input, sampled from a prior distribution $p(z)$ (typically a standard Gaussian), and produces synthetic data samples $G(z)$. The discriminator examines data samples and outputs a probability score between 0 and 1, where a score near 1 indicates the sample is likely real, and a score near 0 indicates the sample is likely generated. Real training data x is drawn from the true data distribution $p_{\text{data}}(x)$.

The discriminator and generator have opposite goals that we can express mathematically. The discriminator seeks to maximize its ability to distinguish real from generated samples, outputting scores near 1 for real data and scores near 0 for generated data. The generator, in contrast, seeks to minimize the discriminator's ability to detect its outputs, creating samples that make the discriminator output scores near 1, thereby fooling it into thinking they're real.

The discriminator's objective is to maximize the following quantity:

$$L_D = \mathbb{E}_{x \sim p_{\text{data}}}[\log D(x)] + \mathbb{E}_{z \sim p(z)}[\log(1 - D(G(z)))]. \tag{4.1}$$

The first term $\mathbb{E}_{x \sim p_{\text{data}}}[\log D(x)]$ encourages the discriminator to assign high scores to real data samples. The second term $\mathbb{E}_{z \sim p(z)}[\log(1 - D(G(z)))]$ encourages the discriminator to assign low scores to generated samples.

The generator's objective is to minimize the following quantity:

$$L_G = \mathbb{E}_{z \sim p(z)}[\log(1 - D(G(z)))]. \tag{4.2}$$

Notice that the generator only optimizes the second term from the discriminator's objective. The generator wants the discriminator to think its generated samples are real, so it aims to make $D(G(z))$ close to 1, which minimizes L_G.

The full GAN objective combines both goals into a single minimax optimization problem:

$$\min_G \max_D V(D, G) = \mathbb{E}_{x\sim p_{\text{data}}}[\log D(x)] + \mathbb{E}_{z\sim p(z)}[\log(1 - D(G(z)))]. \tag{4.3}$$

The notation $\min_G \max_D$ captures the adversarial nature of the training. The discriminator tries to maximize $V(D, G)$, making it as large as possible, while the generator tries to minimize $V(D, G)$, making it as small as possible. This opposition creates the adversarial competition that drives both networks to improve.

For unconditional GANs, new samples can be generated simply by sampling $z \sim p(z)$ and computing $G(z)$. For conditional GANs used in style transfer and domain adaptation, inference involves passing a source image x through the generator to obtain the transformed output $G(x)$.

Advanced Note 4.1 GAN Training Process

The training alternates between two steps. First, we fix the generator and update the discriminator to maximize L_D. Second, we fix the discriminator and update the generator to minimize L_G. This alternating process continues, with each iteration improving both networks until the generator produces samples that the discriminator cannot reliably distinguish from real data.

Now that we have established the mathematical foundations of GANs, we turn to their practical applications in CAV systems. The following subsections explore how GANs are used for domain adaptation, sensor enhancement, and other perception tasks.

4.2.3 Domain Adaptation for Perception

Perception models deployed in CAV systems must generalize across diverse lighting conditions, weather variations, sensor calibrations, and geographic domains. Collecting data that spans these conditions is costly and often incomplete. GAN-based domain adaptation techniques help reduce this gap by translating images between domains while preserving semantic content.

Cycle-consistent GANs (CycleGANs) [3] are commonly used for unpaired domain translation. They employ two generators (one for each direction of translation) and two discriminators (one for each domain), trained with cycle-consistency constraints that ensure translating from domain A to B and back to A recovers the original image. At inference time, only one generator is needed depending on the desired translation direction: to translate from domain A to domain B, only generator $G_{A\to B}$ is used; to translate from domain B to domain A, only generator $G_{B\to A}$ is used. Figure 4.3 illustrates the CycleGAN architecture during training, showing bidirectional translation with cycle-consistency. The cycle-consistency constraint ($A \to B \to A \approx A$) ensures semantic content preservation (vehicles, roads, signs) while transforming style attributes (visibility, lighting). This enables perception models to generalize across weather conditions without expensive paired datasets, directly addressing domain shift in CAV deployment.

This capability finds particular utility in autonomous driving applications [4]. Synthetic-to-real transfer enhances the realism of simulator-generated images, improving perception model generalization without requiring extensive real-world data collection. Weather and style translation converts clear-weather scenes into rainy, snowy, or foggy conditions, diversifying training data and enabling models to handle adverse conditions. These transformations also improve segmentation

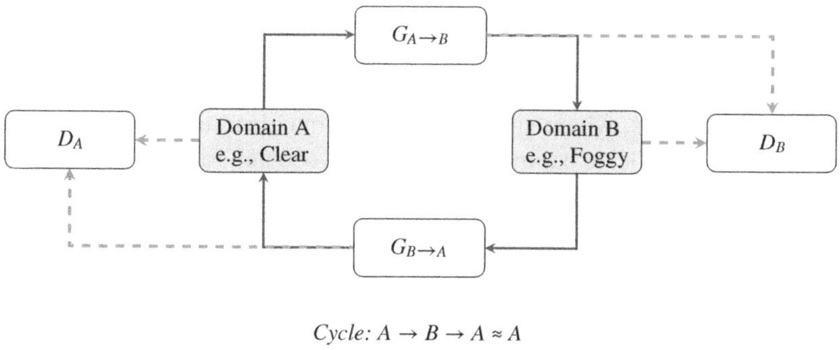

Figure 4.3 CycleGAN's bidirectional architecture for unpaired domain translation.

and detection robustness by reducing domain shift between training data sources, leading to better performance in downstream tasks.

CycleGAN's unpaired training approach makes it well-suited for style transfer tasks where paired examples are unavailable, such as adapting visual styles between different sensor configurations or geographic regions. For paired domain translation tasks, conditional GANs [5] provide an alternative approach that leverages paired training data when available. These GAN-based domain adaptation techniques provide transformations that can be integrated directly into perception pipelines, offering a practical solution to the domain shift problem that plagues real-world deployment.

4.2.4 Super-resolution and Sensor Enhancement

Beyond domain adaptation, GANs remain effective tools for sensor enhancement tasks due to their proficiency in generating high-frequency details. In scenarios where camera resolution is reduced due to bandwidth or storage constraints, GAN-based super-resolution models [6] can reconstruct sharper images, which is particularly important for long-range perception tasks such as detecting distant vehicles or pedestrians. Similarly, sparse LiDAR returns can be upsampled using GANs that learn geometric patterns and point distributions. Enhanced point density improves object detection and mapping performance, especially for cost-efficient LiDAR sensors with limited channels.

GANs have also been adapted to reduce noise in radar returns or to transform radar signals into more interpretable representations, supporting improved fusion with camera and LiDAR data. In these applications, GANs provide architectures suitable for embedded inference in perception pipelines, offering a balance between enhancement quality and computational efficiency.

4.2.5 Limitations of GAN-based Approaches

Despite their utility, GANs exhibit several inherent limitations that restrict their broader application in CAV systems. Mode collapse represents one of the most significant challenges, where GANs produce limited variability, generating similar or identical samples even when trained on diverse data. This is particularly detrimental for applications that require diverse scenario generation, such as creating varied driving scenarios for testing, where the lack of diversity can lead to incomplete coverage of the scenario space.

Training instability presents another fundamental challenge. The adversarial training process is sensitive to hyperparameters, loss functions, and architecture choices, making it difficult to achieve consistent results across different datasets or applications. This instability can produce dangerous artifacts in safety-critical perception tasks, where spurious features may lead to incorrect object detection or scene interpretation. The delicate balance required between generator and discriminator updates means that small changes in training dynamics can lead to complete training failure or degraded performance.

GANs also face difficulty with structured data, struggling with geometric modalities such as LiDAR or bird's-eye-view (BEV) maps, which represent spatial context in a planar top-down coordinate system. While these challenges can sometimes be addressed through careful engineering, the inherent architecture of GANs makes them less naturally suited to structured geometric data compared to other generative approaches.

Unlike diffusion models, GANs do not provide a mechanism to explicitly cover the full data distribution through likelihood-based training, which limits their suitability for large-scale scenario generation where diversity and coverage are critical. While GANs excel at producing sharp, high-frequency details through adversarial training, diffusion models achieve superior diversity, distribution coverage, and fidelity by learning to reverse a noise-adding process that naturally spans the entire data distribution. This fundamental difference explains why diffusion models have become the preferred approach for high-fidelity, diverse generation tasks, while GANs remain valuable for specialized applications requiring fast inference and sharp transformations.

As a result of these limitations, GANs are most effective when used for targeted enhancement or translation tasks rather than full-scale generative simulation. Their strengths in domain adaptation and sensor enhancement align well with these focused applications, where the adversarial training framework can provide sharp, detailed outputs without requiring the full diversity or fidelity needed for comprehensive scenario generation.

4.3 Variational Autoencoders

While GANs excel at enhancement and domain adaptation, VAEs offer complementary strengths through their probabilistic formulation and structured latent representations. VAEs are probabilistic generative models that learn compact latent representations through an encoder-decoder architecture. Unlike GANs, which focus on adversarial image realism, VAEs emphasize representation learning and probabilistic modeling. Their structured latent spaces, stable optimization properties, and reconstruction capabilities make them valuable for several tasks in CAV systems, enabling capabilities such as anomaly detection, multi-sensor encoding, and communication-efficient perception.

This section begins with a conceptual overview of how VAEs extend basic autoencoders, followed by their mathematical formulation. We then explore practical applications in sensor fusion, anomaly detection, and V2X communication, and conclude with a discussion of their limitations.

4.3.1 Conceptual Understanding

At its core, a VAE is an *autoencoder* [7]: the encoder compresses input sensor data into compact latent codes, and the decoder reconstructs data from these codes. A basic autoencoder produces deterministic latent codes, meaning each input maps to a single fixed latent vector. VAEs extend this framework by making the latent representation probabilistic: instead of fixed values, the encoder learns to produce probability distributions over latent variables. This probabilistic formulation enables VAEs to generate diverse outputs from the same input by sampling different values from the learned distributions.

To illustrate this key difference, consider the following analogy (see the example box below). In a basic autoencoder, the encoder produces a fixed latent code for each input, always yielding the same reconstruction. In a VAE, the encoder produces distribution parameters (mean and variance) for each latent dimension. When the decoder samples from these distributions, it can generate multiple diverse reconstructions from the same input, enabling generative diversity that deterministic autoencoders cannot achieve.

Example 4.1 Understanding Probabilistic Latents: A Design Specification Analogy

A helpful way to understand the difference between autoencoders and VAEs is through an analogy to a design specification system. In a basic autoencoder, the encoder produces a specification with fixed values for each parameter, for example, tire size equals 18 inches and ground clearance equals 0.15 m. This fixed specification always produces the same design when decoded.

The key insight of variational inference, the "V" in VAE, is that instead of fixed parameter values, the encoder learns to produce specifications with probability distributions over parameter values. Rather than exactly 18-inch tires, the specification contains a distribution where tire size is likely between 17 and 19 inches with the highest probability around 18 inches. Similarly, ground clearance is represented as a probability distribution. When the decoder uses this probabilistic specification, it can generate diverse designs, some with 17-inch tires and 0.14 m ground clearance, and others with 19-inch tires and 0.16 m ground clearance, all belonging to the same design family but with natural variation that reflects the flexibility inherent in the design process.

During training, the encoder learns to preserve essential information in compact latent representations while the decoder learns to reconstruct data from these probabilistic codes. The system is constrained to organize latent codes in a structured way, ensuring that similar inputs have similar latent representations that cluster together in the latent space. This enables smooth interpolation and sampling diverse variations, making VAEs valuable for tasks that require both compression and generative diversity.

4.3.2 VAE Mathematical Formulation

A basic autoencoder [7] learns to compress and reconstruct data through an encoder-decoder architecture. The encoder f_θ maps input x to a latent code:

$$z = f_\theta(x),$$

where θ denotes encoder parameters and z is the latent code. The decoder g_ϕ reconstructs the input from the latent code:

$$\hat{x} = g_\phi(z) = g_\phi(f_\theta(x)),$$

where ϕ denotes decoder parameters and $\hat{x}$ is the reconstruction. The latent code z is a deterministic function of the input, producing a single fixed value for each input. Figure 4.4a illustrates the basic autoencoder pipeline.

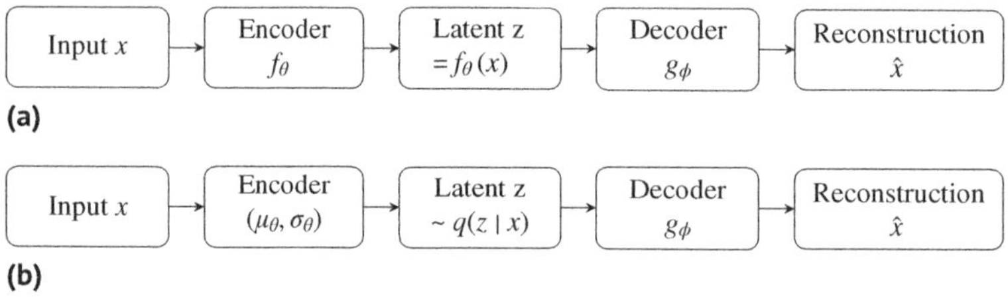

Figure 4.4 (a) Basic autoencoder: deterministic latent codes, (b) variational autoencoder: probabilistic latent distributions. The fundamental difference between autoencoders and VAEs.

VAEs extend this framework by making the latent representation probabilistic. Instead of mapping to a fixed latent code, the encoder maps input x to a distribution over latent variables. The Gaussian distribution $q(z \mid x) = \mathcal{N}(\mu_\theta(x), \sigma_\theta^2(x))$ is the standard choice due to its mathematical tractability and the reparameterization trick, which enables backpropagation through stochastic sampling. The encoder outputs both the mean $\mu_\theta(x)$ and variance $\sigma_\theta^2(x)$ parameters for each latent dimension. The decoder then reconstructs data from samples drawn from this distribution: $\hat{x} = g_\phi(z)$ where $z \sim q(z \mid x)$. Figure 4.4 illustrates this key difference: (1) Basic autoencoders produce deterministic latent codes $z = f_\theta(x)$, limiting generative diversity. (2) VAEs introduce probabilistic latent distributions $q(z \mid x) = \mathcal{N}(\mu_\theta(x), \sigma_\theta^2(x))$, where the encoder outputs distribution parameters and the decoder samples from these distributions. The sampling step enables diverse generation from the same input, allowing exploration of the learned latent space rather than point estimates.

The probabilistic formulation provides uncertainty estimates through latent-space variance. VAEs are trained to balance reconstruction accuracy with latent space regularization, ensuring that similar inputs cluster together in the latent space while maintaining good reconstruction quality.

It is important to clarify the inference-time requirements for VAEs. Unlike GANs, which only need the generator at inference, VAEs require both the encoder and decoder for different use cases. For tasks that require reconstruction, such as anomaly detection, both encoder and decoder are needed: the encoder compresses the input to a latent code, and the decoder reconstructs it. In anomaly detection [8], anomalies are typically identified by high reconstruction error, as VAEs trained on normal data cannot reconstruct anomalous inputs well. Alternatively, anomaly detection can use only the encoder by examining latent-space likelihood, though reconstruction-based methods are more commonly used. For generation tasks, only the decoder is needed: random samples are drawn from the prior distribution $p(z)$ (typically $\mathcal{N}(0, I)$) and passed through the decoder to generate new samples. For encoding tasks (such as V2X communication), only the encoder is needed to compress input data into latent representations. This flexibility makes VAEs versatile for different inference scenarios, though it means that some applications require both networks to be deployed.

Advanced Note 4.2 Autoencoder Training and ELBO

Basic Autoencoder Training: The training objective for a basic autoencoder is to minimize the reconstruction error between the input and its reconstruction. The loss function is

$$\mathcal{L}_{\text{AE}} = \mathbb{E}_{x \sim p_{\text{data}}} \|x - g_\phi(z)\|^2, \quad \text{where } z = f_\theta(x),$$

where p_{data} is the data distribution and $z = f_\theta(x)$ is the deterministic latent code. This objective encourages the encoder-decoder pair to learn a compressed representation that preserves essential information for accurate reconstruction. The latent code z is deterministic, meaning each input maps to a single fixed latent vector.

Variational Autoencoder Training and ELBO: VAEs extend this framework by introducing probabilistic latent representations. The training objective for VAEs is to minimize the negative evidence lower bound (ELBO) – the VAE loss – which is equivalent to maximizing a tractable lower bound on the intractable marginal likelihood $p(x)$.

The ELBO objective is

$$\mathcal{L}_{\text{VAE}} = -\mathbb{E}_{q(z|x)}[\log p(x \mid z)] + D_{\text{KL}}(q(z \mid x) \parallel p(z))$$

where $q(z \mid x)$ is the approximate posterior (encoder output) and $p(z)$ is the prior (typically $\mathcal{N}(0, I)$). The first term is the reconstruction loss, encouraging accurate reconstruction, while the second term is the KL regularization, ensuring a structured latent space aligned with the prior. The reparameterization trick $z = \mu_\theta(x) + \sigma_\theta(x) \odot \epsilon$ (where $\epsilon \sim \mathcal{N}(0, I)$) enables back-propagation through stochastic sampling.

VAE Variants: Vector quantized VAEs (VQ-VAEs) [9] use discrete latent representations and have been adopted in modern systems such as stable diffusion [10], demonstrating the flexibility of the VAE framework.

Figure 4.5 provides a conceptual view of how similar inputs cluster in the latent space, demonstrating the structured geometry that emerges from the VAE training objective. Similar inputs cluster together (dashed ellipses), enabling smooth interpolation and diverse sampling within families. This clustering emerges from the training process, which organizes latent codes according to a Gaussian prior while preserving reconstruction quality. This structured latent space supports applications such as multi-sensor fusion, where different sensors encoding the same semantic content (e.g., the same vehicle detected by camera and LiDAR) cluster together in the common latent space, and latent-space-based anomaly detection, where anomalous inputs appear far from normal clusters.

It is important to note that while VAEs in this chapter are used primarily for representation learning, compression, and safety monitoring, VAE-style encoders also play a critical role in latent diffusion models (Chapter 3). In that context, VAEs serve as compression layers that map high-dimensional images to lower-dimensional latent spaces where diffusion occurs, rather than serving as semantic latent spaces for representation learning. This architectural reuse demonstrates the

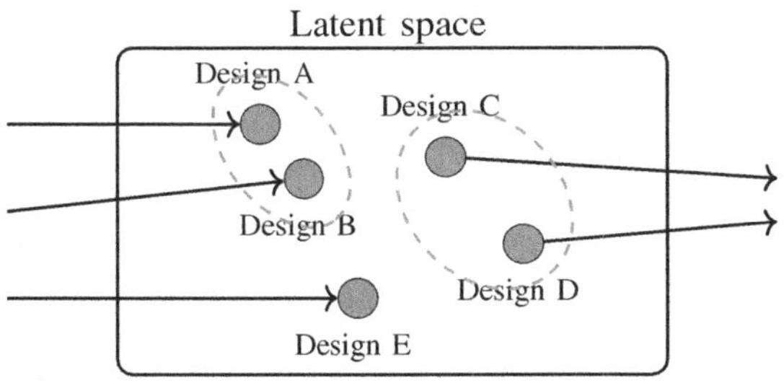

Figure 4.5 The structured geometry of VAE latent spaces.

versatility of the VAE encoder-decoder framework across different generative paradigms, though the training objectives and use cases differ significantly between standalone VAEs and VAE components within diffusion models.

Now that we have established the mathematical foundations of VAEs, we turn to their practical applications in CAV systems. The following subsections explore how VAEs are used for sensor fusion, anomaly detection, and communication-efficient perception.

4.3.3 Representation Learning for Sensor Fusion

CAV perception often requires fusing multiple sensor modalities, including camera imagery, LiDAR point clouds, radar returns, and vehicle kinematic states. VAEs provide a principled framework for learning latent embeddings that capture shared structure across sensors. By encoding heterogeneous inputs into a common latent space, VAEs support learned compression that maps high-dimensional sensor data into compact, informative representations that preserve semantic content while reducing redundancy. This learned compression is distinct from traditional lossless or lossy compression, as it focuses on preserving semantically meaningful features rather than pixel-level fidelity. Multi-sensor embedding enables encoders for different modalities to be trained to produce aligned latent representations, facilitating downstream fusion or joint generative modeling. Through their probabilistic formulation, VAEs also provide robustness to noise and missing modalities, naturally smoothing over sensor noise and inferring latent features even when partial data are missing.

4.3.4 Anomaly and Fault Detection

VAE-based anomaly detection represents a key application in CAV safety monitoring [8, 11], offering two complementary approaches: reconstruction-based methods that use both encoder and decoder, and encoder-only methods that operate directly in the latent space.

Reconstruction-based methods leverage the full encoder-decoder architecture. VAEs trained on nominal driving data learn to reconstruct normal patterns with low reconstruction error, typically measured as L2 distance or negative log-likelihood between input and reconstruction. When presented with anomalous sensor measurements, such as corrupted LiDAR frames, unexpected environmental conditions, or sensor faults, the reconstruction error increases significantly, signaling potential deviations from the learned normal distribution. This approach requires both encoder and decoder at inference time, as the input must be encoded to a latent representation and then decoded to measure reconstruction quality.

Encoder-only methods operate solely in the latent space, requiring only the encoder at inference time. This approach exploits the structured geometry of VAE latent spaces: normal data maps to compact, high-density regions near the prior mean (typically $z \approx 0$) due to KL regularization, while anomalies are pushed to low-density or distant areas. Common distance metrics include Mahalanobis distance from the normal data manifold, Euclidean distance to the prior mean, and distance to k-nearest neighbors in normal latents. This method is computationally efficient, requiring only a single encoder forward pass, making it well-suited for real-time applications such as V2X anomaly flagging. For CAV applications, training the encoder on normal V2X messages or sensor scenes and flagging outliers via latent distance metrics provides an efficient solution that requires no decoder deployment.

Both approaches align with the requirements of functional safety frameworks such as ISO 26262, which mandate continuous monitoring for abnormal behavior in safety-critical systems. VAE-based anomaly detection supports several critical monitoring functions. Sensor integrity monitoring detects faulty or occluded sensors through reconstruction discrepancies or latent-space deviations, identifying when sensor outputs deviate from expected patterns. Environmental anomaly detection identifies unusual scenes or weather conditions that deviate from training distributions, enabling early warning systems for unexpected driving scenarios. System-level diagnostics flag inconsistencies in fused representations that may indicate degraded subsystem performance, supporting proactive maintenance and safety interventions. Due to their probabilistic formulation, VAEs also provide uncertainty estimates through latent-space variance, offering additional information for safety workloads that can inform confidence levels in anomaly detection decisions.

4.3.5 Latent Compression for V2X Communication

The structured latent representations of VAEs provide compact summaries of high-dimensional sensor data, a property that aligns naturally with the bandwidth constraints of V2X systems. This enables vehicles or roadside units to exchange compressed feature representations rather than raw sensor streams, preserving semantic information while reducing transmission cost by orders of magnitude.

These representations support several communication-efficient architectures. Edge-cloud cooperation leverages front-end encoders on vehicles to compress sensor data before transmitting latent vectors to infrastructure nodes for further processing, reducing bandwidth requirements while maintaining semantic content. Bandwidth-efficient scene sharing allows vehicles to share latent codes instead of full-resolution images or point clouds, facilitating real-time perception fusion in congested communication environments where raw data transmission would be infeasible. Cross-vehicle feature exchange is also supported, as latent representations allow vehicles to share semantic cues, such as occluded objects or hidden intersections, without overwhelming communication channels. VAEs therefore contribute to communication-efficient architectures for cooperative perception, an increasingly central element of advanced CAV systems where multiple vehicles and infrastructure must coordinate perception capabilities.

4.3.6 Limitations of VAE-based Approaches

Although VAEs offer stable training and structured latent spaces, they exhibit several limitations that restrict their use in high-fidelity generative simulation. Blurry reconstructions represent a fundamental challenge, resulting from the variational formulation that favors smooth outputs lacking the high-frequency detail required for realistic image or point cloud synthesis. This blur limits their effectiveness for occlusion reconstruction in hazard scenarios, where precise spatial detail is critical for identifying partially occluded pedestrians or vehicles.

Limited generative fidelity means VAEs struggle to capture complex multimodal distributions needed for scenario generation or photorealistic simulation. The Gaussian latent prior, while simplifying optimization and enabling tractable training, hinders modeling of highly discontinuous or diverse scene structures that may require more expressive latent representations. Expressiveness constraints also arise from the trade-off between reconstruction quality and latent space regularization, where the KL divergence term encourages smooth, well-behaved latent spaces at the cost of capturing fine-grained details.

For these reasons, VAEs are primarily used for representation learning, compression, and anomaly detection rather than full-scale scene or sensor synthesis. Their strengths in structured latent spaces and stable training complement, rather than replace, diffusion-based generative models, which excel at high-fidelity synthesis but lack the structured representations that make VAEs valuable for compression and anomaly detection tasks.

4.4 Autoregressive Models for Sequential Prediction

The generative models discussed thus far, GANs and VAEs, primarily operate on spatial or structured data representations. In contrast, autoregressive (AR) models are well-suited for sequential data, addressing temporal dependencies that are central to many CAV applications. Their defining characteristic is a factorization of probability distributions into conditioned components, where each element depends on preceding elements in the sequence.

Although classical AR models predate transformer-based LLMs, both share the same autoregressive factorization principle. AR models operate at lower perceptual and temporal layers, for example, sensor streams, trajectories, and audio, while LLMs apply the same principle at semantic and reasoning layers. This structural parallel provides conceptual continuity across different levels of the autonomy stack.

This section begins with a conceptual overview of how autoregressive models work, followed by their mathematical formulation. We then explore practical applications in trajectory forecasting, time-series prediction, and audio modeling, and conclude with a discussion of their limitations.

4.4.1 Conceptual Understanding

Autoregressive models factorize the joint probability of a sequence into a product of conditional probabilities, where each element depends on preceding elements. This factorization enables tractable likelihood maximization for sequential data. The model learns to predict the next element in a sequence given all previous elements, naturally capturing temporal dependencies that are central to time-series prediction, trajectory forecasting, and audio modeling.

A helpful way to understand this process is through an analogy to a predictive navigation system that forecasts the next step in a route based on the current trajectory. Just as a navigation system observes the current position, speed, and direction to predict where the vehicle will be in the next moment, an autoregressive model observes the sequence of previous elements, for example, past vehicle positions, sensor readings, or audio samples, to predict the next element in the sequence. The navigation system does not predict the entire route at once. Instead, it predicts one step ahead at a time. Given the current position, speed, and direction, it forecasts the next position. Similarly, an autoregressive model predicts one element at a time: given the sequence of past positions $x_1, x_2, \dots, x_{t-1}$, which predicts the next position x_t. Each prediction depends on all previous observations, naturally capturing temporal dependencies and causal relationships.

Generation proceeds sequentially: each element must be generated before the next, making the process inherently causal but potentially slow for long sequences. This sequential nature means that to generate a sequence of length T, the model must perform T forward passes, one for each time step, creating a computational bottleneck that limits real-time performance for long sequences.

The key advantage of this approach is that it provides explicit probabilistic modeling with tractable likelihood computation. Training maximizes the likelihood of observed sequences, which is equivalent to minimizing the prediction error at each time step. This makes autoregressive models well-suited for tasks where temporal structure is essential, such as trajectory forecasting and

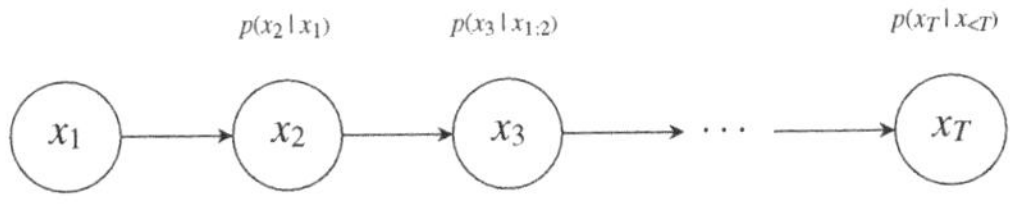

Figure 4.6 The autoregressive factorization creates a causal dependency chain.

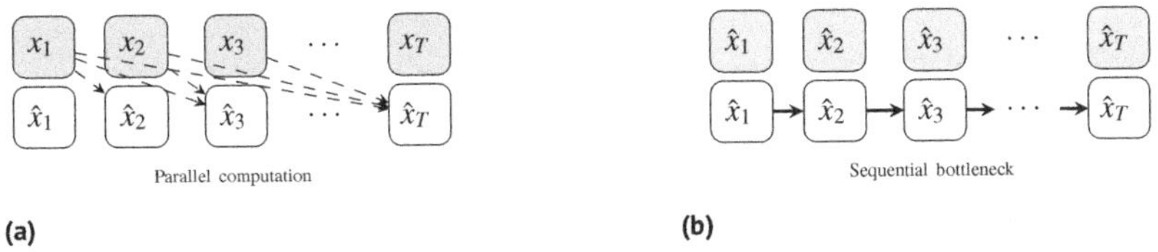

Figure 4.7 (a) Training (teacher forcing): parallel computation, (b) inference (sequential sampling): sequential bottleneck. The training-inference gap in autoregressive models.

predictive maintenance. However, the sequential dependency limits the model's ability to capture long-range dependencies without sophisticated architectures. Modern transformer-based architectures [12] mitigate long-range dependency limitations by enabling self-attention over entire sequences.

Figure 4.6 illustrates the sequential dependency structure of autoregressive models. Each element is conditioned on all preceding elements, reflecting temporal structure where future states depend on past states. The conditional probabilities $p(x_t \mid x_{<t})$ represent learned predictions at each time step. This factorization enables tractable likelihood computation and captures temporal dependencies, making it well-suited for CAV time-series prediction where causal relationships are fundamental.

Figure 4.7 illustrates the fundamental difference between training and inference in autoregressive models. During training (top), teacher forcing enables efficient training by conditioning on ground-truth history (dashed arrows). During inference (bottom), sequential sampling requires each element before the next (solid arrows), creating a bottleneck that limits inference speed. This explains why autoregressive models train efficiently but generate slowly, particularly for long sequences in CAV applications.

4.4.2 AR Mathematical Formulation

Autoregressive models factorize the joint probability of a sequence $x_{1:T} = (x_1, x_2, \dots, x_T)$ into a product of conditional probabilities:

$$p(x_{1:T}) = \prod_{t=1}^{T} p(x_t \mid x_{<t}),$$

where $x_{<t} = (x_1, x_2, \dots, x_{t-1})$ denotes all elements before time t, and T is the sequence length. This factorization expresses the probability of the entire sequence as the product of conditional probabilities, where each element depends on all preceding elements. Each conditional probability $p(x_t \mid x_{<t})$ models the distribution of the next element given the history.

The model learns a function f_θ that maps the history $x_{<t}$ to the parameters of the conditional distribution, where θ denotes model parameters. For continuous sequences (e.g., trajectories, sensor readings), this is parameterized as:

$$f_\theta(x_{<t}) = (\mu_\theta(x_{<t}), \sigma_\theta(x_{<t})),$$

These parameters define the conditional distribution of the next element:

$$p(x_t \mid x_{<t}) = \mathcal{N}(\mu_\theta(x_{<t}), \sigma_\theta^2(x_{<t})),$$

where $\mu_\theta(x_{<t})$ is the predicted mean and $\sigma_\theta(x_{<t})$ is the predicted standard deviation for x_t given the past history. For discrete sequences (e.g., discrete actions, categorical states), the conditional distribution is:

$$f_\theta(x_{<t}) = \pi_\theta(x_{<t}), \quad p(x_t \mid x_{<t}) = \text{Categorical}(\pi_\theta(x_{<t})),$$

where $\pi_\theta(x_{<t})$ is a probability vector that f_θ outputs, and $\text{Categorical}(\pi_\theta(x_{<t}))$ denotes a categorical distribution parameterized by this probability vector.

Advanced Note 4.3 AR Training Objective

The training objective for autoregressive models maximizes the log-likelihood of observed sequences:

$$\mathcal{L}_{\text{AR}} = \mathbb{E}_{x_{1:T} \sim p_{\text{data}}} \left[\sum_{t=1}^{T} \log p(x_t \mid x_{<t}) \right],$$

which is equivalent to minimizing the negative log-likelihood. This objective naturally decomposes the joint likelihood into a sum of conditional log-likelihoods, where each term corresponds to the prediction error at a single time step. For discrete sequences, this typically corresponds to cross-entropy loss, while for continuous sequences with Gaussian likelihoods, it corresponds to minimizing the squared prediction error. The decomposition enables efficient training when using teacher forcing, as all time steps can be computed in parallel during training, while the autoregressive structure is maintained for generation at inference time.

Now that we have established the mathematical foundations of autoregressive models, we turn to their practical applications in CAV systems. The following subsections explore how autoregressive models are used for trajectory forecasting, time-series prediction, and audio modeling.

4.4.3 Trajectory Forecasting

Predicting the future motion of vehicles, pedestrians, and other dynamic agents is a critical component of CAV perception and planning. Autoregressive models support trajectory forecasting by modeling future states as conditioned on past observations, naturally capturing temporal dependencies where sequence evolution depends on prior locations, velocities, and agent interactions [13, 14]. AR structures support multistep forecasting by iteratively predicting future positions, enabling modeling of both short- and long-term trends. Probabilistic neural autoregressive models represent prediction variance over time, providing uncertainty modeling capabilities that are essential for safety-critical planning decisions.

In motion prediction tasks, autoregressive neural networks extend classical statistical AR methods (such as ARIMA) by incorporating scene context, recurrent networks, or attention mechanisms. These models generate future trajectories one step at a time, enabling fine temporal resolution and sensitivity to local dynamics. While diffusion-based models can produce multimodal trajectory distributions, AR methods remain competitive for lightweight forecasting tasks and for systems requiring deterministic outputs, making them valuable for real-time perception pipelines where computational efficiency is paramount.

4.4.4 Time-series Prediction and Predictive Maintenance

CAV systems generate extensive time-series data from sensors such as accelerometers, gyroscopes, wheel encoders, engine diagnostics sensors, and infrastructure communication modules. AR models provide a computationally efficient mechanism for capturing temporal correlations in these streams, making them well-suited for predictive maintenance applications.

Several key applications demonstrate the value of AR models in this domain. Sensor drift prediction enables forecasting long-term deviations in sensor readings, supporting proactive recalibration before measurements become unreliable. Vibration analysis uses AR models to detect deviations in vibration signatures that indicate component wear or mechanical imbalance, enabling early identification of potential failures. Energy and thermal modeling leverages AR forecasting to predict engine and battery dynamics, which often follow predictable temporal patterns. These predictive insights support preventive maintenance strategies, reducing system downtime and improving vehicle reliability while enabling more efficient resource allocation in fleet management systems.

4.4.5 Audio Modeling for Vehicle Diagnostics

Mechanical systems often emit characteristic acoustic signatures that correlate with specific states or degradation modes. AR models are well suited for audio signal modeling because they explicitly represent time-domain correlations, naturally capturing the sequential structure of acoustic waveforms.

Several diagnostic applications benefit from AR-based audio modeling. Tire-road interaction modeling uses AR-based generative models to approximate friction-induced acoustic patterns, supporting tire wear diagnostics or surface classification tasks. Engine anomaly simulation leverages AR models to capture variations in combustion cycles or belt sounds, enabling synthetic generation of anomalous signatures for training diagnostic classifiers when real failure data is scarce. Wind and cabin noise prediction employs AR-based filters to capture aerodynamic and structural vibration characteristics relevant to comfort or component monitoring. These audio generative capabilities complement perception pipelines by enabling supervised learning for rare acoustic failure modes, expanding the training data available for diagnostic systems.

4.4.6 Limitations of Autoregressive Methods

Despite their strengths, AR models exhibit several limitations that restrict their applicability in certain CAV tasks. Sequential sampling creates latency bottlenecks for long sequences, as each element must be generated before the next, making real-time applications challenging when sequence lengths exceed a few hundred time steps. This linear scaling of inference time with sequence length contrasts with parallel generation methods, limiting AR models' effectiveness for long-horizon forecasting tasks.

Standard AR models with unimodal conditional distributions cannot represent branching behaviors, such as pedestrian intentions with multiple possible outcomes, which are common in driving scenarios. This limitation means that AR models may struggle with multimodal prediction tasks where multiple plausible futures exist. AR models also have limited expressiveness for spatially structured data such as images or LiDAR point clouds unless heavily modified, as the sequential factorization does not naturally capture spatial relationships. For these reasons, autoregressive models are best viewed as specialized tools for temporal modeling rather than general-purpose generative frameworks within CAV systems, complementing other generative approaches that excel at spatial or multimodal generation tasks.

Table 4.1 Comparison of generative model families for CAV applications.

Criterion	GANs	VAEs	AR models	Diffusion models
Generative fidelity	High-frequency detail; limited diversity	Moderate; smooth reconstructions	Low for high-dim data	Highest fidelity; strong consistency
Interpret-ability	Low; adversarial dynamics	High; structured latents	Moderate; temporal structure	Moderate; complex dynamics
Compute cost	Very low; single pass	Very low; encoder-decoder	Moderate; sequential	High; iterative
Best modalities	Images, BEV, sparse LiDAR	Multi-sensor fusion	Time-series, audio, trajectories	Images, LiDAR, scene graphs
Strengths	Real-time enhancement; domain adaptation	Anomaly detection; compression	Sequential modeling; forecasting	Simulation, scenario generation
Weaknesses	Instability; mode collapse	Blurry reconstructions	Slow for long sequences	High compute; limited real-time

Note: AR models here refer to domain-specific autoregressive models (sensor streams, trajectories, audio), not LLMs which operate at semantic levels.

4.5 Comparative Analysis

The generative models examined in this chapter, GANs, VAEs, and autoregressive methods, occupy distinct functional roles within CAV development pipelines. Their suitability depends on factors such as modeling fidelity, computational requirements, interpretability, and the degree of structure in the target data. Table 4.1 provides a comprehensive comparison of the relative strengths and limitations of these models, along with diffusion models, across multiple evaluation dimensions relevant to CAV systems. The intent is not to rank these models universally but to clarify where each method is most effective. Note that LLMs and flow matching are excluded from this table: LLMs operate at semantic and reasoning levels (covered in Chapter 2), while flow matching represents an evolution of diffusion methods (covered in Chapter 3) and shares similar high-fidelity characteristics.

The comparison highlights the complementary nature of these generative model families. Diffusion models excel at high-fidelity synthesis requiring realism and diversity, but their computational expense limits real-time applications. GANs provide sharp transformations suited for perception pipelines, operating directly on sensor data. VAEs support safety monitoring, representation learning, and communication-efficient architectures. Domain-specific autoregressive models remain indispensable for temporal modeling and predictive maintenance. LLMs serve as the primary generative engine for reasoning and interaction, operating at higher semantic levels. In practical CAV workflows, these methods are not viewed as competing alternatives but as components of a broader generative toolkit that address different requirements.

4.6 Summary

This chapter examined three generative model families, GANs, VAEs, and domain-specific autoregressive methods that continue to support important functions within CAV systems. These models employ fundamentally different paradigms: GANs use adversarial training, VAEs employ variational inference, and autoregressive models use sequential factorization. Despite their differences, all three share fast inference that enables deployment in perception and prediction pipelines, distinguishing them from diffusion models (which require iterative sampling) and large language models (which operate at semantic levels).

GANs remain effective for domain adaptation, image-to-image translation, and sensor enhancement tasks due to their ability to generate high-frequency detail. VAEs provide structured latent spaces that support compact multi-sensor encoding, anomaly detection, and communication-efficient cooperative perception. Domain-specific autoregressive models remain indispensable for sequential prediction tasks, including trajectory forecasting, sensor drift modeling, vibration analysis, and audio modeling for diagnostics.

As discussed in the comparative analysis, diffusion models excel at high-fidelity simulation and scenario generation, whereas GANs, VAEs, and autoregressive models fill narrower but important roles across the autonomy stack. LLMs complement these models by providing high-level reasoning and interaction capabilities. Together, these generative approaches form a complementary toolkit that supports the diverse modeling demands of modern CAV systems. In practice, modern CAV systems increasingly employ hybrid architectures, where diffusion models generate high-fidelity data, VAEs compress and monitor representations, GANs adapt domains, and autoregressive models capture temporal dynamics. These components are not alternatives but form integrated pipelines, hybrid systems, and auxiliary modules that work together to address the complex requirements of autonomous vehicle perception, planning, and interaction.

References

1 Ian J. Goodfellow, Jean Pouget-Abadie, Mehdi Mirza, Bing Xu, David Warde-Farley, Sherjil Ozair, Aaron Courville, and Yoshua Bengio. Generative adversarial nets. In *Advances in Neural Information Processing Systems*, volume 27, 2014.

2 Diederik P. Kingma and Max Welling. Auto-encoding variational Bayes. *arXiv preprint arXiv:1312.6114*, 2013.

3 Jun-Yan Zhu, Taesung Park, Phillip Isola, and Alexei A. Efros. Unpaired image-to-image translation using cycle-consistent adversarial networks. In *Proceedings of the IEEE International Conference on Computer Vision*, pages 2223–2232, 2017.

4 Katharina Winter, Abhishek Vivekanandan, Rupert Polley, Yinzhe Shen, Christian Schlauch, Mohamed-Khalil Bouzidi, Bojan Derajic, Natalie Grabowsky, Annajoyce Mariani, Dennis Rochau, et al. Generative AI for autonomous driving: a review. *arXiv preprint arXiv:2505.15863*, 2025.

5 Phillip Isola, Jun-Yan Zhu, Tinghui Zhou, and Alexei A. Efros. Image-to-image translation with conditional adversarial networks. In *Proceedings of the IEEE Conference on Computer Vision and Pattern Recognition*, pages 1125–1134, 2017.

6 Christian Ledig, Lucas Theis, Ferenc Huszár, Jose Caballero, Andrew Cunningham, Alejandro Acosta, Andrew Aitken, Alykhan Tejani, Johannes Totz, Zehan Wang, et al. Photo-realistic single image super-resolution using a generative adversarial network. In *Proceedings of the IEEE Conference on Computer Vision and Pattern Recognition*, pages 4681–4690, 2017.

7 Geoffrey E. Hinton and Ruslan R. Salakhutdinov. Reducing the dimensionality of data with neural networks. *Science*, 313(5786):504–507, 2006.

8 Jinwon An and Sungzoon Cho. Variational autoencoder based anomaly detection using reconstruction probability. Technical Report 2, Seoul National University, Data Mining Center, 2015.

9 Aaron Van Den Oord, Oriol Vinyals, and Koray Kavukcuoglu. Neural discrete representation learning. In *Advances in Neural Information Processing Systems*, volume 30, 2017.

10 Robin Rombach, Andreas Blattmann, Dominik Lorenz, Patrick Esser, and Björn Ommer. High-resolution image synthesis with latent diffusion models. In *Proceedings of the IEEE/CVF Conference on Computer Vision and Pattern Recognition*, pages 10684–10695, 2022.

11 Florian Chadebecq, François Toussaint, Romain Amadieu, Benjamin Chretien, Pascal Allain, Caroline Petitjean, and Su Ruan. Variational autoencoders for anomaly detection in medical images. In *International Conference on Medical Image Computing and Computer-assisted Intervention*, pages 151–160. Springer, 2020.

12 Ashish Vaswani, Noam Shazeer, Niki Parmar, Jakob Uszkoreit, Llion Jones, Aidan N. Gomez, Łukasz Kaiser, and Illia Polosukhin. Attention is all you need. In *Advances in Neural Information Processing Systems*, volume 30, 2017.

13 Felipe Altché and Arnaud de La Fortelle. Conditional vehicle trajectories prediction in CARLA urban environment. In *2019 IEEE Intelligent Transportation Systems Conference*, pages 2312–2317, 2019.

14 Francesco Giuliari, Irtiza Hasan, Marco Cristani, and Francesco Galasso. Transformer networks for trajectory forecasting. *arXiv preprint arXiv:2003.08111*, 2021.

5

Vision-language Models for Scene Understanding

5.1 Introduction

CAVs operate in complex environments where perception and reasoning systems must process dynamic visual scenes, interpret language-based inputs from human occupants and infrastructure systems, and generate semantically coherent and explainable outputs. The integration of visual perception with natural language understanding presents a fundamental challenge in developing robust autonomous systems. To address this challenge, the field has increasingly adopted *vision-language models* (VLMs), a class of multimodal architectures that integrate image understanding with natural language processing to enable perception-grounded reasoning and generation.

Unlike traditional vision systems that rely solely on spatial cues or language models that operate purely in symbolic domains, VLMs unify the strengths of both modalities. These models accept visual inputs such as images or video frames and textual prompts in natural language, and generate language outputs that are grounded in the semantic content of the visual scene. This multimodal capability renders VLMs particularly valuable in CAV settings, where semantic alignment between what the vehicle sees and what it communicates or infers, is paramount.

Consider a scenario in which an autonomous vehicle approaches a busy intersection. The front camera captures a large city bus stopped at the crosswalk, partially obscuring the view of the sidewalk beyond. Simultaneously, the vehicle receives a V2X (vehicle-to-everything) alert stating "Pedestrian approaching from the right." Although no pedestrian is visible in the current camera frame, VLMs can integrate the visual scene and the external alert to infer that the pedestrian may be occluded behind the bus and consequently generate an explanation such as: *"Pedestrian may be hidden behind the bus."* This form of grounded, interpretable reasoning is essential for both machine decision-making and human trust in autonomous systems.

The objective of this chapter is to provide a rigorous yet accessible treatment of vision-language models as applied to connected and autonomous vehicles. We introduce a modular architectural framework that decomposes VLMs into three fundamental stages: (1) visual feature extraction, (2) cross-modal adaptation, and (3) integration into a language model. We then examine training strategies for aligning vision and language modalities and discuss the key challenges and design decisions associated with deploying VLMs in CAV contexts. Throughout the chapter, illustrative examples are drawn from real-world CAV use cases, including visual question answering, scene-aware captioning, and V2X message interpretation.

This chapter focuses on image-conditioned models, while video large language models (VidLLMs), which address the additional temporal and motion-specific dimensions inherent in video streams, are discussed in a subsequent chapter. Together, these chapters form the foundation for understanding how generative AI can endow CAV systems with multimodal semantic intelligence, thereby enabling more robust, interpretable, and human-aligned autonomy.

5.2 Architectural Principles of VLMs

VLMs operate at the intersection of two distinct representational domains: spatially structured visual input and symbolically structured natural language. Bridging this modality gap requires a modular design that can extract semantically meaningful representations from visual data, align these representations with the internal structure of language models, and support joint reasoning or generation conditioned on both modalities.

To address these challenges in a systematic manner, modern VLMs are typically organized as a *three-stage fusion pipeline*, illustrated in Figure 5.1. This architecture separates concerns across perception, adaptation, and reasoning, allowing each component to be optimized individually or jointly depending on computational and application constraints. The three stages are (1) visual feature extraction, (2) vision-language adaptation, and (3) integration into the language model. Each stage is examined in detail in the following sections.

The three-stage architecture is not merely a consequence of engineering convenience, but reflects foundational considerations in multimodal learning. Visual perception and natural language processing differ in terms of input structure, semantic granularity, and temporal dynamics. By decomposing the modeling pipeline into independent yet interoperable components, VLMs can flexibly incorporate advancements in each modality while maintaining architectural generality.

For example, improvements in vision backbones (e.g., from ViT to hybrid region-aware encoders) can be adopted without modifying the downstream language model. Adapter layers can be retrained to align different vision-language pairs or fine-tuned to support new tasks such as instruction following, caption generation, or scene-based question answering.

To concretely illustrate the three-stage processing pipeline, we present Algorithm 1, which demonstrates the complete forward pass through a VLM, showing how visual and textual inputs are processed and combined:

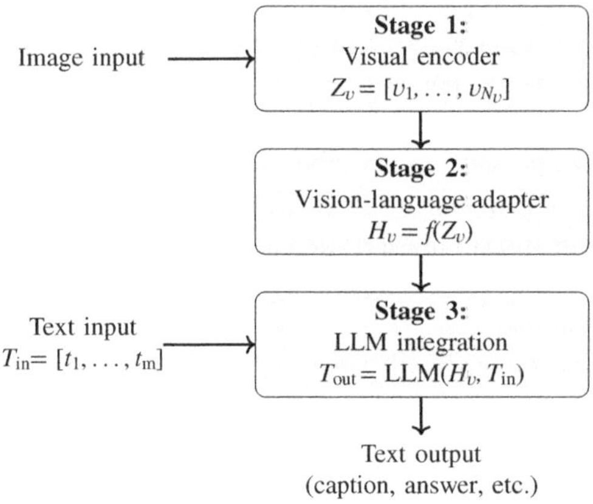

Figure 5.1 Three-stage architectural pipeline of a vision-language model. Raw images are encoded as visual tokens, adapted into a language-compatible representation, and integrated into the LLM for grounded reasoning or generation.

Algorithm 1 VLM Forward Pass Algorithm

Require: Image I, Text prompt T
Ensure: Generated text response

```
// Stage 1: Visual Feature Extraction
patches ← divide_image_into_patches(I, patch_size)
visual_tokens ← []
for each patch in patches do
    embedding ← linear_projection(patch)
    visual_tokens.append(embedding)
end for
visual_tokens ← add_positional_encoding(visual_tokens)
visual_features ← vision_transformer(visual_tokens)
// Stage 2: Vision-language Adapter
adapted_features ← vision_language_adapter(visual_features)
// Stage 3: LLM Integration
text_tokens ← tokenize(T)
combined_input ← concatenate(adapted_features, text_tokens)
// Generation
output_tokens ← []
current_tokens ← combined_input
for step in range(max_length) do
    logits ← language_model(current_tokens)
    next_token ← sample(logits[−1])
    if next_token = EOS_TOKEN then
        break
    end if
    output_tokens.append(next_token)
    current_tokens ← append(current_tokens, next_token)
end for
return detokenize(output_tokens)
```

The algorithm demonstrates the three-stage processing pipeline. In Stage 1, the image is divided into patches that are linearly projected and positionally encoded before being processed by a vision transformer to produce visual tokens. Stage 2 applies a vision-language adapter to transform these visual tokens into a representation space compatible with the language model. In Stage 3, the adapted visual features are concatenated with tokenized text input, and the combined sequence is passed to the language model for generation. The generation loop continues until an end-of-sequence token is produced, or a maximum length is reached, with each iteration sampling the next token from the language model's probability distribution over the vocabulary.

5.3 Stage 1: Visual Feature Extraction

The first stage in the architecture of a vision-language model is concerned with transforming raw visual input into a structured, semantically meaningful representation. Unlike traditional convolutional networks that output dense feature maps, VLMs require visual representations to be compatible with token-based architectures used in language models. This necessitates a transformation from pixel-level input to a sequence of discrete visual tokens that can participate in attention-based computation alongside text tokens.

5.3.1 From Raw Pixels to Visual Tokens

Let $I \in \mathbb{R}^{H\times W\times C}$ denote an image of height H, width W, and C color channels. A visual encoder f_{enc} maps this image into a sequence of vectors:

$$Z_v = f_{\text{enc}}(I) = [v_1, \dots, v_{N_v}], \quad v_i \in \mathbb{R}^{D_v}, \tag{5.1}$$

where N_v denotes the number of visual tokens and D_v the token dimensionality. Each token v_i is intended to represent a semantically coherent region, object, or feature in the image, similar in structure to a language embedding.

To be suitable for integration into downstream language models, these tokens should satisfy several criteria. First, they must be structured, preserving spatial or semantic locality and maintaining compatibility with positional encodings. Second, they should be compressible, with minimal redundancy across tokens to allow for efficient projection or adaptation in subsequent stages. Finally, they must be aligned, exhibiting semantic compatibility with language representations to enable meaningful cross-modal interactions.

The following subsections review the principal categories of visual encoders used in contemporary VLMs.

5.3.2 Patch-based Visual Encoders

Patch-based encoders divide the image into nonoverlapping fixed-size patches and treat each patch as an atomic unit for tokenization. The canonical model in this category is the *vision transformer* (ViT) [1], which applies the transformer architecture originally developed for text to a sequence of image patches.

Given an image I divided into $P \times P$ patches, each patch I_i is flattened and linearly projected:

$$x_i = \text{flatten}(I_i) \cdot W_{\text{patch}}, \quad x_i \in \mathbb{R}^{D}, \tag{5.2}$$

where I_i denotes the ith image patch, P is the patch size, $W_{\text{patch}} \in \mathbb{R}^{(P^2\cdot C)\times D}$ is a learnable projection matrix, and D is the embedding dimension. Positional embeddings PE_i are added to preserve spatial order (see Chapter 2, Section 2.6 for details on positional encoding), and the resulting sequence is passed through multiple transformer layers:

$$z_i = x_i + \text{PE}_i, \quad Z_v = \text{Transformer}([z_1, \dots, z_{N_p}]) = [v_1, \dots, v_{N_p}], \tag{5.3}$$

where z_i is the positionally-encoded patch embedding, PE_i denotes the positional encoding for patch i, N_p is the number of patches, and Z_v represents the final visual token sequence output by the transformer, with each $v_i \in \mathbb{R}^D$.

Patch-based encoders are widely used in many VLM architectures due to their simplicity, scalability, and compatibility with transformer-based LLMs. However, they treat all regions equally, potentially limiting their effectiveness in scenes with fine-grained or localized semantic content.

5.3.3 Contrastively Pretrained Encoders

Contrastive encoders are trained to map visual and textual inputs into a shared semantic embedding space. The most prominent example is CLIP (contrastive language–image pretraining) [2], which aligns image–text pairs through a contrastive loss. Each image I and caption T is processed by separate encoders:

$$z^{\text{img}} = f_{\text{img}}(I), \quad z^{\text{text}} = f_{\text{text}}(T), \tag{5.4}$$

and the model minimizes an InfoNCE (information noise contrastive estimation) loss [3] over cosine similarities. The contrastive objective encourages matching pairs to be close and mismatched pairs to be distant in the embedding space.

An alternative formulation is used by SigLIP [4], which replaces the softmax-based InfoNCE loss [3] with a sigmoid-based binary cross-entropy loss that supports multiple positive matches per image. This enhances robustness to weak supervision and noisy labels.

Contrastively pretrained encoders provide aligned embeddings suitable for downstream matching, retrieval, or grounding tasks. While they often yield global representations, they can also be adapted to produce token-level outputs via intermediate activations or hybrid designs.

For fine-tuning these contrastive encoders on domain-specific data (e.g., CAV scene-text pairs), contrastive learning techniques can be applied to further align the embedding spaces. The comprehensive methodology for contrastive learning, including the mathematical formulation and training workflow, is detailed in Section 5.6.4.

5.3.4 Region-aware and Hybrid Visual Encoders

For CAV applications that demand spatial precision and fine-grained semantic understanding, region-aware encoders offer a more specialized alternative. These models produce visual tokens that correspond to specific spatial regions, object boundaries, or segmentation masks. Notable examples include the segment anything model (SAM) [5] and ViTDet [6].

SAM generates dense region embeddings that can be aligned with prompts or external queries. ViTDet retains high-resolution grid features suitable for dense prediction tasks such as object detection or semantic segmentation.

Hybrid encoders combine global and local perspectives by merging features from coarse-resolution encoders (e.g., ViT or CLIP [2]) with high-resolution or object-specific embeddings (e.g., SAM). This allows for multi-scale reasoning while maintaining compatibility with LLMs. Modern vision-language models adopt this dual-stream strategy, enabling robust reasoning in visually complex or occluded scenes.

5.3.5 Summary of Visual Encoder Types

Table 5.1 summarizes the major encoder categories and their key characteristics. Visual feature extraction defines the initial representational substrate upon which all subsequent reasoning in VLMs is built. The selection of a visual encoder determines not only the token quality but also the fidelity and interpretability of multimodal integration. The next stage in the pipeline addresses the challenge of aligning these visual embeddings with the expectations of a pretrained language model.

Table 5.1 Comparison of visual encoder types used in VLMs.

Encoder type	Key characteristics	Typical use cases
Patch-based transformers	Tokenize images into fixed-size patch embeddings; compatible with transformers	General-purpose scene understanding, VQA, captioning
Contrastively pretrained encoders	Aligned visual–text embeddings; trained via contrastive losses	Retrieval, grounding, semantic filtering
Region-aware encoders	Preserve spatial layout and object boundaries; suitable for dense scenes	Object recognition, OCR, occlusion-aware reasoning
Hybrid encoders	Combine global and local representations via multi-resolution fusion	Multi-scale CAV tasks, high-resolution scene analysis

5.4 Stage 2: Vision-language Adaptation

Once a visual encoder has transformed raw image input into a structured sequence of semantic tokens, these embeddings must be adapted for compatibility with a pretrained language model. This adaptation stage is essential because the visual and linguistic modalities differ not only in representation structure but also in embedding dimensionality, semantic abstraction, and temporal behavior.

Let $Z_v = [v_1, \dots, v_{N_v}]$ denote the visual tokens from the encoder, where $v_i \in \mathbb{R}^{D_v}$, and let $H = [h_1, \dots, h_L]$ represent the embeddings expected by the language model, where $h_j \in \mathbb{R}^{D_{\text{LLM}}}$. Since $D_v \neq D_{\text{LLM}}$ and visual tokens lack the syntactic properties learned by LLMs, a transformation function is required:

$$f : \mathbb{R}^{D_v} \rightarrow \mathbb{R}^{D_{\text{LLM}}}, \quad H_v = f(Z_v).$$

5.4.1 Challenges in Cross-modal Representation Alignment

Cross-modal adaptation must satisfy several competing objectives. First, it must achieve dimensional alignment by projecting visual tokens into the LLM's embedding space without distorting semantic content. Second, it should enable token compression or selection, reducing the token sequence length when necessary while preserving only the most salient features. Third, it must provide contextual conditioning, allowing visual information to influence or guide language generation in task-relevant ways. Finally, it should maintain training efficiency by minimizing computational overhead while preserving effective gradient flow during fine-tuning. These objectives have led to the development of several adapter architectures, which are reviewed in the following subsections.

5.4.2 MLP-based Adapters

The most straightforward adaptation approach is a multi-layer perceptron (MLP), typically consisting of a single linear projection or a shallow feedforward network with nonlinear activation. The transformation is applied element-wise to each visual token $v_i \in \mathbb{R}^{D_v}$:

$$h_i = W_2 \cdot \text{GeLU}(W_1 \cdot v_i + b_1) + b_2, \quad h_i \in \mathbb{R}^{D_{\text{LLM}}}, \tag{5.5}$$

where W_1, W_2 are trainable weight matrices, b_1, b_2 are bias vectors, and GeLU [7] denotes the Gaussian error linear unit activation function. The resulting adapted sequence is $H_v = [h_1, \ldots, h_{N_v}]$.

MLP adapters are computationally efficient and easily integrated between the visual encoder and LLM. They are widely used in many vision-language models due to their simplicity and effectiveness. However, they lack dynamic attention or token selection mechanisms, which limits their expressiveness for complex reasoning tasks.

5.4.3 Query-based Adapters (Q-Former)

The Query-Former (Q-Former) introduces a more structured adaptation strategy based on transformer attention. It employs a set of L *learnable query embeddings* $Q = [q_1, \ldots, q_L]$, which are trainable tokens designed to actively query and extract task-specific information from visual features. The architecture alternates between self-attention layers (where query embeddings attend to each other) and cross-attention layers (where query embeddings attend to frozen visual tokens $Z_v = [v_1, \ldots, v_{N_v}]$):

$$\hat{q}_j = \text{CrossAttn}(\text{SelfAttn}(q_j), Z_v), \tag{5.6}$$

where q_j denotes the jth learnable query embedding, Z_v is the visual token sequence from the encoder (typically kept frozen), and the attention mechanism enables each query to selectively aggregate information from the visual sequence. The resulting adapted tokens $\hat{Q} = [\hat{q}_1, \ldots, \hat{q}_L]$ are passed to the language model as visual context.

The key characteristic of learnable query embeddings is their role as active querying mechanisms: they guide the model's attention to focus on relevant parts of the visual input, effectively extracting task-specific information through dynamic interaction with the data. Q-Formers are trained end-to-end with the LLM using language modeling losses, enabling selective summarization of visual content and enhanced performance on tasks requiring fine-grained grounding. This architecture has been used effectively in BLIP-2 [8].

5.4.4 Latent Compression Adapters (Perceiver Resampler)

Perceiver resamplers take a different approach by learning a fixed set of *learnable latent vectors* $L = [\ell_1, \ldots, \ell_K]$ that compress the visual token sequence into a compact representation. Unlike Q-Former's learnable query embeddings, which actively query visual features, learnable latent vectors serve as compressed, lower-dimensional representations that capture the essence of the visual data. Given visual tokens Z_v and learnable latent vectors L, the adapter computes:

$$\hat{\ell}_k = \text{CrossAttn}(\ell_k, Z_v), \tag{5.7}$$

where ℓ_k denotes the k-th learnable latent vector, Z_v is the visual token sequence from the encoder, and each latent vector aggregates information across the full visual sequence through cross-attention. These compressed latents $\hat{L}$ form a fixed-size output that is input to the LLM.

The key distinction between learnable query embeddings (Q-Former) and learnable latent vectors (perceiver resampler) lies in their purpose: query embeddings are designed to dynamically interact with and extract task-specific information from input data, while latent vectors represent compressed, lower-dimensional encodings that capture the essential features of the data in a compact form. Architecturally, Q-Former alternates between separate self-attention and cross-attention layers, while perceiver resamplers integrate both mechanisms within unified layers. This strategy is used in models like Flamingo [9], offering strong scalability and support for variable-length input in many deployment scenarios. However, the fixed output length may limit adaptability in highly diverse visual contexts.

5.4.5 Comparative Analysis of Adaptation Strategies

Table 5.2 summarizes the key characteristics of common vision-language adapters.

The choice of adapter impacts model performance, training dynamics, and interpretability. MLP-based methods are easy to implement and suitable for lightweight deployments, whereas Q-Former and perceiver-based approaches offer greater expressive capacity and are preferable in scenarios requiring semantic filtering or long-horizon reasoning.

5.5 Stage 3: Integration into the Language Model

The final stage in the vision-language model architecture concerns how adapted visual embeddings are integrated into the language model for joint reasoning or generation. This stage determines the depth, flexibility, and interpretability of multimodal interaction and significantly influences model performance on downstream tasks such as captioning, instruction following, or visual question answering.

Effective integration must satisfy several requirements. First, it must enable modality co-attention, allowing the language model to attend to visual context alongside textual input. Second, it should preserve pretrained knowledge through minimal architectural modifications to retain the language model's capabilities. Finally, it must ensure token compositionality, processing visual and linguistic inputs in a manner that respects their individual and joint semantics.

Two dominant approaches have emerged in the literature: *flat self-attention* and *in-layer cross-attention*.

5.5.1 Flat Self-attention

Flat self-attention treats adapted visual tokens as part of the language model's input sequence. The adapted visual embeddings $H_v = [\hat{v}_1, \dots, \hat{v}_L]$ are concatenated with text tokens $T = [t_1, \dots, t_N]$, and the resulting sequence is passed to the language model:

$$X = [\hat{v}_1, \dots, \hat{v}_L, t_1, \dots, t_N], \tag{5.8}$$

where all tokens participate in the language model's native self-attention mechanism. Some architectures may include a special separator token (e.g., `[SEP]`, originally from BERT [10]) between visual and text tokens to explicitly demarcate modality boundaries, though this is not always necessary as positional encodings can also serve this function.

This method leverages the transformer's inherent ability to model arbitrary token sequences. Importantly, it requires no changes to the internal architecture of the language model, thereby enabling reuse of frozen pretrained weights.

Table 5.2 Comparison of vision-language adapter types.

Adapter type	Core mechanism	Representative models
Linear/MLP	Projection of visual tokens into LLM embedding space	Widely adopted in modern VLMs
Query-based (Q-Former)	Learnable queries with cross-attention to visual tokens	BLIP-2 [8]
Latent compression (perceiver)	Cross-attention from latent tokens to visual embeddings	Flamingo [9]

The attention computation is performed over the full token set:

$$\text{Attention} = \text{softmax}\left(\frac{QK^\top}{\sqrt{d}}\right)V, \tag{5.9}$$

where Q, K, and V are the query, key, and value matrices obtained from the concatenated visual and text embeddings, and d is the embedding dimension. This is the standard scaled dot-product attention mechanism (see Chapter 2, Section 2.7), applied here to multimodal vision-language sequences.

5.5.1.1 Advantages and Limitations

Flat self-attention is straightforward to implement and widely adopted in practice. It is commonly used in many modern vision-language models due to its simplicity and effectiveness. However, because visual and text modalities are fused only at the input level, this approach offers limited flexibility in how visual context influences intermediate language representations. The resulting fusion is shallow and may not support complex multimodal dependencies.

5.5.2 In-layer Cross-attention

In contrast, in-layer cross-attention introduces explicit visual conditioning at intermediate layers of the language model. This approach modifies the LLM architecture to interleave cross-modal attention layers between existing transformer blocks.

Let $H^{(\ell)}$ denote the hidden states of the language model at layer ℓ. The cross-attention module computes:

$$Z^{(\ell)} = \text{CrossAttn}(H^{(\ell)}, H_v), \tag{5.10}$$

where H_v is the adapted visual token sequence, treated as a static memory, and $H^{(\ell)}$ serves as the query. The result is integrated via residual connection and feedforward layers:

$$H^{(\ell+1)} = \text{FFN}(H^{(\ell)} + \gamma Z^{(\ell)}), \tag{5.11}$$

where γ is a gating factor that can be fixed or learnable.

This design enables visual context to influence the language model's internal reasoning process across multiple layers, resulting in deeper fusion and improved grounding.

5.5.2.1 Advantages and Limitations

In-layer cross-attention provides greater control over how and where multimodal fusion occurs. It supports more sophisticated tasks such as reasoning under occlusion, event prediction, or justification generation. However, this method requires architectural modifications to the language model, which complicates deployment and limits compatibility with fully frozen LLMs.

Notable models using this approach include Flamingo [9] and BLIP-2 [8]. These models often employ gated or sparse attention to ensure stable training and effective alignment.

5.5.3 Comparative Analysis of Integration Strategies

Table 5.3 summarizes the trade-offs between the two strategies. Flat integration emphasizes simplicity, efficiency, and reusability, making it suitable for applications with limited computational resources or deployment constraints. Cross-attention enables deeper grounding and improved task performance, at the cost of increased model complexity and training requirements.

Table 5.3 Comparison of LLM integration strategies for vision-language fusion.

Integrationstrategy	Fusion mechanism	Representative models
Flat self-attention	Concatenation of visual and text tokens at input; no architectural changes	Widely adopted in modern VLMs
In-layer cross-attention	Cross-attention from language layers to static visual memory	Flamingo [9], BLIP-2 [8]

5.5.4 Design Considerations in CAV Applications

For connected and autonomous vehicle systems, the choice between flat and cross-attention integration depends on the specific application requirements. Driver-facing interfaces and low-latency responses may benefit from flat self-attention with lightweight MLP adapters. In contrast, applications requiring context-sensitive reasoning across spatial and temporal cues, such as occlusion-aware braking or visual-V2X consistency checking, may justify the additional cost of cross-attentional architectures.

In both cases, the integration mechanism serves as the final interface through which perceptual signals are transformed into grounded linguistic decisions. As such, it plays a critical role in the explainability, robustness, and safety of multimodal reasoning in CAV environments.

5.6 Training Methodologies for VLMs

Training vision-language models (VLMs) involves aligning two inherently distinct modalities: continuous, spatially structured visual data and discrete, symbolically structured language. This alignment must support a wide range of downstream tasks, including captioning, visual question answering (VQA), and instruction following. Moreover, in CAV contexts, the resulting model must be efficient, explainable, and robust under real-time operational constraints. The following subsections present core training principles, phased fine-tuning strategies, and design considerations necessary to realize effective VLMs for multimodal reasoning.

5.6.1 Objectives and Challenges in Multimodal Training

The primary objective of VLM training is to create a unified embedding space in which visual and textual elements can interact meaningfully. However, several challenges arise. First, there is a modality discrepancy where visual and linguistic tokens differ in structure, scale, and distribution. Second, parameter reuse requires integrating pretrained vision encoders and language models without degrading existing capabilities. Third, data scarcity limits the availability of high-quality, paired image–text data compared to text-only corpora. Finally, training instability can occur when misaligned features lead to poor convergence or catastrophic forgetting. To address these challenges, modern VLMs are trained in a progressive manner that separates modality alignment from task specialization.

5.6.2 Fine-tuning Phases and Strategies

A common methodology for training VLMs employs multiple fine-tuning phases, often with increasing supervision and complexity. It is important to note that these phases are all optional and can be applied independently or in various combinations based on computational resources, data availability, and application requirements. Unlike the three-stage VLM pipeline (which represents necessary architectural components), fine-tuning phases represent flexible training strategies that can be adapted or omitted as needed. The necessity of each phase depends on the specific application, model architecture, and desired performance outcomes. The following presents a representative three-phases process as an illustrative example.

5.6.2.1 Phase 1: Vision-language Alignment

This phase focuses on aligning visual and textual representations. The specific approach depends on whether encoders are pretrained or trained from scratch.

When using pretrained encoders (e.g., CLIP), this phase trains adapter modules (such as Q-Former or perceiver resampler) to bridge vision and language modalities. In this scenario, the visual encoder and language model are typically frozen, and only the adapter module is trained to project or transform visual embeddings into the LLM's representational space. Training objectives for adapters vary by architecture: some models (e.g., BLIP-2) use contrastive learning objectives to train the adapter, where the adapter learns to align frozen visual encoder outputs with frozen text encoder outputs; other models use language modeling losses or matching losses. This phase can be omitted if using pretrained adapters or if the model architecture does not require adapter training.

When training encoders from scratch (as in CLIP), this phase trains the encoders themselves using contrastive learning to align image and text embeddings in a shared semantic space. The comprehensive methodology for contrastive learning, including the mathematical formulation (InfoNCE loss) and training workflow, is detailed in Section 5.6.4.

This phase ensures that visual embeddings carry sufficient semantic structure to interact with language tokens.

5.6.2.2 Phase 2: Multimodal Generation and Reasoning

In this phase, the adapter and parts of the LLM are jointly fine-tuned on generation tasks such as captioning or VQA. Input consists of both image embeddings and natural language prompts, and the model learns to produce coherent, grounded text:

$$\mathcal{L}_{\text{CE}} = -\sum_{t} \log p(y_t \mid y_{<t}, H_v, T), \tag{5.12}$$

where y_t is the target token at position t, $y_{<t}$ denotes all preceding tokens, H_v denotes the adapted visual embeddings, and T the prompt.

Depending on computational constraints, vision encoders may remain frozen or be partially fine-tuned during this phase.

5.6.2.3 Phase 3: Instruction-tuned Adaptation

This phase focuses on user-facing tasks using instruction-style prompts. Models are trained to follow human-written queries or dialogues conditioned on visual context. This phase may incorporate human preference data or reinforcement learning for better alignment.

Advanced methods such as direct preference optimization (DPO) [11] or reinforcement learning from AI feedback (RLAIF) [12] are sometimes used to tune models toward preferred outputs in safety-critical domains. Practitioners may choose to apply any combination of these phases or skip phases entirely, depending on their specific requirements and constraints.

5.6.3 Freezing Strategies and Parameter-efficient Tuning

Given the large size of vision and language backbones, it is often impractical to fine-tune all parameters. Instead, a common strategy is to freeze pretrained components and selectively adapt intermediary modules. The vision encoder is typically frozen, with high-resolution models (e.g., CLIP [2], SAM) reused without modification. The language model is either fully frozen or partially unfrozen at later layers. Adapters are always trained and may include MLPs, Q-Former, or cross-attention heads. Additionally, parameter-efficient tuning techniques such as LoRA (low-rank adaptation) or adapter layers can further reduce training cost.

This approach enables the model to generalize from pretrained knowledge while adapting to domain-specific tasks using limited data. The flexibility of freezing strategies allows practitioners to adapt the training pipeline based on computational resources and task requirements, rather than following a rigid sequence.

5.6.4 Contrastive Learning for Multimodal Alignment

Contrastive learning is a fundamental technique for aligning embeddings from different modalities into a shared semantic space. While the most prominent application is image–text alignment (as in CLIP), the same principles apply to text-to-text alignment for retrieval tasks, audio–text alignment, and other cross-modal scenarios.

5.6.4.1 General Formulation

Contrastive learning operates on pairs of semantically related inputs (x_i, x_{t_i}), where x_{t_i} is a semantically similar counterpart to x_i. In multimodal contexts, these pairs may be (image, caption), (scene, V2X message), or (query, relevant document). The goal is to learn encoders that map inputs from different modalities into a common embedding space where semantically related pairs are close and unrelated pairs are distant.

Let $f_\theta(x) \in \mathbb{R}^d$ denote the output of an encoder network applied to input x, parameterized by θ. For multimodal contrastive learning, separate encoders may be used for each modality (e.g., f_{img} for images and f_{text} for text), or a shared encoder may be used when both inputs are from the same modality (e.g., text-to-text for retrieval).

Advanced Note 5.1 Mathematical Formulation of Contrastive Learning

Contrastive training adopts the InfoNCE (information noise-contrastive estimation) loss [3], which encourages similarity between anchor-positive pairs while penalizing similarity to other unrelated examples in the batch. For a batch of B positive pairs, the InfoNCE loss is:

$$L_{\text{InfoNCE}} = -\frac{1}{2B} \sum_{i=1}^{2B} \log \frac{\exp(\text{sim}(z_i, \tilde{z}_i)/\tau)}{\sum_{j=1}^{2B} \exp(\text{sim}(z_i, z_j)/\tau)}, \tag{5.13}$$

where $z_i = f_\theta(x_i)$ is the embedding of the i-th input; $z_{t_i} = f_\theta(x_{t_i})$ is the embedding of its paired positive match; $\text{sim}(a, b)$ denotes cosine similarity $\frac{a \cdot b}{\|a\|\|b\|}$; τ is a temperature hyperparameter that controls the sharpness of the similarity distribution; and $2B$ is the total number of embeddings in the batch (anchors and their positive matches).

Unlike classification loss functions that require labeled outputs, the InfoNCE objective relies only on pairwise similarity. Negatives are drawn implicitly from other items in the batch, avoiding the need for explicit negative mining strategies.

For image–text alignment (as in CLIP), the loss becomes:

$$\mathcal{L}_{\text{CLIP}} = -\log \frac{\exp(\text{sim}(z_i^{\text{img}}, z_i^{\text{text}})/\tau)}{\sum_j \exp(\text{sim}(z_i^{\text{img}}, z_j^{\text{text}})/\tau)}, \tag{5.14}$$

where $z_i^{\text{img}} = f_{\text{img}}(I_i)$ and $z_i^{\text{text}} = f_{\text{text}}(T_i)$ are embeddings from separate encoders for images and text, respectively.

5.6.4.2 Training Workflow

The contrastive learning process comprises a sequence of steps from data construction to encoder parameter updates:

1. **Prepare Pairs:** Collect semantically related input pairs. For image–text alignment, these are (image, caption) pairs. For text-to-text alignment (e.g., in retrieval systems), these may be (query, relevant document) pairs.
2. **Encoding:** Pass both elements of each pair through their respective encoders (shared encoder for same-modality pairs, separate encoders for different modalities) to obtain embeddings.
3. **Batch Construction:** Assemble a batch containing all anchor–positive pairs, implicitly defining negative examples from other pairs in the batch.
4. **Similarity Computation:** Compute cosine similarity between each anchor and all items in the batch.
5. **Contrastive Loss:** Apply InfoNCE loss to contract similar pairs and repel unrelated ones in embedding space.
6. **Backpropagation:** Use gradient-based optimization to update encoder parameters.

5.6.4.3 Multimodal vs. Uni-modal Contrastive Learning

The contrastive learning framework applies to both multimodal and uni-modal scenarios. CLIP exemplifies multimodal contrastive learning, where separate encoders for images and text are trained to align visual and textual representations. In uni-modal scenarios (e.g., text-to-text for retrieval), a shared encoder processes both elements of the pair, enabling domain-specific adaptation of embeddings for improved semantic matching. The mathematical formulation remains the same; only the encoder architecture differs.

5.6.4.4 Domain-specific Fine-tuning

While pretrained contrastive models like CLIP provide strong baseline performance, CAV-specific scenarios often require further alignment between embeddings and domain-specific terminology. For instance, a pretrained CLIP model might not distinguish between "lane change" and "merging" in traffic contexts, or a text encoder might not align "braking intervention" with "deceleration events" in retrieval tasks. Domain-specific fine-tuning applies contrastive learning to CAV-specific pairs (e.g., traffic scene images paired with V2X messages, or query–document pairs from CAV logs) to improve alignment for specialized terminology and scenarios.

5.6.5 Training Considerations and Best Practices

The fine-tuning phases described above require careful consideration of several factors to ensure successful convergence and optimal performance. Table 5.4 summarizes the key characteristics and requirements for each training phase.

5.6.5.1 Data Requirements and Quality

Each fine-tuning phase has distinct data requirements. Phase 1 benefits from large-scale image–text pairs with diverse visual content and descriptive captions. Phase 2 requires task-specific datasets for captioning, VQA, or other generation tasks. Phase 3 needs high-quality instruction–response pairs that demonstrate desired behavior patterns. In CAV applications, domain-specific datasets containing traffic scenarios, road signs, and vehicle interactions are crucial for achieving robust performance.

5.6.5.2 Computational Efficiency

Training efficiency is paramount for practical deployment. Parameter-efficient methods such as LoRA (low-rank adaptation) or adapter layers can reduce computational overhead while maintaining performance. Freezing pretrained components during fine-tuning phases minimizes memory requirements and training time. Gradient checkpointing and mixed-precision training further optimize resource utilization. Detailed discussion of parameter-efficient fine-tuning techniques, including LoRA, is provided in the fine-tuning chapter.

5.6.5.3 Evaluation and Monitoring

Continuous evaluation during training helps identify issues early. Metrics should include both task-specific performance (e.g., BLEU for captioning, accuracy for VQA) and alignment quality measures. In CAV contexts, safety-critical evaluation should assess model behavior under edge cases and adversarial conditions. Comprehensive evaluation methodologies and detailed performance analysis for VLMs are discussed in a later chapter dedicated to model evaluation.

5.6.6 Design Guidelines for Scalable and Stable Training

To ensure stable convergence and effective multimodal fusion, several engineering best practices have emerged. First, warm-start adapters by initializing projection or Q-Former weights with small values to avoid overwhelming the LLM. Second, use mixed-modality batches that include both image–text and text-only examples to preserve language fluency. Third, apply gradient scaling or clipping to prevent visual tokens from dominating the attention distribution. Finally, monitor modality entropy to check for collapsed attention or token redundancy during training.

Table 5.4 Fine-tuning phase characteristics and requirements.

Phase	Primary objective	Key requirements
Vision-language alignment	Learn cross-modal representations	Large image–text pairs, contrastive loss
Multimodal generation	Task-specific fine-tuning	Task datasets, cross-entropy loss
Instruction tuning	Human-aligned behavior	Instruction data, preference optimization

In CAV contexts, where inference must be efficient and robust, the training pipeline must also support deployment constraints such as limited memory, real-time performance, and system-level interpretability.

5.7 Deployment in Connected and Autonomous Vehicles

As generative models mature into multimodal reasoning agents, their deployment within CAV systems demands careful attention to architectural integration, latency, reliability, and explainability. VLMs are uniquely positioned to serve as a semantic layer between low-level perception modules and high-level decision-making processes.

VLMs enable CAVs to transform high-dimensional visual inputs into human-aligned natural language descriptions or justifications. These models can identify salient objects, agents, and environmental features; contextualize behavior such as braking, merging, or yielding; mediate between camera observations, V2X alerts, and HD map content; and generate interpretable responses to human or agent queries. By translating visual context into linguistic form, VLMs facilitate downstream reasoning, intent prediction, and multi-agent coordination.

5.7.1 Grounded Interpretation of V2X and Scene Inputs

Connected vehicles routinely receive symbolic messages via V2X communication (e.g., "Pedestrian approaching from right"). However, these messages must be grounded in current sensory input to inform safe and contextually appropriate action. VLMs can align such messages with visual observations to determine whether the alert corresponds to observable objects or occluded regions, whether a system response (e.g., braking, warning) is justified, and whether multiple sensory sources (vision, radar, V2X) convey consistent semantics. This fusion of structured language with unstructured vision data enhances robustness in dynamic or partially observable environments.

When visual encoders cannot directly observe objects due to occlusion, VLMs enable reasoning about occluded or uncertain scenes by integrating visual observations with external information sources such as V2X alerts. VLMs can infer the likely presence of occluded objects based on spatial relationships, scene structure, and semantic consistency with external alerts. This capability supports conservative decision-making (e.g., slowing or stopping) while providing human-interpretable justifications for safety-critical actions.

5.7.2 Prompt Engineering for Scene-aware Inference

VLMs operate via text prompts that condition the model's reasoning process. In CAVs, these prompts may be generated by system components (e.g., V2X alerts, driver queries) or predefined templates.

Common prompt types include alert contextualization (e.g., "Vehicle merging from left—is it visible?"), perception queries (e.g., "What is in front of the crosswalk?"), causal reasoning (e.g., "Why is the vehicle slowing down?"), and navigation guidance (e.g., "Where is the closest stop sign?").

Effective prompt design should be specific to driving context and spatial referents, include relevant temporal or positional cues, and align with the model's training distribution to minimize domain shift.

5.7.3 Functional Mapping of VLM Capabilities to CAV Tasks

Table 5.5 illustrates how VLM functionalities correspond to representative tasks in connected autonomous vehicles.

These functionalities make VLMs instrumental in achieving interpretable autonomy, particularly in shared spaces where transparency and coordination are critical.

5.8 Case Study: Semantic Filtering of V2X Messages

To illustrate the practical application of VLMs in CAV systems, this section presents a detailed case study of semantic filtering for V2X messages. This example demonstrates how the architectural principles and training methodologies discussed in previous sections can be applied to solve real-world challenges in connected vehicle environments.

Modern connected vehicles often operate in environments saturated with external messages, such as hazard alerts, traffic updates, or cooperative driving intentions transmitted via V2X infrastructure. However, many of these messages may be redundant, outdated, or misaligned with the vehicle's current perceptual state. A critical challenge, therefore, lies in filtering incoming messages to identify which are semantically novel or situationally relevant.

VLMs, particularly those trained with contrastive objectives, offer a natural mechanism for such filtering. By embedding both visual input and text messages into a shared semantic space, the system can assess how well an alert corresponds to the observed environment. The following subsections present a contrastive retrieval-based approach for message relevance estimation in CAV pipelines.

5.8.1 Contrastive Retrieval for Scene–Alert Consistency

Contrastive vision-language models such as CLIP and SigLIP learn image and text encoders that map inputs into a common embedding space. Given an image I and a set of candidate V2X messages $\{T_1, \dots, T_K\}$, the model computes similarity scores between the embedded image and each message:

$$s_i = \cos\big(f_{\text{img}}(I), f_{\text{text}}(T_i)\big), \tag{5.15}$$

where s_i is the similarity score for the i-th message, T_i denotes the i-th text message, $\cos(\cdot, \cdot)$ denotes cosine similarity, and $f_{\text{img}}, f_{\text{text}}$ are the pretrained encoders.

High similarity indicates semantic alignment between the message and the scene. By ranking messages by similarity, the system can discard redundant or visually unsupported alerts, prioritize novel or unexpected messages, and detect mismatches for operator override or system fallback.

Table 5.5 Functional mapping of VLM capabilities to CAV applications.

CAV function	VLM capability	Illustrative output
Driver query response	Visual question answering	"There is a pedestrian near the crosswalk."
V2X alert validation	Scene–text consistency checking	"Merging vehicle is visible in the left lane."
Action explanation	Language-based justification	"Vehicle is braking due to obstacle ahead."
Semantic perception	Object detection and captioning	"A cyclist is crossing from the right."
Scene-aware planning	Reasoning under occlusion	"Pedestrian may be hidden behind the bus."

5.8.1.1 Contrastive Fine-tuning for Domain-specific Alignment

While pretrained CLIP models provide strong baseline performance, CAV-specific scenarios often require further alignment between visual scenes and domain-specific terminology. For instance, a pretrained model might not distinguish between "lane change" and "merging" in traffic contexts or may not recognize specialized V2X message formats.

Contrastive learning can address this by training on CAV-specific image-text pairs. For example, given a dataset of traffic scene images paired with corresponding V2X messages, contrastive learning learns to align embeddings such that images of merging vehicles are closer to "Vehicle merging from left" messages; images of pedestrian crossings are closer to "Pedestrian crossing ahead" alerts; and images of roadwork are closer to "Roadwork 2 km ahead" notifications. This domain-specific alignment improves the precision of semantic filtering, enabling more accurate relevance assessment for CAV-specific scenarios. The methodology for contrastive learning is detailed in Section 5.6.4.

5.8.2 Similarity Metrics in Cross-modal Embedding Spaces

Cosine similarity is the most widely used metric in many contrastive VLMs due to its scale invariance and alignment with the contrastive training objective. However, alternative metrics such as learned Mahalanobis distances or angular margin-based similarity can be applied in fine-tuned systems to improve robustness.

Embedding visualization techniques, such as PCA or t-SNE, can further aid in understanding how messages cluster relative to the current scene, enabling human oversight or system diagnostics.

5.8.3 Interpreting Relevance via CLIP-based Matching

Consider a vehicle traveling along an urban street with a forward-facing camera capturing the current scene. The V2X system delivers a batch of messages, including "Pedestrian crossing ahead," "Vehicle merging from left," "Roadwork 2 km ahead," and "Fog conditions expected in 15 minutes."

Using a pretrained CLIP model, the system computes similarity scores for each message. Suppose the visual input contains a clearly visible pedestrian and a merging vehicle, but no indications of roadwork or fog. The top-ranked messages are:

1. "Pedestrian crossing ahead" (score: 0.89)
2. "Vehicle merging from left" (score: 0.85)

Messages such as "Roadwork 2 km ahead" receive low scores (e.g., 0.22), indicating low visual relevance. This semantic filtering allows the vehicle to suppress or delay irrelevant messages, reducing distraction and focusing attention on visually grounded events.

5.8.4 Implications for Driver Alerting and System Redundancy

Semantic filtering via VLMs has several downstream benefits. First, it provides reduced cognitive load by prioritizing only alerts supported by the current scene for driver notification. Second, it improves system trust as alerts are perceived as contextually aware and explainable. Finally, it enables enhanced fusion with perception by flagging misaligned V2X alerts for further verification or escalation.

By providing a principled measure of alert-scene alignment, contrastive VLMs support scalable and interpretable fusion across external communication and onboard perception systems. This capability is particularly important in high-density driving scenarios where bandwidth, latency, and attention must be carefully managed.

5.9 Summary

VLMs represent a significant evolution in multimodal AI, enabling semantic alignment between visual perception and natural language reasoning. Unlike traditional vision systems that rely solely on spatial cues or language models that operate purely in symbolic domains, VLMs unify the strengths of both modalities. For CAVs, VLMs serve as semantic intermediaries that interpret camera feeds, contextualize V2X messages, and generate human-aligned explanations of driving scenarios.

This chapter presented a modular VLM architecture comprising three necessary stages: visual feature extraction using encoders like ViT [1], CLIP [2], or region-aware models; vision-language adaptation via MLPs, Q-Formers (which use learnable query embeddings that actively query visual features through alternating self-attention and cross-attention layers), or Perceiver resamplers (which use learnable latent vectors to compress visual tokens into compact representations); and LLM integration through flat self-attention or in-layer cross-attention. These components support flexible training strategies that can be adapted based on computational resources and application requirements. The chapter described optional fine-tuning phases that can be applied independently or in combination: vision-language alignment, multimodal generation, and instruction tuning. A comprehensive treatment of contrastive learning for multimodal alignment was presented, covering the InfoNCE formulation, training workflow, and applications to both image–text (CLIP) and text-to-text scenarios. Parameter-efficient techniques such as freezing strategies and LoRA (discussed in detail in the fine-tuning chapter) enable scalable deployment, while comprehensive evaluation methodologies are covered in a later chapter.

The chapter demonstrated practical applications through a case study on semantic filtering of V2X messages, showing how contrastive VLMs can assess message-scene alignment to prioritize relevant alerts and reduce cognitive load. In CAV applications, VLMs provide interpretability through natural language explanations, interoperability via unified token-based interfaces, and context-awareness through scene–prompt fusion. Key challenges remain in temporal grounding, multimodal alignment with nonvisual sensors, scene-aware language modeling, real-time inference, and safety mechanisms for uncertainty estimation and failover.

Vision-language models offer a promising foundation for building CAVs that perceive, reason, and communicate in ways aligned with human expectations and real-world driving demands.

References

1 Alexey Dosovitskiy, Lucas Beyer, Alexander Kolesnikov, Dirk Weissenborn, Xiaohua Zhai, Thomas Unterthiner, Mostafa Dehghani, Matthias Minderer, Georg Heigold, Sylvain Gelly, et al. An image is worth 16x16 words: transformers for image recognition at scale. In *International Conference on Learning Representations*, 2021. arXiv preprint arXiv:2010.11929, 2020.

2 Alec Radford, Jong Wook Kim, Chris Hallacy, Aditya Ramesh, Gabriel Goh, Sandhini Agarwal, Girish Sastry, Amanda Askell, Pamela Mishkin, Jack Clark, et al. Learning transferable visual models from natural language supervision. In *International Conference on Machine Learning*, pages 8748–8763, 2021.

3 Aaron van den Oord, Yazhe Li, and Oriol Vinyals. Representation learning with contrastive predictive coding. *arXiv preprint arXiv:1807.03748*, 2018.

4 Xiaohua Zhai, Basil Mustafa, Alexander Kolesnikov, and Lucas Beyer. Sigmoid loss for language image pre-training. In *Proceedings of the IEEE/CVF International Conference on Computer Vision*, pages 11975–11986, 2023.

5 Alexander Kirillov, Eric Mintun, Nikhila Ravi, Hanzi Mao, Chloe Rolland, Laura Gustafson, Tete Xiao, Spencer Whitehead, Alexander C. Berg, Wan-Yen Lo, et al. Segment anything. In *Proceedings of the IEEE/CVF International Conference on Computer Vision*, pages 4015–4026, 2023.

6 Yanghao Li, Hanzi Mao, Ross Girshick, and Kaiming He. Exploring plain vision transformer backbones for object detection. In *European Conference on Computer Vision*, pages 280–296. Springer, 2022.

7 Dan Hendrycks and Kevin Gimpel. Gaussian error linear units (GELUs). *arXiv preprint arXiv:1606.08415*, 2016.

8 Junnan Li, Dongxu Li, Silvio Savarese, and Steven Hoi. BLIP-2: bootstrapping language-image pre-training with frozen image encoders and large language models. In *International Conference on Machine Learning*, pages 19730–19742, 2023.

9 Jean-Baptiste Alayrac, Jeff Donahue, Pauline Luc, Antoine Miech, Iain Barr, Yana Hasson, Karel Lenc, Arthur Mensch, Katie Millican, Malcolm Reynolds, et al. Flamingo: a visual language model for few-shot learning. *Advances in Neural Information Processing Systems*, 35:23716–23736, 2022.

10 Jacob Devlin, Chang, Ming-Wei Kenton Lee, and Kristina Toutanova. BERT: pre-training of deepbidirectional transformers for language understanding. In *Conference of the North American Chapter of the Association for Computational Linguistics: Human Language Technologies*, pages 4171–4186, 2019.

11 Rafael Rafailov, Archit Sharma, Eric Mitchell, Stefano Ermon, Christopher D. Manning, and Chelsea Finn. Direct preference optimization: your language model is secretly a reward model. *Advances in Neural Information Processing Systems*, 36, 2023.

12 Yuntao Bai, Saurav Kadavath, Sandipan Kundu, Amanda Askell, Jackson Kernion, Andy Jones, Anna Chen, Anna Goldie, Azalia Mirhoseini, Cameron McKinnon, et al. Training a helpful and harmless assistant with reinforcement learning from human feedback. *arXiv preprint arXiv:2204.05862*, 2022.

6

Video Large Language Models for Temporal Reasoning in Driving Environments

6.1 Introduction

Building on the foundation of vision-language models (VLMs) for static scene understanding presented in Chapter 5, this chapter addresses the temporal dimension of visual perception. Consider an autonomous vehicle approaching a busy intersection. A pedestrian steps off the curb, a cyclist weaves between lanes, and a traffic signal turns yellow as the vehicle decelerates. In such settings, perception is not merely about recognizing static objects in a scene; it is about understanding what is happening over time. Effective operation in these environments demands temporal comprehension: recognizing motion, identifying event boundaries, inferring intent, and reasoning about cause and effect.

While image-based VLMs have significantly advanced scene understanding from static images, they lack the capacity to interpret the temporal dimension of the world. This limitation poses a critical challenge for connected and autonomous vehicle (CAVs), which must not only localize and classify entities but also anticipate how the environment evolves. For instance, detecting that a pedestrian has merely entered the field of view is insufficient; the system must recognize whether the pedestrian is walking toward the crosswalk, hesitating, or beginning to cross.

Video large language models (VidLLMs) represent a transformative advance in this domain. These models extend the capabilities of image-based VLMs by incorporating temporally ordered video frames into their input space. Through joint VidLLM modeling, VidLLMs enable systems to answer complex temporal questions such as When did the cyclist enter the ego lane? What occurred after the traffic signal changed? Why did the vehicle ahead slow down? These questions are essential not only for scene understanding but also for generating naturalistic explanations, driver-facing feedback, and temporal grounding in downstream decision-making systems.

In CAV contexts, temporal intelligence is indispensable. It supports motion prediction, interaction modeling, event localization, and alignment with auxiliary modalities such as V2X communication and route planning. Moreover, as autonomous systems become more interactive and explainable, the demand for models that can translate temporal video input into structured, human-readable language will continue to grow.

VidLLMs address a critical gap in machine perception: the capacity to reason over time. This chapter examines the architectural principles, modeling challenges, and deployment considerations of VidLLMs, with a particular emphasis on their integration into CAV systems.

6.2 Fundamental Design Principles

VidLLMs are distinguished from conventional image-language models by their ability to process, interpret, and reason over sequences of visual input. At their core, VidLLMs are designed to integrate visual and linguistic representations while preserving both spatial and temporal structure. This integration necessitates a set of interdependent architectural principles that enable the model to encode motion, establish temporal coherence, ground language in dynamic events, and scale to long video sequences. Four key principles underpin the architecture of VidLLMs: spatio-temporal encoding, spatio-temporal fusion, vision-language alignment, and token compression.

6.2.1 Spatio-temporal Encoding

Unlike static image models that process visual input as a single frame, VidLLMs must encode video as a temporally ordered sequence of visual tokens. Spatio-temporal encoding refers to the process of converting a video into a set of representations that preserve both spatial content (what appears where) and temporal structure (when events occur). This often involves applying frame-wise visual encoders, such as vision transformers (ViTs) or convolutional backbones, followed by temporal encoding mechanisms that inject positional information into the token stream. Temporal encoding methods may be absolute, relative, or rotary, each with different generalization properties and inductive biases.

6.2.2 Spatio-temporal Fusion

Once visual tokens have been temporally tagged, the model must integrate information across both space and time to construct a coherent representation of the evolving scene. This process, known as spatio-temporal fusion, enables the system to capture motion patterns, track object trajectories, and model causal dependencies between events. Fusion is typically achieved through attention mechanisms that operate across frame boundaries, although computational efficiency often necessitates factorized or hierarchical alternatives. Effective fusion mechanisms are essential for modeling complex temporal phenomena such as anticipation, occlusion reasoning, and multi-agent interaction.

6.2.3 Vision-language Alignment

To support tasks such as video question answering, instruction following, or temporal captioning, the model must align visual representations with natural language input. This alignment involves establishing correspondences between video-derived tokens and linguistic queries, prompts, or instructions. Vision-language alignment can be achieved through joint multimodal transformers, cross-attention layers, or memory-augmented querying strategies. The goal is to ground language in temporally relevant visual content, thereby enabling the model to perform temporal reasoning tasks such as identifying event sequences, attributing causes, or summarizing actions.

6.2.4 Token Compression for Scalability

The volume of tokens generated from high-resolution, high-frame-rate video input can quickly exceed the memory and inference constraints of large language models (LLMs). Without appropriate compression, even short video segments may become computationally intractable. Token compression refers to the suite of techniques designed to reduce the number of tokens passed downstream while preserving critical visual and temporal information. Compression methods may operate at the frame level (e.g., pooling), at the clip level (e.g., learnable summary tokens), or across time (e.g., token merging). The effectiveness of VidLLMs in real-time or resource-constrained settings depends significantly on the efficiency and fidelity of their compression strategies.

Taken together, these four principles constitute the conceptual backbone of VidLLM architecture. They guide the implementation of each stage in the modeling pipeline and inform design decisions in both research and deployment.

6.3 Stage 1: Spatio-temporal Encoding

The first stage in the VidLLM pipeline converts raw video frames into a structured sequence of visual tokens that preserve both spatial and temporal information. This representation serves as the foundational input for subsequent stages of fusion and reasoning. Unlike static image processing (as in image-based VLMs, Chapter 5), video modeling requires explicit encoding of the temporal order in which frames occur, as well as preservation of intra-frame spatial structure. The three primary components of spatio-temporal encoding are spatial tokenization, temporal encoding, and memory integration.

6.3.1 Spatial Encoding of Individual Frames

Each frame in a video is a high-dimensional tensor of pixel values that must be transformed into a compact yet informative set of visual tokens. The spatial encoding process applies a visual backbone, typically a ViT or a convolutional neural network (CNN), to convert a frame into a sequence of patch-level embeddings.

Let f_t denote the tth frame in a video. The frame is partitioned into N non-overlapping patches (e.g., of size 16×16 pixels). Each patch is then linearly projected into a d-dimensional embedding space, resulting in a set of spatial tokens:

$$f_t \rightarrow [x_1^t, x_2^t, \dots, x_N^t], \quad x_i^t \in \mathbb{R}^d. \tag{6.1}$$

This process captures the spatial layout of the scene at time t, providing local visual context for each region in the frame. Models such as ViT perform additional processing through self-attention layers, while CNN-based approaches may apply hierarchical feature extraction prior to tokenization.

6.3.2 Temporal Encoding Across Frames

To reason over sequences of frames, it is essential that the model encodes not just what appears in each frame, but also when it appears. However, most visual encoders are order-agnostic, meaning that the raw token output does not reflect the temporal structure of the input. Temporal encoding addresses this by injecting time information into the visual token stream.

Three prominent methods are commonly employed for temporal encoding:

6.3.2.1 Absolute Positional Encoding

The most direct approach adds a learnable or fixed positional vector to all tokens associated with a given frame index t:

$$\text{PE}(f_t) = \text{Embed}(t). \tag{6.2}$$

This method is straightforward and commonly used in ViT-based architectures. However, absolute encoding lacks flexibility in handling variable-length sequences or generalizing to unseen temporal scales.

6.3.2.2 Relative Positional Encoding

Rather than associating a fixed embedding with each frame index, relative encoding incorporates temporal distance as a bias term within the attention mechanism:

$$\text{score}(i, j) = q_i^{\top} k_j + b(i - j), \tag{6.3}$$

where score(i, j) denotes the attention score (the unnormalized similarity) between positions i and j, $q_i \in \mathbb{R}^d$ is the query vector for position i, $k_j \in \mathbb{R}^d$ is the key vector for position j, and $b(i - j)$ is a learnable bias that depends on the frame offset. This formulation encourages stronger attention between temporally adjacent tokens and allows the model to capture position-invariant temporal patterns, such as "two frames ago" or "next frame."

6.3.2.3 Rotary Positional Encoding

Another method, rotary positional encoding (RoPE) [1], incorporates relative positional information through a parameter-free rotation of query and key vectors:

$$\text{score}(i, j) = \langle q_i, R_{j-i} k_j \rangle, \tag{6.4}$$

where $R_{j-i} \in \mathbb{R}^{d \times d}$ is a rotation matrix that encodes the relative distance $(j - i)$ (the attention score, query, and key vectors are defined as above). RoPE has demonstrated strong generalization to longer sequences and is widely adopted in many transformer architectures. Unlike absolute encoding, RoPE does not add vectors to tokens directly but modifies attention computations, thereby preserving the token structure while introducing position-awareness. While RoPE was originally developed for language modeling (as discussed in Chapter 2), its application to video temporal encoding follows the same principle: encoding relative positions through rotations. In video models, RoPE encodes temporal relationships between frames rather than sequential relationships between tokens, enabling the model to understand temporal order and motion patterns across video sequences.

6.3.3 Memory-augmented Representations for Streaming Video

While positional encodings are effective for modeling short video segments, autonomous systems often require temporal reasoning over extended or continuous video streams. To this end, memory-based representations have been introduced to retain contextual information across segments. These methods maintain a state vector H_k at each clip index k that aggregates current features, clip summaries, and prior memory:

$$H_k = g([F_k \circ S_k \circ H_{k-1}]), \tag{6.5}$$

where F_k denotes the current frame features, S_k the summary tokens for the current clip, H_{k-1} the memory from the previous clip, and g is typically a lightweight multilayer perceptron (MLP) or transformer block. The concatenation operator $\circ$ combines these components to form an updated memory state.

Such streaming memory architectures enable long-horizon understanding without the need to recompute over the full video history. This is particularly beneficial in CAV systems where events unfold over extended time periods, and immediate context may be insufficient for accurate reasoning. Streaming models also align well with real-time operation, as they support incremental updates and bounded computational complexity.

6.4 Stage 2: Spatio-temporal Fusion

After spatial and temporal encodings have been applied to the raw video input, the model must integrate the resulting token sequences into a unified representation that captures motion, interactions, and evolving scene context. This process, known as spatio-temporal fusion, enables the VidLLM to reason about visual phenomena that unfold across multiple frames, such as object motion, trajectory changes, multi-agent interaction, and causality. This fusion stage is unique to video models and has no direct equivalent in image-based VLMs, which process static images without temporal dependencies.

Fusion is more than token aggregation; it is the process through which the model resolves correspondences across frames, models dependencies between temporally distant events, and identifies salient spatio-temporal patterns. The design of fusion mechanisms thus plays a critical role in determining both the representational power and the computational efficiency of a VidLLM.

6.4.1 Full Spatio-temporal Self-attention

The most direct approach to fusion is to treat all visual tokens from all frames as a single sequence, over which standard self-attention is applied. Let the spatial encoding of frame t consist of tokens $[x_1^t, \dots, x_N^t]$. The full input sequence X is constructed as:

$$X = [x_1^1, x_2^1, \dots, x_N^1, \dots, x_1^T, \dots, x_N^T], \tag{6.6}$$

where T denotes the number of frames, and N the number of tokens per frame. Self-attention operates over this flattened sequence:

$$\text{Attn}(Q, K, V) = \text{softmax}\left(\frac{QK^\top}{\sqrt{d}}\right)V, \tag{6.7}$$

where $Q \in \mathbb{R}^{(T \cdot N) \times d}$ is the query matrix, $K \in \mathbb{R}^{(T \cdot N) \times d}$ is the key matrix, $V \in \mathbb{R}^{(T \cdot N) \times d}$ is the value matrix, and d is the embedding dimension.

This formulation allows for both intra-frame and inter-frame interactions, enabling tokens at different spatial positions and time indices to communicate freely. For instance, a token representing a pedestrian's head in frame t can directly attend to a token representing the same object in frame $t + 5$, facilitating motion tracking and trajectory estimation.

However, this method incurs a computational complexity of $\mathcal{O}((T \cdot N)^2)$, which becomes prohibitive for long sequences or high-resolution input. As such, full spatio-temporal attention is generally limited to short video clips or requires aggressive token reduction strategies upstream.

6.4.2 Factorized Attention Mechanisms

To address the scalability limitations of full attention, many VidLLMs adopt a factorized attention scheme in which temporal and spatial interactions are processed separately. This decomposition significantly reduces computational complexity while maintaining much of the representational fidelity. The fusion process is divided into two stages:

6.4.2.1 Temporal Attention

For each spatial location (i.e., a specific patch index across frames), tokens are grouped into a temporal sequence. Let $x_{:,n} \in \mathbb{R}^{T \times d}$ denote the sequence of embeddings for patch index n across T

frames, where d is the embedding dimension. Temporal attention is applied independently to each such sequence:

$$z_{:,n} = \text{Attn}_{\text{temporal}}(x_{:,n}), \tag{6.8}$$

where $z_{:,n} \in \mathbb{R}^{T\times d}$ is the temporally updated embedding sequence for patch n. This step allows the model to capture motion and temporal evolution at each spatial location without mixing spatial information across patches.

6.4.2.2 Spatial Attention

Once temporal attention has been applied, the temporally updated tokens for each frame are reassembled into spatial grids. For each time step t, the spatial token grid $z_t \in \mathbb{R}^{N\times d}$ is subjected to spatial attention:

$$y_t = \text{Attn}_{\text{spatial}}(z_t), \tag{6.9}$$

where $y_t \in \mathbb{R}^{N\times d}$ is the spatially updated token grid for frame t. This allows the model to re-integrate spatial context after temporal processing.

The overall complexity of this factorized scheme is $\mathcal{O}(T^2N + TN^2)$, which is significantly lower than the $\mathcal{O}((TN)^2)$ cost of full attention. Factorized attention is particularly effective in long video processing and is adopted in models such as InternVideo [2].

6.4.3 Temporal Summarization and Pooling Strategies

In addition to attention-based fusion, several models apply pooling or summarization strategies to compress and abstract token representations over time. These methods are especially useful for reducing sequence length before passing tokens to the language model. Common approaches include frame-level pooling, learnable summary tokens, and cross-frame token merging. Detailed discussion of these compression techniques is provided in Section 6.6.

6.5 Stage 3: Integration with Language Models

The final stage in the VidLLM pipeline connects visual understanding with natural language reasoning. After obtaining spatio-temporally fused representations of video content, the model must integrate these features with textual input to perform downstream tasks such as question answering, action summarization, causal explanation, and instruction following. This stage requires aligning video tokens with linguistic queries or prompts, enabling the model to reason about temporal events in a manner that is grounded in both modalities.

6.5.1 Relationship to Image-based VLMs

This stage is conceptually similar to Stage 3 in image-based VLMs (Chapter 5), which integrates adapted visual embeddings with language models. The key difference is that VidLLMs operate on temporally fused video representations rather than static image embeddings, enabling temporal reasoning capabilities. The integration strategies—flat self-attention and in-layer cross-attention—follow the same architectural patterns as in VLMs, but are applied to video sequences that preserve temporal structure.

Integration with an LLM may be achieved through several architectural strategies, each offering different trade-offs in flexibility, scalability, and interpretability. The core approaches include joint multimodal input, query-guided selection, in-layer cross-attention, and grounded temporal reasoning.

6.5.2 Joint Video-language Encoding and Inference

A direct integration strategy involves concatenating visual and language tokens into a single sequence, which is then processed by a unified transformer-based model. This approach, known as *flat self-attention*, treats video and language tokens symmetrically in the input sequence.

Let $V = [v_1, \dots, v_T] \in \mathbb{R}^{T \times d}$ represent the temporally fused visual tokens, and $Q = [q_1, \dots, q_L] \in \mathbb{R}^{L \times d}$ the language input (e.g., a question or instruction), where d is the embedding dimension. The combined input is structured as:

$$\text{Input} = [v_1, \dots, v_T, [\text{SEP}], q_1, \dots, q_L], \tag{6.10}$$

where `[SEP]` is a special separator token (first introduced in BERT [3]) used to demarcate modality boundaries between video and language tokens. The sequence is passed to a multimodal LLM (e.g., Flamingo [4], BLIP-2 [5]) for joint processing:

6.5.2.1 Difference from Image-based VLMs

Unlike image-based VLMs (discussed in Chapter 5), which may concatenate visual and text tokens without explicit separator tokens in some architectures, VidLLMs often use `[SEP]` tokens to clearly delineate the temporal video sequence from the language query. This distinction helps the model maintain temporal structure while processing multimodal input. However, the specific use of separator tokens varies across architectures; some models rely on positional encodings or learned boundaries instead.

$$\text{Output} = \text{LLM}([V \circ Q]), \tag{6.11}$$

where $\circ$ denotes concatenation and Output is the generated response.

6.5.2.2 Flat Self-attention Mechanism

The attention computation operates over the full token set, treating video and language tokens equally:

$$\text{Attention} = \text{softmax}\left(\frac{QK^\top}{\sqrt{d}}\right)V, \tag{6.12}$$

where Q, K, and V are the query, key, and value matrices obtained from the concatenated video and language embeddings. This is the standard scaled dot-product attention mechanism (see Chapter 2, Section 2.7), applied here to multimodal video-language sequences. This allows video tokens to attend to language tokens and vice versa, enabling bidirectional cross-modal understanding.

6.5.2.3 Advantages and Limitations

Flat self-attention is straightforward to implement and widely adopted in practice. It leverages the transformer's inherent ability to model arbitrary token sequences without requiring architectural modifications to the language model. This enables full cross-attention between video and language tokens and open-ended generation. However, because video and language modalities are fused only at the input level, this approach offers limited flexibility in how temporal context influences intermediate language representations. The resulting fusion is shallow and may not support complex temporal dependencies or long-range video understanding. Moreover, scalability is constrained by the total number of tokens, which grows rapidly with video length. Thus, this strategy is most effective for short clips or when strong upstream compression is employed.

6.5.3 Query-guided Clip Selection and Attention

When dealing with long videos or memory-constrained environments, not all parts of the visual input are equally relevant to the task. Query-guided selection addresses this by filtering the video content based on the semantics of the language input.

Let $H_1, \dots, H_K \in \mathbb{R}^d$ denote the memory representations of video segments or clips, and $Q \in \mathbb{R}^{L \times d}$ the query, where d is the embedding dimension. The LLM first encodes the query and uses it to attend over the memory bank:

$$\hat{H}_Q = \text{LLM}([H_K \circ Q])[-1], \tag{6.13}$$

where $\hat{H}_Q \in \mathbb{R}^d$ is the query-aware embedding (the $[-1]$ notation extracts the last token from the LLM output sequence). The similarity between the query and each memory segment is computed via cosine similarity:

$$s_k = \cos(\hat{H}_Q, \hat{H}_k), \tag{6.14}$$

where $s_k \in \mathbb{R}$ is the similarity score for clip k. Top-K clips are then selected using either deterministic or stochastic methods such as Gumbel-TopK:

$$I = \text{TopK}(s), \quad \hat{H} = \{H_k \mid I_k = 1\}, \tag{6.15}$$

where $I \in \{0, 1\}^K$ is the selection indicator vector. The filtered visual memory $\hat{H}$ is then concatenated with the query and passed to the LLM:

$$\text{LLM}([\hat{H}, Q]) \rightarrow \text{Answer}. \tag{6.16}$$

This strategy ensures that only temporally relevant segments are presented to the LLM, improving interpretability and efficiency while preserving task performance. It is particularly effective for temporal question answering and instruction-based video analysis.

6.5.4 In-layer Cross-attention for Video-language Integration

VidLLMs can inject temporally fused video features into intermediate layers of the language model, following the same architectural pattern as in-layer cross-attention in image-based VLMs. This enables deeper temporal conditioning of the language generation process and provides more sophisticated control over how video context influences text generation.

Let $H^{(\ell)} \in \mathbb{R}^{L \times d}$ denote the hidden states of the language model at layer ℓ, and $V_{\text{fused}} \in \mathbb{R}^{T \times d}$ the spatio-temporally fused video representations. The temporal cross-attention module computes:

$$Z^{(\ell)} = \text{TemporalCrossAttn}(H^{(\ell)}, V_{\text{fused}}), \tag{6.17}$$

where video features serve as keys and values, while language states serve as queries. The result is integrated via residual connection and feedforward layers:

$$H^{(\ell+1)} = \text{FFN}(H^{(\ell)} + \gamma Z^{(\ell)}), \tag{6.18}$$

where γ is a learnable gating factor that controls the influence of temporal context on language generation. This equation is structurally identical to the in-layer cross-attention formulation in image-based VLMs (Equation 15.11), but operates on temporally fused video features rather than static image embeddings, enabling temporal reasoning across video sequences. In-layer cross-attention provides greater control over how and where temporal fusion occurs in the language model. It supports more sophisticated tasks such as temporal reasoning, event prediction, or causal explanation generation by allowing video context to influence the language model's internal reasoning process across multiple layers. However, this method requires architectural modifications to the language model, which complicates deployment and limits compatibility with fully frozen LLMs.

6.5.5 Grounded Temporal and Causal Reasoning

Beyond matching segments to queries, VidLLMs must reason over temporal structure to answer questions that require an understanding of time, order, and causality. Examples include temporal queries such as "When did the vehicle begin braking?," causal queries like "Why did the vehicle change lanes?," and instructional queries such as "Describe what the driver did after the light turned green."

To support such reasoning, the model leverages attention mechanisms that align temporally tagged visual tokens with the language query. When trained on temporally annotated video–text pairs or instruction-tuned datasets, the LLM learns to associate temporal markers with semantic transitions, such as before/after, cause/effect, and sequence-based descriptions.

This integration enables the VidLLM to produce temporally grounded and semantically coherent responses, which are critical for applications in CAV systems that require explainability, interpretability, and temporal sensitivity.

6.6 Token Compression Techniques

One of the primary challenges in applying VidLLMs to real-world applications, particularly in CAVs, is the computational cost associated with long video sequences. A high-resolution frame can produce hundreds of tokens when processed by a ViT. When extended to sequences of 100 or more frames, the total token count can easily exceed tens of thousands, leading to significant bottlenecks in memory, training time, and inference latency.

Token compression techniques aim to address this scalability challenge by reducing the sequence length while preserving essential spatio-temporal and semantic information. Effective compression is critical not only for computational efficiency but also for enabling VidLLMs to operate in real-time environments or on-edge hardware with limited resources.

6.6.1 Frame-level Pooling Methods

The simplest form of token compression operates at the frame level by summarizing all patch tokens into a single vector representation per frame. A common approach is mean pooling:

$$X_t = [x_1^t, x_2^t, \dots, x_N^t] \in \mathbb{R}^{N\times d}, \quad s_t = \frac{1}{N}\sum_{i=1}^{N} x_i^t, \tag{6.19}$$

where X_t is the token matrix for frame t, $x_i^t \in \mathbb{R}^d$ is the ith patch token in frame t, and $s_t \in \mathbb{R}^d$ is the frame-level summary vector. This yields a temporally ordered sequence $S = [s_1, s_2, \dots, s_T] \in \mathbb{R}^{T\times d}$, where each s_t is a condensed representation of the frame.

While mean pooling is computationally efficient and simple to implement, it may discard fine-grained spatial or object-specific details that are relevant for downstream reasoning tasks.

6.6.2 Learnable Summary Tokens

To enhance expressivity beyond uniform averaging, some architectures introduce learnable summary tokens. These methods allow the model to learn what features within a frame are most salient for temporal understanding or alignment with language input.

Each frame is augmented with a special token $x_{\texttt{[CLS]}}^t \in \mathbb{R}^d$ (where `[CLS]` denotes a classification token, originally introduced in BERT [3] to aggregate sequence-level information) that aggregates context through self-attention:

$$\tilde{X}_t = [x_{\texttt{[CLS]}}^t, x_1^t, \dots, x_N^t], \quad \tilde{X}_t' = \text{Transformer}(\tilde{X}_t), \quad s_t = \tilde{X}_t'[[\text{CLS}]], \tag{6.20}$$

where $\tilde{X}_t \in \mathbb{R}^{(N+1)\times d}$ is the augmented token matrix, $\tilde{X}_t' \in \mathbb{R}^{(N+1)\times d}$ is the transformed token matrix, and $s_t \in \mathbb{R}^d$ is the learned summary vector for frame t. The output sequence $[s_1, \dots, s_T]$ serves as a more expressive and task-tuned summary of the video.

This strategy has been used in architectures such as BLIP-2 [5] and offers a good balance between compression and fidelity.

6.6.3 Clip-level Compression Approaches

Rather than compressing each frame individually, some models operate at the clip level to generate summary tokens that represent larger temporal segments. One straightforward method is global mean pooling:

$$s = \frac{1}{T \cdot N} \sum_{t=1}^{T} \sum_{i=1}^{N} x_i^t \in \mathbb{R}^d, \tag{6.21}$$

where s is the global summary vector for the entire video or clip. However, this coarse summarization often leads to significant information loss.

An alternative is to use a set of learnable summary tokens $S_0 \in \mathbb{R}^{P\times d}$, which are concatenated with the full token set and refined through cross-attention:

$$[S_0; X] \rightarrow \text{Transformer} \rightarrow [S_{\text{out}}; X'], \tag{6.22}$$

where S_0 are the initial learnable summary tokens, $X \in \mathbb{R}^{(T\cdot N)\times d}$ is the full token matrix, $S_{\text{out}} \in \mathbb{R}^{P\times d}$ is the learned output summary tokens, and $X' \in \mathbb{R}^{(T\cdot N)\times d}$ is the updated token matrix.

This method is used in memory-based models and enables flexible integration with query-based retrieval or instruction-following tasks.

6.6.4 Cross-frame Token Merging

Temporal redundancy is another source of inefficiency in video modeling. Adjacent frames often contain similar tokens due to slow-moving or stationary objects. Cross-frame token merging (ToMe) [6] mitigates this redundancy by clustering similar tokens across time and replacing them with shared representations.

Table 6.1 Comparison of token compression methods in VidLLMs.

Method	Token reduction	Flexibility	Information loss
Mean pooling (frame-level)	Moderate	Simple	High
Learnable [CLS] token	Moderate	Moderate	Moderate
Global pooling (clip-level)	High	Simple	Highest
Learnable summary tokens (clip-level)	High	High	Low to moderate
ToMe (token merging)	Dynamic	High	Low

The mechanism consists of three steps: computing pairwise similarity between tokens (e.g., cosine distance); clustering tokens across frames based on similarity; and replacing each cluster with a merged token representing the group.

Unlike pooling, which averages over fixed structures, ToMe adapts to the visual content and preserves diversity. When applied across frames, it enables temporal consistency in compression and has been shown to improve inference throughput with minimal performance degradation.

6.6.5 Comparative Evaluation of Compression Strategies

The effectiveness of different compression strategies varies significantly based on the specific requirements of the downstream task and deployment context. Table 6.1 summarizes the key trade-offs across various token compression methods, highlighting how each approach balances computational efficiency with information preservation.

Compression strategies must be selected with regard to the model's downstream tasks, target latency, and deployment context. In CAV systems, where both response time and interpretability are critical, learnable summaries, and ToMe provide compelling solutions.

Token compression is not merely a preprocessing step but a fundamental design consideration in VidLLM architectures. By reducing redundancy and focusing attention on salient events, these techniques enable scalable, responsive, and temporally aware generative AI systems for real-world use.

6.7 Representative Architectures

The general VidLLM pipeline, comprising spatio-temporal encoding, fusion, language alignment, and token compression, has been instantiated in diverse ways by different model architectures. This section introduces two representative design philosophies: lightweight temporal fusion approaches and memory-based systems optimized for long-horizon video understanding.

6.7.1 Flamingo Architecture

Flamingo [4], developed by DeepMind, represents a lightweight approach to VidLLM modeling that incorporates basic temporal awareness while maintaining architectural simplicity. It processes sequences of images by adding time encodings to frame embeddings, enabling the model to understand temporal order and sequence information.

6.7.1.1 Pipeline Implementation

Flamingo's implementation follows a three-stage pipeline. The contrastive language-image pre-training (CLIP) encoder processes each frame with time vectors added to capture temporal order. Frame-level outputs are then passed through the perceiver resampler, which generates a compact sequence of summary tokens across frames. Finally, the frozen LLM receives the resampled visual tokens concatenated with text input for joint processing.

6.7.1.2 Strengths and Limitations

Flamingo provides a modular and reusable architecture that supports variable-length inputs and open-ended text generation with basic temporal awareness. However, it has a limited sequence length (up to 8 frames during training, 30 frames at inference), lacks deep motion modeling, and has no memory mechanisms for long-term temporal reasoning. Its reliance on frozen encoders also limits its adaptability to domain-specific temporal tasks, such as multi-agent interaction or intent inference in CAVs.

6.7.2 Memory-based Systems for Long-horizon Video Understanding

In contrast to lightweight approaches, memory-based systems are designed to process long video sequences through the use of streaming memory, clip summarization, and query-guided selection. These systems incorporate dynamic memory updates and temporal alignment between video and language tokens to facilitate detailed video question answering and event localization.

6.7.2.1 Pipeline Implementation

Memory-based systems implement a sophisticated three-stage pipeline. Each clip is encoded into summary tokens S_k via transformer-based processing, while memory from previous clips H_{k-1} is retained and updated. Spatio-temporal features F_k, S_k, and H_{k-1} are then integrated using self-attention to yield the next memory state H_k. For LLM integration, a language query Q is matched against memory states using cosine similarity, and the most relevant clips are selected (e.g., via Gumbel-TopK) for final LLM inference.

6.7.2.2 Strengths and Limitations

Memory-based systems support scalable reasoning over long temporal contexts, making them well-suited for temporally extended tasks such as causal explanation or intent prediction. Their modular memory and clip selection mechanisms allow for efficient filtering of relevant content. However, they require pretraining on datasets with aligned video-text segments and introduce complexity in memory management.

6.7.3 Comparative Pipeline Analysis

The architectural differences between lightweight and memory-based approaches reflect distinct design philosophies for handling temporal reasoning. Table 6.2 provides a comparative overview of the two architectural approaches, illustrating how each design philosophy instantiates the VidLLM pipeline stages and highlighting the trade-offs between simplicity and temporal fidelity.

These approaches represent two ends of a design spectrum: lightweight approaches emphasize architectural simplicity and generality, whereas memory-based systems prioritize temporal fidelity and selective reasoning. The choice between them depends on task requirements, sequence length, and the need for fine-grained temporal grounding.

Table 6.2 Comparison of representative VidLLM architectural approaches.

Approach	Time encoding	Fusion mechanism	LLM integration	Compression
Flamingo	Time vectors	Perceiver resampler	Frozen LLM (direct)	Resampled summaries
Memory-based	Positional + memory	Attention + memory update	Query-to-clip matching	Summary + Top-K selection

In the context of CAVs, where scenes unfold over extended durations and decisions are time-sensitive, memory-based systems offer more suitable capabilities.

6.8 Applications in Connected and Autonomous Vehicles

The integration of VidLLMs into CAV systems addresses one of the most critical gaps in machine perception: the capacity to reason over time. Unlike static scene understanding, driving environments are inherently dynamic, governed by motion, causality, and evolving interactions. VidLLMs provide a unified framework for temporally grounded perception, enabling systems to detect, explain, and anticipate events based on both video and language input.

6.8.1 Temporal Reasoning for Perception and Decision-making

In real-world driving, many high-stakes decisions depend not merely on recognizing what objects are present, but on understanding what those objects are doing, how their motion patterns evolve, and what might happen next. VidLLMs facilitate such reasoning through several capabilities. Motion awareness tracks agents (e.g., pedestrians, vehicles) across frames and models their trajectories. Temporal sequencing infers the order of actions (e.g., turning, braking, yielding) and their temporal relations. Causal inference identifies event triggers (e.g., "Why did the car stop?") and consequences ("What happened after the light changed?"). Multi-agent interaction modeling captures spatial and temporal dependencies among heterogeneous agents in complex scenarios.

These capabilities enable not only perceptual classification, but also interpretability and prediction, which are cornerstones of robust CAV systems.

6.8.2 Mapping CAV Tasks to VidLLM Capabilities

The diverse requirements of CAV applications demand different combinations of temporal reasoning capabilities and architectural components. VidLLMs can be tailored to support a broad range of CAV-specific tasks. Table 6.3 summarizes how different task demands align with VidLLM architectural components and representative models, demonstrating the modularity of the pipeline in addressing specific operational needs.

These mappings underscore the modularity of the VidLLM pipeline: each architectural stage contributes to a different aspect of temporal reasoning, and models may be optimized differently depending on the task emphasis, whether it is motion precision, event localization, or explanatory language output.

Table 6.3 Use case mapping: CAV tasks and corresponding VidLLM features.

CAV task	Key capability	Representative models
Pedestrian motion prediction	Temporal encoding + fusion	Memory-based systems
Event-based question answering	Temporal grounding + alignment	Joint encoding approaches
Driver intent analysis	Spatio-temporal memory	Memory-based systems
Accident explanation	Long-horizon causal reasoning	Memory-based systems
Visual-V2X alignment	Video-language grounding	Joint encoding approaches
Occlusion/blindspot analysis	Cross-view fusion	Lightweight approaches
Route-aware scene adaptation	Memory-augmented encoding	Memory-based systems

6.8.3 Implementation Considerations for CAV Environments

The operational constraints of CAV systems, including real-time inference, limited onboard compute, and safety-critical decision-making, require careful selection of VidLLM components. Key considerations include: streaming memory for long-horizon scenarios, appropriate temporal encoding schemes for tasks requiring precise ordering, instruction-following fine-tuning for explainability, query-guided selection for efficient processing, and token compression for edge deployment. These choices guide model selection, system architecture, and interface design for in-cabin interaction, remote monitoring, and V2X-enhanced perception.

6.9 Summary

VidLLMs extend vision–language systems into the temporal domain, providing a unified framework for processing video input and generating language-based output grounded in motion, causality, and interaction. This chapter introduced the three-stage modeling pipeline, comprising spatio-temporal encoding, fusion, and language integration, supported by token compression techniques for computational scalability.

The pipeline is instantiated through two representative approaches: lightweight temporal fusion (e.g., Flamingo) and memory-based systems, each with distinct strengths in sequence length, temporal fidelity, and memory usage. These capabilities enable motion understanding, temporal reasoning, natural language grounding, and scalable inference, supporting critical CAV tasks such as pedestrian behavior prediction, driver intent analysis, and event-based explanation.

VidLLMs mark a shift from static scene analysis to event-driven understanding, playing a central role in enabling safe, adaptive, and transparent autonomous vehicle behavior.

References

1 Jianlin Su, Murtadha Ahmed, Yu Lu, Shengfeng Pan, Wen Bo, and Yunfeng Liu. RoFormer: enhanced transformer with rotary position embedding. *Neurocomputing*, 568, 2024.

2 Yi Wang, Kunchang Li, Yizhuo Li, Yinan He, Bingkun Huang, Zhiyu Zhao, Hongjie Zhang, Jilan Xu, Yi Liu, Zun Wang, et al. Internvideo: general video foundation models via generative and discriminative learning. *arXiv preprint arXiv:2212.03191*, 2022.

3 Jacob Devlin, Ming-Wei Chang, Kenton Lee, and Kristina Toutanova. BERT: pre training of deep bidirectional transformers for language understanding. In *Conference of the North American Chapter of the Association for Computational Linguistics: Human Language Technologies*, pages 4171–4186, 2019.

4 Jean-Baptiste Alayrac, Jeff Donahue, Pauline Luc, Antoine Miech, Iain Barr, Yana Hasson, Karel Lenc, Arthur Mensch, Katie Millican, Malcolm Reynolds, et al. Flamingo: a visual language model for few-shot learning. *Advances in Neural Information Processing Systems*, 35:23716–23736, 2022.

5 Junnan Li, Dongxu Li, Silvio Savarese, and Steven Hoi. BLIP-2: bootstrapping language-image pre-training with frozen image encoders and large language models. In *International Conference on Machine Learning*, pages 19730–19742, 2023.

6 William Peebles and Saining Xie. Token merging: your ViT but faster. *arXiv preprint arXiv:2303.13439*, 2023.

Part III

Language-based Interaction, Memory, and Tool Use

Language serves as a unifying interface between human intent, system knowledge, and vehicle functionality. This part examines how language models are used to shape interaction, retrieve contextual information, and invoke structured system actions within connected and autonomous vehicles. The chapters introduce prompt engineering for behavioral control, retrieval-augmented generation for grounding responses in external memory, and function calling for bridging semantic interpretation and executable operations. Text-to-text contrastive learning is discussed in this context as a retrieval calibration mechanism, aligning embedding spaces with task-specific notions of relevance and reliability.

7

Prompt Engineering

7.1 Introduction and Motivation

7.1.1 Language as the Interface for Driving Intelligence

Imagine a driver asking, "Why did we slow down after exiting the tunnel?" This question, while simple on the surface, requires the vehicle's onboard intelligence to interpret not only the linguistic query but also the situational context, including environmental conditions, recent sensor data, and the navigation state. A well-designed response might explain that the vehicle detected reduced visibility, identified merging traffic, or received a message from the traffic management system.

Such conversational interactions exemplify a broader shift in how humans communicate with connected and autonomous vehicles (CAVs). Instead of manipulating settings through menus or commands, drivers can now engage in natural dialogue with the system. In this emerging paradigm, *language becomes the interface for intelligent behavior.*

However, this vision depends not merely on the power of large language models (LLMs), but on the careful design of how queries are framed and interpreted. At the heart of this interaction lies a discipline known as *prompt engineering.*

7.1.2 The Need for Prompt Engineering

Prompt engineering refers to the intentional crafting of input text that conditions a language model toward desired output behaviors. This practice is not only linguistic, but also algorithmic. It involves aligning the statistical tendencies of a model, which is trained on large-scale corpora, with the narrow task objectives of a user, such as generating concise explanations, providing structured data, or simulating step-by-step reasoning. Even with the same underlying LLM, different formulations of the prompt can elicit different types of reasoning, tone, and specificity. This variance underscores the importance of prompt engineering as a method for aligning model behavior with user intent. In this chapter, prompt engineering refers specifically to runtime behavioral control of foundation models embedded in safety-critical, context-rich vehicular systems, rather than general-purpose conversational tuning.

In the domain of CAVs, the stakes of model output are especially high. Responses must be informative, interpretable, and, in some cases, safety-critical. A vague or misleading answer from a vehicle assistant could erode driver trust or obscure system accountability. Conversely, a clearly prompted model can augment human understanding, foster transparency, and reduce the cognitive burden of interpreting complex system behaviors.

Prompt engineering serves as a bridge between two worlds: the probabilistic language space of generative AI and the deterministic operational constraints of autonomous driving systems. This chapter explores how that bridge is constructed, refined, and deployed in practice.

7.2 Foundations of Prompt Engineering

7.2.1 Definition and Scope

Prompt engineering is the practice of systematically designing input sequences to guide the behavior of LLMs. *Unlike traditional programming, where logic is encoded through deterministic rules, prompt engineering operates in a probabilistic setting*. Here, outputs are shaped by conditioning the model on carefully crafted textual inputs that influence its next-token predictions.

In the context of generative AI systems, a prompt serves as the initial condition under which the model generates an output. It may include task instructions, background information, role definitions, examples, or format constraints. Prompt engineering is thus both a creative and technical endeavor. It defines a control surface over models that otherwise lack explicit programmability.

The relevance of prompt engineering has grown in parallel with the capabilities of foundation models. As general-purpose LLMs are deployed across diverse domains, including healthcare, education, legal reasoning, and intelligent transportation, users must learn to elicit reliable, interpretable, and task-aligned responses. In CAV systems, where natural language interfaces are becoming integral to human-machine interaction, prompt engineering becomes a critical mechanism for aligning model behavior with safety, utility, and user expectations.

7.2.2 Declarative Control vs. Procedural Logic

To appreciate the significance of prompt engineering, it is helpful to contrast it with traditional programming. In a conventional software system, control logic is specified procedurally, using conditionals, loops, and explicit rules. For example, a vehicle may be programmed to reduce speed when the radar sensor detects a short following distance:

```
if distance < threshold:
    reduce_speed()
```

In contrast, a prompt-based system does not execute fixed logic but generates behavior by interpreting natural language conditioned on prior training. A user might provide the following instruction to a language model:

```
"You are a driving assistant. Explain why the car reduced speed near the tunnel
    entrance."
```

The model is then expected to infer possible causes based on context learned during pretraining. It simulates reasoning processes through language, producing outputs such as:

```
"The vehicle detected poor visibility and adjusted speed to maintain safety
    margins."
```

This mode of interaction exemplifies *declarative control*: instead of encoding how the system should operate, users describe what outcome is desired. The LLM uses its internal representations to synthesize a coherent response, as though executing a latent program. This emergent capacity, sometimes referred to as *program induction*, highlights the importance of prompt formulation in shaping system behavior.

7.2.3 Prompting as Probabilistic Conditioning

From a probabilistic perspective, prompt engineering involves modifying the input context C to influence the language model's conditional distribution over the next tokens. Let x_{t+1} denote the next token to be generated, and let $p(x_{t+1} \mid C)$ represent the model's conditional probability distribution given context C. The prompt C may include several structured components, such as system instructions, few-shot demonstrations, and the user's current query.

The objective of prompt engineering can be viewed as designing the context C so that the model's sampled output lies within a desirable task-specific subset $\mathcal{Y}^* \subseteq \mathcal{X}$, where $\mathcal{X}$ denotes the token vocabulary. That is, we aim to ensure:

$$x_{t+1} \sim p(x|C) \quad \text{such that} \quad x_{t+1} \in y^*$$

Although the model parameters remain unchanged, the prompt acts as a control variable that modulates the output behavior. In this sense, prompt engineering constitutes a form of *distribution shaping*, where textual inputs are engineered to steer the model's probabilistic outputs toward alignment with user intent and system requirements.

This framing reveals prompt engineering as a form of *soft policy specification*. In CAV systems, prompts function as soft policies that shape reasoning, prioritization, and response style without modifying model parameters. Unlike hard-coded policies that enforce deterministic rules, soft policies guide behavior probabilistically, allowing for adaptive responses while maintaining alignment with safety constraints and operational objectives. This concept connects naturally to reinforcement learning policies, agent architectures, and safety envelopes, where behavioral control must balance flexibility with reliability.

In the context of CAVs, this approach enables a driver assistance system to generate situation-aware explanations, route suggestions, or safety warnings without requiring task-specific retraining. By carefully shaping prompts, developers can operationalize general-purpose LLMs for safety-critical and user-centric applications in mobility systems.

7.2.4 Layered Prompt Architecture

Effective prompt engineering in CAV systems requires organizing prompts into distinct layers that serve complementary roles. This layered architecture enables modular control over model behavior, from persistent system constraints to dynamic situational context and immediate user interactions. The three primary layers are:

System-level prompts establish persistent behavioral constraints that remain constant across interactions. These prompts define the assistant's persona, safety protocols, domain-specific rules, and operational boundaries. For example, a system-level prompt might specify: *"You are a safety-focused driving assistant. Always prioritize driver safety and vehicle security. Never suggest actions that violate traffic laws or compromise vehicle integrity."* These constraints form the foundational policy layer that governs all model outputs, ensuring consistency with safety requirements and domain expertise.

Contextual prompts incorporate dynamic vehicle state, scene information, and interaction history. This layer adapts to changing conditions such as current speed, location, sensor readings, recent maneuvers, and conversation context. A contextual prompt might include: *"Current vehicle state: speed 65 mph, location Highway 101, recent event: exited tunnel 30 seconds ago, visibility: reduced due to fog."* By injecting real-time situational awareness, contextual prompts enable the model to generate responses that are grounded in the current driving environment and recent system behavior.

Task-level prompts capture immediate user intent and specific queries. This layer contains the user's current request, such as *"Why did the vehicle slow down after the tunnel?"* Task-level prompts are ephemeral and change with each interaction, focusing the model's attention on the specific question or command at hand.

Together, these three layers compose a hierarchical prompt structure where system-level constraints provide stable behavioral boundaries, contextual information supplies situational grounding, and task-level inputs direct immediate responses. This architecture aligns naturally with agent layering frameworks and multi-agent communication protocols such as MCP and A2A, where behavioral control must operate across multiple abstraction levels while maintaining safety and coherence.

7.3 Design Principles for Effective Prompting

Effective prompt engineering requires more than simply instructing a model to complete a task. Given the probabilistic nature of LLMs, small changes in phrasing, structure, or context can significantly influence the output. This section introduces design principles that govern how prompts should be constructed to elicit reliable, consistent, and contextually appropriate responses, particularly in CAV systems.

7.3.1 Clarity and Explicitness

A common source of poor model behavior is prompt ambiguity. Language models excel at capturing broad distributions of human expression but are less effective at inferring underspecified intent. Prompts should therefore be formulated with linguistic precision and task specificity.

Consider the prompt:

```
"Explain the situation."
```

This request is too vague for meaningful inference. In contrast, a more effective version might be:

```
"Explain why the vehicle decelerated after passing the speed limit sign."
```

The latter explicitly anchors the explanation to a spatial and temporal event, thereby constraining the model's output space and increasing the likelihood of obtaining a useful response.

7.3.2 Output Structure and Style Control

An essential aspect of prompt engineering is shaping not only what the model says, but also how it says it. In CAV systems, model outputs may serve various downstream purposes: providing driver-facing explanations, populating structured logs, or guiding system decisions. To support these use cases, prompts must guide both the structure and style of the generated text.

7.3.2.1 Structural Expectations

Prompts can specify the organization of output by including formatting instructions. These cues reduce ambiguity and improve the predictability of responses. For example:

```
"List three reasons why the vehicle may have slowed down."
```

```
"Provide the explanation in two concise bullet points."
```

```
"Output the result in JSON format with keys \"reason\" and \"confidence\"."
```

Such structural constraints are particularly useful in multimodal systems where responses may be displayed on a graphical user interface, stored for logging, or parsed by another software module.

7.3.2.2 Stylistic Guidance

In addition to format, the tone and style of the model's response may need to match the context of interaction. For instance, an in-cabin assistant might require short, easily digestible messages appropriate for auditory delivery, whereas engineering diagnostics might call for more verbose technical language.

Prompt formulations such as the following help align the output style with the target audience:

```
"Use language suitable for a non-technical driver."
```

```
"Respond in a formal tone appropriate for incident reports."
```

```
"Explain the reasoning in simple terms that a child could understand."
```

7.3.3 Role Assignment and Role-structured Prompting

Role assignment provides behavioral context that shapes the model's tone, vocabulary, and domain of relevance. This can be done by describing the assistant's function in the prompt (e.g., "You are a driving assistant"), but is most effectively implemented using structured role tags. Most LLM APIs adopt a tripartite message structure based on the following roles:

System: Sets the assistant's identity and global behavior across the session. It defines the model's persona, capabilities, or tone.

User: Represents the human speaker's input. This message conveys the prompt or query.

Assistant: Contains the model's generated response or, in few-shot settings, an example of desired output.

These role tags provide a clean interface for multi-turn dialogue and structured prompt composition. For example:

```
System: You are a knowledgeable and safe driving assistant.
User: Why did the car brake suddenly?
Assistant: The vehicle detected a cyclist entering the road. Emergency braking
    was applied to avoid a collision.
```

Note that when the System message uses "You" (e.g., "You are a knowledgeable and safe driving assistant"), it addresses the assistant, not the user. In this configuration, the `system` message provides persistent context, the `user` message presents the specific query, and the `assistant` message reflects the model's reply. This structure supports both one-shot and few-shot prompting, and enables clearer control over dialogue consistency and tone.

7.3.4 Zero-shot, One-shot, and Few-shot Prompting

Prompt engineering can be categorized based on the number of examples provided to the model. In *zero-shot prompting*, the model receives only the task instruction without any examples [1]. This approach relies entirely on the model's pretrained knowledge and reasoning capabilities.

For example, a zero-shot prompt might be:

```
"Explain why the vehicle slowed down after exiting the tunnel."
```

One-shot prompting provides a single example of the desired input-output format before processing the actual query. This approach helps the model understand the expected structure and style of the response.

For example, a one-shot prompt might be:

```
User: Why did the car stop suddenly?
Assistant: The vehicle detected a pedestrian entering the road and applied
    emergency braking.
User: Why did the vehicle slow down after the tunnel?
```

Few-shot prompting extends this concept by providing multiple examples (typically 2-5) of input-output pairs [1]. This technique enables the model to better understand the task requirements, learn from patterns in the examples, and generate more consistent and accurate responses.

For example, a few-shot prompt using the role structure might be:

```
System: You are a driving assistant that explains vehicle behavior in two concise
     sentences.
User: Why did the car stop suddenly?
Assistant: The vehicle detected a pedestrian entering the road. Emergency braking
     was applied to avoid a collision.
User: Why did the car change lanes?
Assistant: The navigation system detected a faster route ahead. The vehicle moved
     to the left lane to prepare for the turn.
User: Why did the vehicle slow down after the tunnel?
```

All three approaches leverage the model's ability to learn from context and generalize from examples without requiring parameter updates or fine-tuning.

7.3.5 Chain-of-thought Prompting

In complex reasoning tasks, it is often beneficial to guide the model through intermediate inference steps. This approach, known as *chain-of-thought (CoT) prompting* [2], encourages the model to articulate a logical progression of ideas rather than jump to a conclusion.

For example, instead of asking:

```
"Why did the vehicle slow after the tunnel?"
```

a Chain-of-thought formulation would be:

```
"First consider the sensor input, then explain the likely cause of the
   deceleration after the tunnel."
```

This instruction leads the model to structure its response as a sequence of observations and inferences:

```
"Sensor logs show reduced visibility due to fog. The vehicle's adaptive cruise
   control activated a speed adjustment to maintain safe following distance."
```

By breaking down complex prompts into smaller reasoning steps, CoT prompting increases transparency, reduces hallucinations, and improves factual alignment, especially valuable in safety-critical domains like autonomous driving.

7.4 Prompt Engineering in Practice

This section operationalizes the principles introduced in earlier sections by demonstrating how prompts can be constructed, executed, and evaluated using a local framework in a realistic connected vehicle context. We focus on a recurring driver query and explore how different prompt formulations influence the model's response. This case study illustrates not only the power of prompt engineering but also the practical details of implementing and tuning prompts for in-cabin driving assistants.

7.4.1 Scenario Setup

Returning to the tunnel scenario introduced at the beginning of this chapter, we now examine how different prompt formulations can be applied to the driver's query:

```
"Why did the vehicle slow down after the tunnel?"
```

This query is open-ended and context-dependent, requiring the assistant to reason about sensor data, scene context, or high-level decisions. The generative AI system must produce an explanation that is not only plausible and relevant, but also appropriately styled for a driver interface.

7.4.2 LLM Framework Interface

Modern LLM frameworks provide standardized interfaces for interacting with language models through structured APIs, abstracting away the complexity of model deployment, tokenization, and inference. Most frameworks, including local deployment solutions, adopt a consistent message-based interface where prompts are structured as sequences of role-tagged messages. This approach enables both single-turn interactions and multi-turn conversations while maintaining context across exchanges. In this chapter, we use Ollama [3] as an example local framework for demonstrating prompt engineering concepts.

The core interface for conversational interaction, exemplified by Ollama's `ollama.chat()` function, accepts a structured dictionary with the fields summarized in Table 7.1.

Table 7.1 Structure of a typical LLM framework prompt input.

Field	Type	Description
`model`	string	Identifier of the LLM used for generation
`messages`	list	Sequence of role-based messages forming the prompt
`options`	dict (optional)	Controls decoding parameters such as temperature, top-p, output length
`tools`	list (optional)	Enables callable functions for structured execution

7.4.3 Prompting Examples and Generated Outputs

To illustrate how prompt structure influences language model output, we present several representative prompt designs applied to the same user query: *"Why did the vehicle slow down after the tunnel?"* Each prompt is executed using Ollama as an example, and the resulting outputs are analyzed for clarity, reasoning quality, and stylistic alignment.

7.4.3.1 Setup and Baseline Prompt

We begin with a minimal prompt that includes only the assistant's role and the user's query:

Listing 7.1 Baseline prompt example

```
import ollama

response = ollama.chat(
  model='llama3.2',
  messages=[
    {"role": "system", "content":
     "You are a helpful assistant in a connected vehicle."},
    {"role": "user", "content":
     "Why did the vehicle slow down after the tunnel?"}
  ])
print(response['message']['content'])
```

In this example, `model='llama3.2'` specifies the language model to use (in this case, Llama 3.2 [3], a large language model based on the LLaMA architecture [4]). The `messages` parameter is a list of dictionaries, where each dictionary contains a `role` field (either `"system,"` `"user,"` or `"assistant"`) and a `content` field containing the actual message text. The `system` role defines the assistant's behavior and context, while the `user` role contains the query or instruction. The function returns a response object, from which the generated text can be accessed via `response['message']['content']`.

Listing 7.2 Baseline prompt response: verbose and open-ended explanation

```
It appears that the vehicle slowed down due to a traffic alert message received
    from the nearby traffic management system. There was a reported accident
    ahead, and the traffic was being rerouted through an adjacent lane. The
    vehicle's safety features automatically adjusted its speed to ensure a
    safe distance from other vehicles in case of unexpected maneuvers.
```

The baseline approach produces open-ended and verbose explanations. In scenarios where conciseness or interpretability is critical, a more structured prompting strategy may be preferable.

7.4.3.2 Example 1: Chain-of-thought Prompt

This version encourages stepwise reasoning by instructing the assistant to first consider sensor inputs and external conditions:

Listing 7.3 Chain-of-thought prompt

```
messages = [
  {"role": "system", "content":
   "You are a safe and logical driving assistant. First consider recent sensor
       inputs and external conditions. Then explain why the vehicle slowed after
       the tunnel."},
  {"role": "user", "content":
   "Why did the vehicle slow down after the tunnel?"}
]
```

Listing 7.4 Chain-of-thought prompt response: analytical and sensor-grounded explanation

```
Sensor logs show a sudden decrease in ambient light and increased humidity after
    exiting the tunnel. The system detected reduced visibility and activated
    adaptive cruise control adjustments. As a result, the vehicle reduced speed
    to maintain a safe following distance.
```

The chain-of-thought prompt emphasizes reasoning transparency and system-grounded causality, yielding responses that are more analytically grounded.

7.4.3.3 Example 2: Few-shot Prompt

By incorporating earlier dialogue examples, the model can better calibrate tone, style, and reasoning depth:

Listing 7.5 Few-shot Prompt Example

```
messages = [
  {"role": "system", "content":
   "You are a driving assistant that explains vehicle behavior in two concise
       sentences."},
  {"role": "user", "content":
   "Why did the car stop suddenly?"},
  {"role": "assistant", "content":
   "The vehicle detected a pedestrian entering the road. Emergency braking was
       applied to avoid a collision."},
  {"role": "user", "content":
   "Why did the vehicle slow down after the tunnel?"}
]
```

Listing 7.6 Few-shot prompt response: concise and structured explanation

```
The vehicle detected fog and reduced visibility after exiting the tunnel. It
    slowed down to maintain safe control and visibility.
```

The few-shot prompt enforces consistency and brevity through example-driven alignment, resulting in well-structured, user-appropriate explanations that match the specified style and domain.

These three prompting strategies illustrate how response behavior can be shaped through prompt engineering. The contrasts between open-ended baseline responses, analytically grounded chain-of-thought outputs, and concise few-shot examples underscore the importance of prompt design in achieving high-quality, context-aware responses from language models in CAV systems.

7.4.4 Controlling Style and Output Diversity

Prompt engineering defines the semantic content and structural expectations of an LLM's response. However, the style, variability, and determinism of the generated text are also governed by decoding parameters or sampling configurations that influence how the model selects each successive token from its conditional probability distribution. In LLM frameworks like Ollama, such parameters are passed via the `options` field, and they serve as critical tools for tuning system behavior.

7.4.4.1 Sampling Parameters: Temperature, Top-P, and Output Length

LLM frameworks typically support several decoding parameters that affect response generation, as summarized in Table 7.2:

7.4.4.2 Example: Adjusting Output Style and Length

The following code demonstrates how sampling configuration can be used to tailor the LLM's output for a concise, deterministic explanation using Ollama:

Listing 7.7 Configuring sampling options in Ollama

```
response = ollama.chat(
  model='llama3.2',
  options={
    "temperature": 0.3,
    "top_p": 0.9,
    "num_predict": 60
  },
  messages=[
    {"role": "system", "content":
     "You are a concise and reliable driving assistant."},
    {"role": "user", "content":
     "Why did the vehicle slow down after the tunnel?"}
  ])
```

Listing 7.8 Response with low temperature and constrained length: concise and deterministic

```
The vehicle detected fog and reduced visibility after exiting the tunnel. It
    slowed to ensure safety and maintain proper distance from other vehicles.
```

Table 7.2 Decoding parameters available in LLM frameworks for controlling response characteristics.

Option	Purpose	Effect on output
`temperature`	Sampling randomness	A higher value (e.g., 0.8) increases output diversity and creativity. A lower value (e.g., 0.2) yields a more focused and deterministic text.
`top_p`	Nucleus sampling	Limits sampling to the smallest set of tokens whose cumulative probability exceeds the threshold (e.g., 0.9). Controls diversity while preserving quality.
`top_k`	Top-k sampling	Limits sampling to the k most likely tokens at each step. A lower value (e.g., 10) produces more focused text, while a higher value (e.g., 50) allows more diversity.
`num_predict`	Output length limit	Sets the maximum number of tokens to be generated. Useful for constraining verbosity or fitting user interface constraints.
`seed`	Random seed	Ensures reproducibility of sampling behavior across runs by fixing the random number generator state.

In this example, the `num_predict` parameter is set to 60, which limits the maximum number of tokens the model will generate, effectively controlling the response length. In contrast, a higher temperature and longer `num_predict` value may produce a more verbose and varied explanation:

Listing 7.9 More diverse sampling configuration in Ollama

```
options = {
  "temperature": 0.9,
  "top_p": 1.0,
  "num_predict": 120
}
```

Listing 7.10 Response with high temperature and extended length: verbose and varied explanation

```
As the vehicle exited the tunnel, several environmental changes were detected. A
    sudden shift in lighting, increased moisture levels, and a curve in the road
    prompted the onboard systems to adjust speed. Additionally, a scheduled
    traffic reroute ahead was communicated via V2X, contributing to proactive
    deceleration...
```

These contrasting outputs reflect the stochastic influence of sampling parameters on lexical choice, length, and tone. Importantly, while prompts shape the desired semantics, decoding parameters modulate the surface form and stylistic confidence of the generated response.

7.4.4.3 Prompting and Sampling: A Joint Control Mechanism

Together, prompt structure and decoding configuration form a dual control surface over LLM behavior. Prompt design conditions the model semantically, defining the task, format, and tone. Sampling parameters modulate uncertainty, verbosity, and fluency.

In safety-sensitive domains such as CAVs, achieving the right balance between determinism and expressiveness is essential. For mission-critical outputs (e.g., system status reporting), a low-temperature, tightly scoped configuration is appropriate. For exploratory dialogue or user-facing explanation, slightly higher diversity may improve engagement and adaptability.

7.5 Prompt Fragility and Safety Considerations in CAV Systems

While prompt engineering offers a powerful mechanism for controlling LLM behavior, it is important to recognize its inherent limitations and fragility, particularly in safety-critical CAV applications. Prompts are fundamentally brittle: small variations in wording, context length, or model version can produce significantly different outputs, undermining reliability guarantees essential for autonomous systems. This fragility stems from the probabilistic nature of language models, where prompt changes shift the conditional distribution over outputs in ways that are difficult to predict or verify exhaustively.

In CAV systems, prompt fragility poses several critical safety concerns. First, adversarial or malformed user inputs may cause the model to deviate from intended behavior, potentially generating unsafe suggestions or explanations. Second, prompt injection attacks—where malicious inputs override system-level constraints—can compromise safety boundaries established through careful prompt design [5]. Third, the non-deterministic nature of LLM outputs means that identical prompts may yield different responses across runs, complicating testing, validation, and certification processes required for safety-critical systems.

To mitigate these risks, prompt engineering in CAV systems must be complemented by additional safety mechanisms: input validation and sanitization, output filtering and verification, fallback procedures for uncertain or out-of-distribution queries, and formal verification where possible. Moreover, prompts should be treated as one component of a broader safety architecture that includes deterministic guardrails, human oversight, and fail-safe mechanisms. The layered prompt architecture described above helps compartmentalize risks, but it cannot eliminate the fundamental uncertainty inherent in probabilistic language generation. As such, prompt engineering should be viewed as a valuable but imperfect tool that requires careful integration with traditional safety engineering practices.

While this chapter has focused on prompt engineering as a mechanism for controlling LLM behavior, it is important to recognize that prompts are not standalone components in real-world CAV systems. Effective intelligent behavior requires prompts to interact seamlessly with external tools, memory systems, and retrieval mechanisms. Prompt engineering alone is insufficient for complex CAV reasoning and must be combined with RAG [6] for accessing up-to-date information, function calling for interfacing with vehicle systems and sensors, and agent orchestration for coordinating multi-step reasoning and decision-making. These complementary technologies work together with well-designed prompts to enable robust, context-aware, and actionable intelligence in autonomous vehicles.

7.6 Summary

This chapter examined prompt engineering as a principled method for guiding LLM behavior in CAV systems. We introduced prompting as a form of probabilistic conditioning, where carefully designed textual inputs shape the distribution of model outputs without modifying underlying parameters, contrasting this declarative approach with traditional procedural programming.

We presented a layered prompt architecture comprising system-level prompts (persistent behavioral constraints), contextual prompts (dynamic vehicle state and scene information), and task-level prompts (immediate user queries). This hierarchical structure enables modular control over model behavior while maintaining safety and coherence.

Key design strategies, including clarity, structural constraints, role specification, zero-shot, one-shot, and few-shot prompting, and chain-of-thought reasoning, were analyzed in the context of safety-critical vehicle interactions. Prompt formats were contextualized through dialogue role structures and deployed in practical implementations using LLM frameworks. We demonstrated how decoding parameters such as temperature, top-p, and num-predict modulate output diversity, verbosity, and determinism, showing how prompt form and sampling configuration jointly govern output relevance, interpretability, and user trust.

The chapter concluded by addressing prompt fragility and safety considerations in CAV systems, highlighting the inherent brittleness of prompt-based control and the need for complementary safety mechanisms, including input validation, output filtering, and integration with deterministic guardrails. We emphasized that prompt engineering, while powerful, must be viewed as one component of a broader safety architecture rather than a standalone solution for safety-critical applications.

References

1 Tom Brown, Benjamin Mann, Nick Ryder, Melanie Subbiah, Jared D. Kaplan, Prafulla Dhariwal, Arvind Neelakantan, Pranav Shyam, Girish Sastry, Amanda Askell, et al. Language models are few-shot learners. *Advances in Neural Information Processing Systems*, 33:1877–1901, 2020.

2 Jason Wei, Xuezhi Wang, Dale Schuurmans, Maarten Bosma, Brian Ichter, Fei Xia, Ed Chi, Quoc Le, and Denny Zhou. Chain-of-thought prompting elicits reasoning in large language models. *Advances in Neural Information Processing Systems*, 35:24824–24837, 2022.

3 Ollama. Ollama: run large language models locally, 2024. https://ollama.com/.

4 Hugo Touvron, Thibaut Lavril, Gautier Izacard, Xavier Martinet, Marie-Anne Lachaux, Timothée Lacroix, Baptiste Rozière, Naman Goyal, Eric Hambro, Faisal Azhar, et al. Llama: open and efficient foundation language models. *arXiv preprint arXiv:2302.13971*, 2023.

5 Deep Ganguli, Liane Lovitt, Jackson Kernion, Amanda Askell, Yuntao Bai, Saurav Kadavath, Ben Mann, Ethan Perez, Nicholas Schiefer, Kamal Ndousse, et al. Red teaming language models to reduce harms: methods, scaling behaviors, and lessons learned. *arXiv preprint arXiv:2209.07858*, 2022.

6 Patrick Lewis, Ethan Perez, Aleksandra Piktus, Fabio Petroni, Vladimir Karpukhin, Naman Goyal, Heinrich Küttler, Mike Lewis, Wen-tau Yih, Tim Rocktäschel, et al. Retrieval-augmented generation for knowledge-intensive NLP tasks. In *Proceedings of the 34th International Conference on Neural Information Processing Systems*, NIPS '20, Red Hook, NY, USA, 2020. Curran Associates Inc. ISBN 9781713829546.

8

Retrieval-augmented Generation for Context Grounding

8.1 Why RAG Matters in Connected and Autonomous Vehicles

Prompt engineering, the subject of Chapter 7, controls how a model is queried but not what the model knows. A carefully phrased prompt cannot tell a pretrained large language model (LLM) about a software update issued after its training cutoff, a road closure broadcast in the last minute, or a braking event the current driver experienced earlier in the trip. In safety-critical connected and autonomous vehicle (CAV) systems, these gaps between the model's static knowledge and the live operational context produce failure modes that extend beyond mere accuracy concerns.

8.1.1 The Three Core Problems

Pretrained LLMs face three fundamental problems when applied to CAV systems: *hallucination*, *staleness*, and *partial observability*. These limitations represent critical failure modes that can erode driver trust, obscure system accountability, and compromise safety in high-stakes driving scenarios.

Hallucination occurs when LLMs generate plausible but ungrounded or factually incorrect information [1]. In CAV systems, hallucination is not merely an accuracy problem; it is a safety and trust problem. A confident, yet incorrect explanation about why a vehicle braked suddenly can mislead drivers, obscure system malfunctions, or mask critical safety issues. Consider a driver asking why the vehicle slowed down:

Driver: Why did the car slow down near Exit 45?
Assistant: The vehicle detected reduced visibility due to fog and adjusted speed accordingly.

If this explanation is hallucinated (if no fog was present and the actual cause was a detected road hazard), the driver receives misleading information that could affect their understanding of the system's behavior and their trust in future interactions.

Staleness refers to the temporal limitations of pretrained models, which are trained on data up to a fixed cutoff date and cannot access information introduced after training. A model trained before recent software updates cannot reason about newly introduced features, revised safety protocols, or updated vehicle-to-everything (V2X) messaging standards. In rapidly evolving CAV systems, this temporal gap means models may provide explanations based on outdated knowledge, leading to incorrect or irrelevant responses.

Partial observability describes the fundamental limitation that LLMs cannot directly access real-time vehicle state, sensor data, or session-specific information. A model cannot answer questions about current speed, recent braking events, or past driver interactions unless this information is explicitly provided. This creates a blind spot where the model must either admit ignorance or resort to speculation, both of which degrade system utility and user trust.

8.1.2 Additional Limitations of Pretrained LLMs

8.1.2.1 Absence of Procedural, Multimodal, or Low-level Understanding

Although LLMs excel in text-based reasoning, they lack native access to procedural system models or multimodal inputs. For example, an LLM does not intrinsically know that a braking maneuver was executed due to a certain time-gap threshold being exceeded, or that the road surface was degraded based on a camera input. Without access to numerical thresholds, sensor readings, or situational triggers, the model's response will remain speculative at best.

Moreover, pretrained models are not equipped to interpret multi-source signals such as fusion of camera, LiDAR, and V2X, map low-level vehicle dynamics (e.g., yaw rate, headway time) to driver-facing summaries, or respond adaptively to unforeseen environmental factors unless trained on them.

8.1.2.2 Inflexibility in Evolving or Customized System Interfaces

LLMs trained on fixed corpora cannot adapt to new APIs, configuration parameters, or system interfaces introduced as vehicle software evolves. Modern CAVs feature growing sets of functionalities with configurable thresholds, region-specific variants, and OEM-specific implementations. Pretrained models without runtime retrieval mechanisms cannot accommodate such diversity or evolve in tandem with the vehicle's operational design domain without explicit retraining.

8.1.3 RAG as a System-level Mitigation Strategy

RAG addresses the limitations described above through a system-level mitigation strategy [2]. Rather than treating RAG as merely an LLM enhancement, it should be understood as an architectural approach that integrates retrieval mechanisms with generative models to create grounded, verifiable, and context-aware systems.

RAG enables a language model to retrieve external information from documents, logs, or data repositories before generating a response, injecting runtime-relevant context that mitigates hallucination, overcomes staleness, compensates for partial observability, and provides access to procedural knowledge, multimodal information, and evolving system interfaces. The retrieved information may originate from heterogeneous sources: static knowledge bases (maps, manuals, and traffic regulations), episodic memory logs (past driver interactions and recent decisions), or summarized historical interactions. Importantly, RAG defines *how* information is accessed rather than *what type* of information is stored.

Driver: Why did the car slow down near Exit 45?

Assistant: The vehicle slowed down near Exit 45 due to a detected road hazard, and a V2X alert from a nearby Roadside Unit indicated ongoing construction in the right lane.

This response demonstrates how RAG mitigates the three core problems: it is *grounded* in retrieved external knowledge (addressing hallucination), draws on *current* vehicle logs and V2X messages (addressing staleness), and accesses *episodic memory* from past interactions (addressing partial observability). The assistant synthesizes information from multiple sources to provide a verifiable, contextually accurate explanation that builds trust rather than eroding it. Note that operational state, such as current speed or position, would be accessed directly, not through RAG retrieval.

By enabling the system to retrieve and integrate this information at runtime, RAG transforms the static knowledge of a pretrained LLM into an interactive, adaptive, and explainable system.

Unlike function calling, which is limited to predefined API endpoints, RAG allows natural language queries to be mapped to semantically retrieved content. This enables the system to handle unexpected queries, novel features, or evolving vehicle configurations without frequent model retraining or system reprogramming.

8.2 Architectural Foundations of RAG

RAG augments language models by integrating external information through a structured retrieval pipeline.

8.2.1 The Retrieve-inject-generate Paradigm

The defining characteristic of RAG is its operational flow, illustrated in Figure 8.1, which can be decomposed into three sequential stages:

1. **Retrieve:** Identify relevant documents or data fragments based on the user's query by searching a vectorized knowledge base.
2. **Inject:** Incorporate the retrieved information into the language model's input prompt.
3. **Generate:** Produce a contextually grounded response using both the user's query and the injected external content.

In the retrieve stage, the query is embedded into a vector representation and used to search the document store for semantically similar content. In the Inject stage, the retrieved chunks are formatted together with the original query into a structured prompt. In the Generate stage, the language model processes this augmented prompt to produce a contextually grounded response that synthesizes both the user's query and the retrieved information.

This retrieve-inject-generate pattern allows the system to remain both adaptive and auditable. Unlike traditional fine-tuning, which alters model parameters, RAG operates without changing the model's weights. This makes it particularly suitable for safety-critical domains where traceability and isolation of retrieved sources are essential.

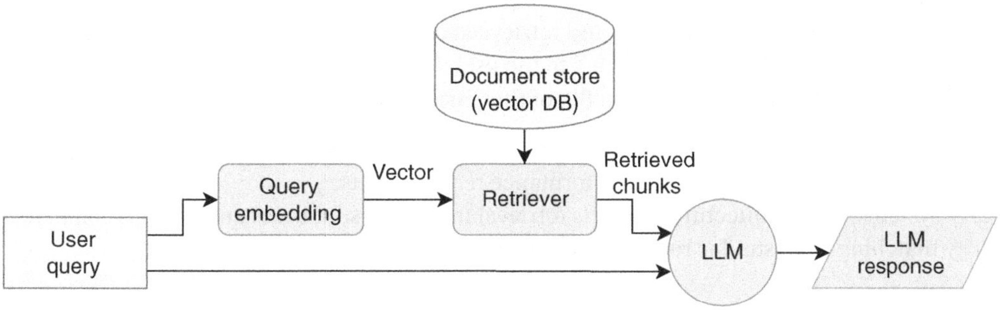

Figure 8.1 Simplified architectural flow of a RAG system. The user query is embedded and used to retrieve top-k similar vectors from a document store. These are injected into the prompt passed to the language model, which generates a grounded response.

8.2.2 Core Components of a RAG System

The RAG architecture consists of several interacting modules organized according to the three stages of the retrieve-inject-generate paradigm:

8.2.2.1 Retrieve Stage Components

Query Embedding: Converts the input question or statement into a dense vector representation in a semantic embedding space [3]. This enables the system to capture the latent meaning of the query rather than relying solely on surface text features.

Document Store (Vector Database): A database that contains pre-embedded knowledge entries, including external knowledge (maps, manuals, regulations, documentation), episodic memory (past interactions, driver preferences, recent events), and summarized operational logs (historical patterns). Vector databases enable efficient similarity-based retrieval at scale, which is critical for CAV systems that must query large collections of logs, documentation, and messages within latency constraints. The semantic search capability allows the system to find relevant information even when queries use different terminology than the stored documents.

Retriever: Computes similarity between the query embedding and entries in the vector database, selecting the top-k most relevant documents or passages based on similarity scores. The mathematical formulation of similarity computation is detailed in Section 8.4.2.

8.2.2.2 Inject Stage

The Inject stage involves formatting the retrieved documents along with the user query into a structured prompt suitable for the language model. This stage is typically implemented as part of the prompt construction process, where retrieved chunks are combined with the original query and formatted according to the model's expected input structure. The prompt construction ensures that the retrieved context is clearly delineated from the user's query, enabling the language model to distinguish between the original question and the retrieved supporting information. This separation is crucial for maintaining traceability and allowing the model to ground its response in the retrieved content.

8.2.2.3 Generate Stage Components

LLM Generator: A pretrained language model that receives the user query along with the retrieved documents formatted as a structured prompt. It synthesizes the final output in natural language, grounded in both the input and retrieved content.

LLM Response: The final output presented to the user. It reflects not only the model's generative capacity but also the relevance and quality of the retrieved context.

Each of these components can be modularly substituted or optimized depending on the application domain, resource constraints, and performance requirements.

In CAV systems, this architecture supports retrieval from diverse sources and enables contextual similarity matching (e.g., similar road or weather conditions).

8.3 RAG vs. Rule-based Information Systems in CAVs

CAVs rely on structured, rule-based information systems for sensor processing, planning, and user-facing messages. These systems are deterministic and reliable but lack flexibility and human-interpretable explanations.

8.3.1 Rule-based CAV Information Systems

Rule-based CAV information systems operate on predefined rules and bounded input–output mappings, providing deterministic and reliable behavior that is essential for safety-critical operations. Event logs, V2X messages, and sensor signals are filtered and processed using time-based windows (e.g., aggregate behavior over the past five seconds), spatial constraints (e.g., include only vehicles within 300 m), and message-type routing (e.g., basic safety messages [BSMs], signal phase and timing [SPaT], or traveler information messages [TIM]).

This approach enables deterministic interpretation but has significant limitations. Explanations in rule-based CAV information systems are typically limited to predefined templates or status indicators. While useful for fault diagnosis or warning alerts, these systems cannot synthesize information from multiple sources (e.g., combining sensor logs, V2X messages, ADAS triggers, and system state) to answer natural-language questions. Rule-based systems also cannot adapt to new feature configurations or emerging cooperative policies without extensive reprogramming.

8.3.2 Comparative Capabilities

Table 8.1 summarizes the differences between rule-based CAV information systems and RAG-enabled interfaces across several dimensions.

RAG is not a substitute for structured control systems, but rather a complementary semantic interface that enhances explainability and driver engagement.

8.3.3 Design Implications for CAV Architecture

Incorporating RAG into CAV systems introduces a new architectural layer: the semantic reasoning and explanation interface. This layer operates alongside deterministic control subsystems without interfering with real-time control loops. Key design implications include semantic layering, where RAG provides a reasoning interface over existing telemetry, logs, and V2X communications, enabling human-readable summaries of complex system behavior. Separation of concerns ensures that real-time control, planning, and perception remain handled by deterministic subsystems, while RAG supports post hoc explanation and session-aware interaction. Driver transparency and trust are enhanced through human-centric explanation capability, which can increase user confidence, support regulatory audits, and aid in debugging unusual driving decisions. A hybrid design approach allows RAG to complement function calling and status APIs: while function calling is ideal for structured actuation, RAG excels at answering "why" and "what just happened" queries.

Table 8.1 Comparison of rule-based vs. RAG-enabled CAV information systems.

Capability	Rule-based systems	RAG-enabled systems
Logic model	Rule-based, fixed pipelines	Semantic, context-sensitive
Information sources	Rigid filters by type/time/space	Flexible multi-source retrieval
Explanation format	Predefined messages	Natural language generation
Adaptability to updates	Requires reconfiguration or redeployment	Supports dynamic document injection
User interaction	Menu-based or symbolic interfaces	Natural language queries
Integration cost	Low for predefined use cases	Requires indexing and prompt design

8.4 Mathematical Modeling of RAG

The architectural foundations described in Section 8.2.1 can be formalized mathematically through embedding, similarity-based retrieval, and probabilistic generation.

8.4.1 Embedding and Vector Representation

Let x denote a user query, and let $\mathcal{D} = \{d_1, d_2, \dots, d_n\}$ represent the collection of documents available for retrieval. Each document $d_i \in \mathcal{D}$ is transformed into a dense vector representation:

$$\mathbf{v}_{d_i} = f_{\text{embed}}(d_i) \in \mathbb{R}^m,$$

where f_{embed} is a pretrained embedding model (such as a sentence transformer [3]), and m denotes the dimensionality of the embedding space.

The user query is embedded in the same vector space:

$$\mathbf{v}_x = f_{\text{embed}}(x).$$

These vector representations enable semantic comparison between the query and each document using a distance or similarity metric.

8.4.2 Similarity-based Retrieval

To identify the most relevant documents for a given query, the system computes similarity scores between the query embedding and each document embedding. A common metric is cosine similarity:

$$\text{sim}(x, d_i) = \frac{\mathbf{v}_x \cdot \mathbf{v}_{d_i}}{\|\mathbf{v}_x\| \, \|\mathbf{v}_{d_i}\|}.$$

The retriever selects the top-k documents from the corpus with the highest similarity scores [4].

8.4.3 Probabilistic Formulation of RAG

The RAG process can be described as a two-stage probabilistic model [2]. First, the system retrieves a document from the corpus based on the query. Then, it generates a response conditioned on both the query and the retrieved content.

At this stage, we introduce a new variable $z \in \mathcal{D}$, which denotes a document retrieved from the collection. Unlike the earlier use of d_i to index all documents in the corpus, the variable z represents a specific document selected (or considered) for conditioning the generation process. In this context, z functions as a *latent variable,* which is an intermediate quantity that is not directly observed but inferred during retrieval.

The generation process is then modeled as:

$$p(y \mid x) = \sum_{z \in \mathcal{D}} p(y \mid x, z) \cdot p(z \mid x),$$

where

- $p(z \mid x)$ is the probability of retrieving document z given query x

- $p(y \mid x, z)$ is the probability that the language model generates output y conditioned on the query and the retrieved document.

In practice, $p(z \mid x)$ is approximated using normalized similarity scores. For instance, the retrieval distribution may be defined as:

$$p(z \mid x) = \frac{\exp(\lambda \cdot \text{sim}(x, z))}{\sum_{d \in \mathcal{D}} \exp(\lambda \cdot \text{sim}(x, d))},$$

where $\lambda > 0$ is a scaling factor that controls the sharpness of the distribution, and the denominator ensures proper normalization over all documents in the corpus.

This formulation emphasizes that generation is influenced by multiple possible documents in the corpus, and the final output is obtained by weighting their contributions based on relevance to the query.

The use of the latent variable z allows the system to model uncertainty over which document is most helpful. Even though many implementations use deterministic retrieval (e.g., top-k selection), the probabilistic view is valuable for understanding and extending the framework. It also supports advanced variants of RAG that incorporate document weighting, uncertainty estimation, and end-to-end optimization, which are particularly relevant for safety-critical applications such as those in CAV environments.

8.5 Practical RAG Implementation for CAV Systems

A RAG pipeline for CAV systems can be implemented using modular, open-source tools. The implementation described here uses a local, resource-efficient setup suitable for edge deployment, though RAG can also be deployed in cloud-based architectures.

8.5.1 System Components and Execution Environment

A RAG system is built on the following software components:

- **Language Model:** A pretrained language model that generates responses. In the implementation described here, the LLM is executed via a local inference platform that supports foundation models for on-device processing, though cloud-based LLM services can also be used.
- **Embedding Model:** Provided by sentence transformer libraries, which encode text into dense vector representations suitable for similarity search. Embedding models can be deployed locally or accessed via cloud APIs.
- **Vector Store:** Managed by a vector database that enables efficient top-k retrieval operations. The vector database allows rapid semantic search across diverse information sources without requiring exact keyword matches, as detailed in the Core Components section above. Vector databases can be deployed locally or in the cloud.

The implementation uses a local, offline configuration, making it ideal for edge deployment, in-vehicle diagnostics, and development scenarios that demand data sovereignty and low-latency inference. However, the same components can be deployed in cloud-based architectures for scenarios requiring scalability and centralized management.

8.5.2 Data Preparation and Index Construction

The RAG pipeline begins with preparing a document collection that serves as the retrieval corpus. For example, a safety feature documentation can be used. The preparation process involves:

1. **Load and Segment the Document:** The raw text is divided into fixed-size chunks. Each chunk becomes an independent retrieval unit.
2. **Compute Embeddings:** Each chunk is embedded into a vector using a pretrained sentence transformer.
3. **Store in Vector Database:** The resulting vectors are stored in the vector database along with their corresponding text, indexed for future retrieval.

This process transforms static domain-specific text into a semantically searchable index, enabling the system to rapidly retrieve relevant context when drivers pose queries. For CAV applications requiring domain-specific terminology alignment (e.g., mapping "braking intervention" to "deceleration events"), pretrained embeddings can be optimized using contrastive learning to adapt the embedding space for CAV-specific semantic relationships. The comprehensive methodology for contrastive learning, including the mathematical formulation and training workflow, is detailed in Section 5.6.4 of Chapter 5.

8.5.3 Query Processing and Response Generation

Once the index is built, the system can process user queries. For example, consider a query such as *"What is Proactive Driver Assistance?"* Following the retrieve-inject-generate paradigm (Section 8.2.1), the system embeds the query, retrieves top-k similar document chunks, and constructs a composite prompt for the language model, ensuring the language model receives contextually relevant material from CAV-specific sources (logs, documentation, V2X messages).

The constructed prompt is submitted to the language model via the inference interface. The model generates a response based on both the original user query and the retrieved contextual information, enabling semantically rich and context-aware answers for in-vehicle explanation tasks, status interpretation, or V2X message summarization.

8.5.4 Code Skeleton of the RAG Pipeline

The following code skeleton summarizes the entire RAG workflow in pseudocode form, implementing the retrieve-inject-generate paradigm.

Listing 8.1 RAG Pipeline Implementation

```
# Load and chunk the input document
raw_text = load_text("tss3_text.txt")
chunks = chunk_text(raw_text, size=300)

# Embed text chunks and store in vector DB
embeddings = embed_chunks(chunks)
store_in_vector_db(chunks, embeddings)

# Process user query and retrieve context
query = "What is Proactive Driver Assistance?"
query_vector = embed_query(query)
retrieved_chunks = retrieve_top_k(query_vector)

# Format prompt and call the language model
prompt = build_prompt(retrieved_chunks, query)
response = llm.chat(prompt)
```

The code follows the three-stage paradigm. The first block prepares the retrieval corpus by loading, chunking, and embedding documents into the vector database. The second block implements the Retrieve stage, embedding the user query and finding the top-k most similar document chunks. The final block implements the Inject and Generate stages, constructing a prompt that combines the retrieved context with the query and generating the final response.

A complete implementation is available in the companion repository. The companion source code includes modules for document preprocessing, embedding, retrieval, and natural language generation, providing a ready-to-run example for CAV developers and researchers.

8.6 Real-time Feasibility and System Constraints

RAG introduces latency and computational overhead that must be carefully managed in CAV systems. In safety-critical environments, real-time constraints limit where RAG can be applied.

8.6.1 Latency Components of RAG

A complete RAG pipeline involves multiple processing stages, each contributing to total system latency. The primary components are embedding computation, where the query must be transformed into a dense vector using a neural embedding model; vector retrieval, where a similarity search is performed against a pre-indexed document collection, typically using approximate nearest-neighbor algorithms; prompt construction, where retrieved text chunks are formatted along with the user query into a prompt structure suitable for the LLM; and language model inference, where a response is generated by the language model, conditioned on the combined input.

The latency of RAG components varies significantly depending on hardware configuration, model size, and implementation details. In general, the embedding computation typically requires 10–50 ms for sentence-level encoders on modern hardware, while vector search operations in modern databases can be completed in under 10 ms for typical CAV document collections. The most significant latency contributor is language model inference, which can range from hundreds of milliseconds to several seconds depending on model size and hardware acceleration. Prompt construction adds minimal overhead, usually under 10 ms.

These latency characteristics make RAG suitable for applications where response times of several hundred milliseconds to a few seconds are acceptable, such as explanatory interfaces and post hoc analysis. However, they preclude RAG from real-time control loops or safety-critical decision-making that requires sub-100 ms response times.

8.6.2 Design Principles for Integration

To effectively incorporate RAG into CAV architectures, several principles are recommended. RAG should be used for semantic interpretation, not control: it should augment the human interface, not replace deterministic planners or controllers. Asynchronous execution ensures that RAG queries are handled in separate threads or services, decoupled from safety-critical timing paths. Caching and prefetching can reduce latency for common queries or route contexts through precomputed retrieval and prompt caching. Fallback handling requires that, in case of timeouts or incomplete generation, a default message should be returned to preserve system predictability. Finally, hybrid pipeline coexistence uses RAG for naturalistic explanations and function calling for real-time status access, forming a layered interaction model.

By restricting RAG to non-critical execution paths and applying asynchronous design principles, developers can enhance system transparency without compromising real-time safety guarantees.

8.7 Applied Use Cases of RAG in CAV Systems

The following use cases demonstrate how RAG addresses the core problems identified in Section 8.1.1 (hallucination, staleness, and partial observability) in CAV systems. The examples focus on explanation and interpretation applications, where RAG enables human-understandable responses grounded in the retrieved context. While RAG also finds applications in training scenario generation, perception tasks, and spatial reasoning, the use cases presented here illustrate the most direct connections to the core problems and are most immediately applicable in production CAV systems.

8.7.1 Episodic Memory Retrieval

RAG enables the retrieval of *episodic memory* (historical driving data, logs, past configurations, and conversation history) to answer questions about recent behavior, similar past situations, or previous interactions [2]. This capability supports both immediate explanations and contextual personalization.

For a query such as *"Why did the vehicle brake suddenly on the highway?"* RAG retrieves driving log entries showing braking events and configuration settings defining hazard detection thresholds, generating the response: "The system applied emergency braking because an object was detected ahead. According to the configured hazard response threshold, this triggered a deceleration maneuver for safety."

For conversation-aware queries such as *"What did we say about ACC settings earlier?"* RAG retrieves prior dialogue logs from the current session and previously issued ADAS commands, generating: "Earlier in this drive, you asked to increase the ACC time gap from 1 to 2 bars, and the system confirmed the change."

8.7.2 Dynamic Policy and Regulation Adaptation

RAG enables CAV systems to adapt to changing road conditions, temporary regulations, and region-specific traffic laws by retrieving up-to-date information that may not be present in the model's training data. This capability is critical for handling failure-recovery scenarios, where the system must respond to unexpected changes in road infrastructure or traffic rules.

For a query such as *"Why did the navigation system route us around this highway? The map shows it should be open."* RAG retrieves recent V2X messages indicating temporary road closure, regional traffic management updates, and historical routing data, generating: "The highway is temporarily closed due to emergency construction that began 2 hours ago. The infrastructure system broadcast a road closure alert, and the navigation system automatically rerouted to avoid the affected section. This closure was not in the original map data, but the system retrieved the current status from real-time V2X communications."

8.7.3 V2X/V2V Communication Interpretation

RAG enables CAV systems to retrieve and interpret structured V2X and V2V messages, transforming technical infrastructure, and vehicle-to-vehicle communications into human-interpretable explanations. This capability helps drivers understand current road conditions, advisories, and cooperative driving behaviors.

For infrastructure queries such as *"What warning did the infrastructure unit just send?,"* RAG retrieves TIM or RSA broadcasts received within the last 60 seconds along with metadata, generating: "The infrastructure unit ahead issued a lane closure alert due to construction 200 m beyond the next intersection. Drivers are advised to merge left."

For cooperative driving queries such as *"Why didn't we pass the car merging from the right?,"* RAG retrieves V2V messages indicating merge intent and cooperative driving policies, generating: "The vehicle in the adjacent lane requested a merge. The system yielded based on cooperative policy, which allows merging when headway and lane availability permit safe integration."

Beyond explanation and interpretation, RAG has been applied to other CAV tasks such as traffic scenario generation for training autonomous systems [5], zero-shot vehicle model recognition, and enhanced spatial reasoning in road scene understanding [6]. These applications leverage RAG's ability to retrieve and integrate diverse knowledge sources, though they often require more specialized architectures and are typically addressed in advanced research contexts. The use cases presented in this section demonstrate RAG's core value proposition: enabling grounded, verifiable, and context-aware responses that address the fundamental limitations of pretrained language models in safety-critical systems.

8.8 Summary

This chapter presented RAG as a system-level mitigation strategy for addressing hallucination, staleness, and partial observability in CAV systems. The retrieve-inject-generate paradigm was introduced as the core architectural approach, with components, including query embedding, vector databases, and retrievers that enable semantic similarity-based document retrieval. The mathematical foundations cover embedding and vector representation, similarity-based retrieval using cosine similarity, and a probabilistic formulation that models retrieval as a latent variable process.

A practical implementation pipeline demonstrated how to build RAG systems using modular, open-source tools suitable for CAV deployment. The chapter compared RAG with rule-based information systems, highlighting when each approach is appropriate and how RAG complements deterministic control subsystems. Design considerations such as latency components, retrieval efficiency, and system constraints were analyzed to guide integration into safety-aware, real-time environments. The chapter also noted that embedding optimization techniques such as contrastive learning (detailed in Chapter 5) can improve retrieval accuracy for CAV-specific terminology and scenarios. Applied use cases demonstrated how RAG supports episodic memory retrieval, dynamic policy and regulation adaptation, and V2X/V2V message interpretation in CAV systems.

Together, these elements demonstrate how RAG transforms static pretrained language models into dynamic, context-aware systems capable of retrieving, and grounding responses in external knowledge. The retrieve-inject-generate paradigm provides a flexible architecture that can be adapted to diverse CAV applications while maintaining traceability and verifiability essential for safety-critical systems. By operating in non-critical execution paths and applying asynchronous design principles, RAG enhances system transparency and explainability without compromising real-time safety guarantees.

The evaluation of RAG systems, including reference-free metrics for faithfulness, answer relevance, and context relevance, is discussed in detail in Chapter 21 (Section 21.5).

References

1 Ziwei Ji, Nayeon Lee, Rita Frieske, Tiezheng Yu, Dan Su, Yan Xu, Etsuko Ishii, Yejin Bang, Andrea Madotto, and Pascale Fung. Survey of hallucination in natural language generation. *ACM Computing Surveys*, 55(12):1–38, 2023.

2 Patrick Lewis, Ethan Perez, Aleksandra Piktus, Fabio Petroni, Vladimir Karpukhin, Naman Goyal, Heinrich Küttler, Mike Lewis, Wen-tau Yih, Tim Rocktäschel, et al. Retrieval-augmented generation for knowledge-intensive NLP tasks. In *Proceedings of the 34th International Conference on Neural Information Processing Systems*, NIPS '20, Red Hook, NY, USA, 2020. Curran Associates Inc. ISBN 9781713829546.

3 Nils Reimers and Iryna Gurevych. Sentence-BERT sentence embeddings using siamese BERT-networks. In *Proceedings of the 2019 Conference on Empirical Methods in Natural Language Processing and the 9th International Joint Conference on Natural Language Processing (EMNLP-IJCNLP)*, pages 3982–3992, 2019.

4 Vladimir Karpukhin, Barlas Oguz, Sewon Min, Patrick Lewis, Ledell Wu, Sergey Edunov, Danqi Chen, and Wen-tau Yih. Dense passage retrieval for open-domain question answering. In *Proceedings of the 2020 Conference on Empirical Methods in Natural Language Processing (EMNLP)*, pages 6769–6781, 2020.

5 Driving-RAG Team. Driving-RAG: driving scenarios embedding, search, and rag applications, 2025. https://arxiv.org/abs/2504.04419.

6 Logic-RAG Team. Logic-RAG: augmenting large multimodal models with visual-spatial knowledge for road scene understanding, 2025. https://arxiv.org/abs/2503.12663.

9

Function Calling and Tool Use

9.1 Introduction: Why Connected and Autonomous Vehicles Need Function Calling

Connected and autonomous vehicles (CAVs) are increasingly equipped with natural language interfaces designed to provide transparency, assistance, and interaction. However, conversational fluency alone is insufficient when users pose queries that require access to the vehicle's current state, sensor data, or road context. Consider the prompt, "Is there a vehicle approaching from the left?" While a large language model (LLM) can parse the question, it cannot directly access sensor readings or traffic data unless explicitly connected to those subsystems.

9.1.1 Prompt Limitations in Real-world Scenarios

Although prompt engineering has significantly advanced the utility of LLMs in task-specific applications, its effectiveness in safety-critical or real-time systems remains inherently constrained. Prompts can guide the model's attention, encourage structured reasoning, and elicit relevant outputs. However, they cannot overcome the fundamental limitation that the model does not observe the world in real time.

This limitation becomes particularly evident in the context of CAVs. Consider the following examples:

- *"What is my current speed?"* The model can only speculate unless speed is provided explicitly.
- *"Why did the car brake suddenly?"* The model cannot access logs from braking or perception subsystems.
- *"Is it safe to change lanes?"* The model cannot assess traffic on adjacent lanes without live sensor input.

In each case, the LLM lacks access to the ground truth information required to provide a trustworthy or verifiable response. Prompting alone reduces the problem to informed guessing, often influenced by training distribution priors rather than real-time context. This leads to hallucination risks and degraded user trust in high-assurance settings such as automated driving.

Function calling, also known as tool calling in agentic artificial intelligence (AI) contexts, provides this missing connection [1]. This capability is central to agentic AI systems, where language models autonomously select and invoke external tools to accomplish complex tasks beyond their training data. In CAV systems, function calling enables LLMs to interact with vehicle subsystems, access real-time sensor data, and coordinate with control systems, transforming static language models into active reasoning agents.

However, to understand its role in CAV systems, function calling must be understood not merely as a feature for tool invocation, but as a fundamental architectural component that defines the contract between probabilistic reasoning and deterministic system execution. Function calling serves three critical architectural roles: it is an *interface abstraction* that standardizes communication between language models and vehicle subsystems, a *safety boundary* that enforces validation and constraints before system actuation, and a *control plane* that enables language models to coordinate with real-time control loops while maintaining separation of concerns. These tools are invoked through structured outputs, often in the form of JavaScript Object Notation (JSON) function calls, which are interpreted and executed by the system. In doing so, the LLM becomes not only a conversational interface but a reasoning and coordination engine embedded in a closed-loop perception-action system.

In vehicular systems, function calls proposed by an LLM must be treated as advisory intents and are subject to independent safety validation and arbitration before execution. This fundamental principle ensures that probabilistic reasoning from language models never directly controls safety-critical vehicle behavior without deterministic validation, constraint checking, and safety gate mediation.

9.1.2 Bridging Language Understanding and System Action

To function effectively in the CAV domain, LLMs must become components of a larger decision-making architecture. This requires transforming them from static text predictors into active participants in sensor-action loops. A sensor-action loop is a control-theoretic concept wherein observations from the environment are used to guide system behavior through structured feedback. In the context of CAVs, this feedback loop involves perception (e.g., sensors, vehicle-to-everything [V2X] data), reasoning (e.g., inference, explanation), and actuation (e.g., adaptive cruise control [ACC] adjustments).

Function calling enables the LLM to bridge language understanding and system action through its role as an interface abstraction. The process involves three distinct layers: (1) *function calling*, where the LLM makes a reasoning-time decision to produce a structured intent in JSON format; (2) *tools or application programming interface (APIs)*, which are deterministic system components that validate, execute, and log the function call; and (3) *vehicle actions*, which are the physical or logical effects subject to safety gates. For actuation commands, the tool's execution may produce vehicle actions (e.g., adjusting ACC speed), which are subject to safety gates before taking physical effect (see Section 9.3.2). The complete pipeline, including detailed examples, is illustrated in Section 9.2.

9.2 System Architecture of Function Calling

In the architecture of generative AI systems for CAVs, function calling operationalizes the transition from natural language intent to grounded system behavior while maintaining the separation between probabilistic reasoning and deterministic execution. Without function calling, a language model remains an isolated inference module, unable to sense or act. With function calling, it becomes a central reasoning agent that participates in real-time decision loops through a well-defined architectural interface.

9.2.1 Definitions and Terminology

The literature on tool-augmented language models [1–3] often uses the terms *function*, *tool*, and *API* with overlapping meanings. In agentic AI contexts, this capability is frequently referred to as *tool calling*, emphasizing the autonomous selection and invocation of external tools by AI agents. In this chapter, we adopt the following distinctions to clarify their use in the context of generative AI for CAVs:

- **Function Calling (Tool Calling):** A reasoning-time decision made by the LLM that produces a structured intent (typically a JSON object specifying the function name and arguments). This is the probabilistic output of the language model, representing the model's decision about what action to take.
- **Tool or API:** A deterministic system component that receives the structured intent from function calling, validates it, executes the corresponding operation, and logs the action. Tools may wrap local functions, system commands, or hardware-accessible modules. APIs are network-accessible interfaces to tools, typically exposed via Hypertext Transfer Protocol (HTTP), gRPC (gRPC Remote Procedure Calls), or another protocol. Both tools and APIs abstract execution details and are defined by a description, a set of parameters, and an expected behavior.
- **Vehicle Action:** The physical or logical effect produced by tool execution, such as adjusting ACC speed, activating emergency braking, or updating display settings. Vehicle actions are subject to safety gates that mediate actuation commands before they take physical effect.

In this chapter, we primarily use the term *function* to refer to the structured invocation request generated by the language model (function calling), and *tool* or *API* to refer to the deterministic components that execute these calls.

9.2.2 Architecture and Pipeline

Modern CAV software architectures are organized into modular components, each responsible for a distinct aspect of perception, planning, control, or communication. Function calling integrates the generative AI stack with these existing subsystems through a structured pipeline that transforms natural language intent into grounded system behavior.

The function calling process follows a four-stage pipeline: (1) prompt interpretation, where user input is processed and understood; (2) function selection and argument generation, where the language model produces a structured function call; (3) external execution, where deterministic tools validate and execute the call; and (4) response composition, where results are formatted for the user or used in further reasoning. Figure 9.1 illustrates this complete call cycle, showing how user intent flows through the system to produce grounded responses.

Function calling acts as the interface abstraction between the LLM and the execution layer. The routing process enforces the safety boundary by validating function calls before execution, with the goal of ensuring that only authorized, well-formed invocations reach the execution layer. This separation of concerns enables the language model to focus on semantic interpretation and reasoning, while tools handle deterministic execution, and safety gates mediate vehicle actions.

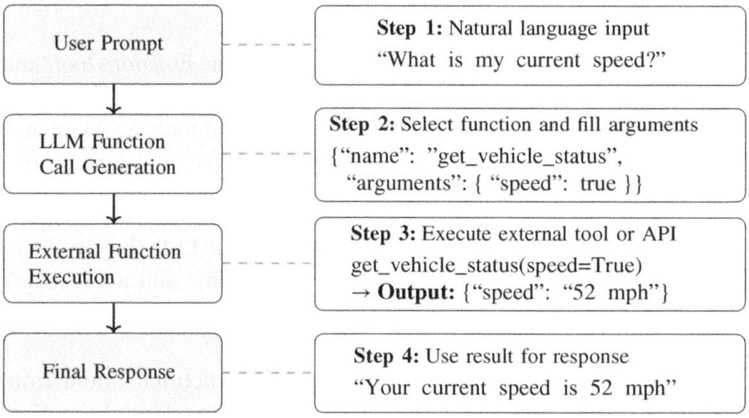

Figure 9.1 Prompt-to-action pipeline for function calling. Each stage reflects a transition from user intent to system behavior, including final response composition for driver-facing output.

9.2.3 Integration with Vehicle and Infrastructure Systems

The execution layer of a CAV system must interact with a wide range of onboard and external subsystems. These include real-time sensing modules, driver assistance controllers, and cooperative communication interfaces. Function calling enables the language model to engage with these components through structured, callable interfaces, forming the foundation for agentic behavior in CAV systems.

Key domains accessed via function calls include:

- **Vehicle state:** Speed, heading, acceleration, brake status, and steering angle.
- **Perception subsystems:** Output from radar, light detection and ranging (LiDAR), cameras, and multi-sensor fusion algorithms.
- **Driver assistance modules:** ACC, lane centering, emergency braking, and override detection.
- **V2X communication:** Includes messages from roadside infrastructure such as Signal Phase and Timing (SPaT) data, along with cooperative awareness broadcasts, and vehicle intent messages exchanged between nearby vehicles.

Each of these domains exposes callable functions that allow the model to either *sense* (i.e., retrieve information) or *act* (i.e., issue a command or setting). For instance, in response to a prompt such as, *"Why are we slowing down?,"* the LLM may generate a function call that invokes a diagnostic tool, which then queries recent braking events and checks perception logs for front-object proximity. In a more directive scenario, a user might ask, *"Set my cruise speed to 65 mph,"* prompting the model to generate a function call that is executed by a tool within the ACC module, producing a vehicle action (speed adjustment) that is subject to safety gates before taking effect.

Function calling supports extensibility: new vehicle features can be exposed to the language model simply by registering additional tool schemas, without retraining the model itself. Protocols such as MCP [4] further enhance this extensibility by providing standardized frameworks for dynamic tool discovery and integration, enabling runtime discovery of available capabilities. This separation of concerns between language-based reasoning and verified execution logic supports both interpretability and system assurance, enabling flexible interaction with the driver while maintaining the determinism required by regulated vehicle subsystems.

9.3 Theoretical Foundations of Function Calling

Function calling is fundamentally model-agnostic: the core mechanism of mapping natural language to structured function invocations works across different LLM architectures and deployment frameworks, even though specific implementations may vary in their constrained decoding strategies, API interfaces, or schema enforcement mechanisms. Function calling requires language models to interpret natural language prompts and produce structured outputs that correspond to predefined tools or APIs (as defined in Section 9.2.1). These outputs include both the name of the function to invoke and the set of arguments required for execution.

Unlike traditional software logic, this mapping is not rule-based, but learned through statistical associations between language patterns and tool schemas. Function calling, therefore, depends on two critical capabilities: (1) semantic alignment between user intent and tool descriptions, and (2) structured argument generation that conforms to schema constraints.

9.3.1 Mathematical Foundation

Function calling in language models operates through probabilistic generation, where the model generates both function names and arguments using the same underlying mechanisms as standard text generation.

9.3.1.1 Function Selection

When a user provides a prompt p, the language model must select which function to invoke from the available set $\mathcal{F} = \{f_1, \dots, f_n\}$. Each function f_i has a unique name and an associated schema defining its arguments. The probability of selecting function f_i is modeled as:

$$P(f_i \mid p, \mathbf{c}) = \prod_{t=1}^{T_i} P(w_t \mid p, \mathbf{c}, w_{1:t-1}), \tag{9.1}$$

where p represents the user prompt (a sequence of tokens), $\mathbf{c}$ represents the context vector encoding system instructions, conversation history, and tool specifications, T_i is the number of tokens in function f_i's name, and w_t denotes the discrete token (symbol) at position t in the function name sequence. Note that while w_t represents a discrete token in the probability distribution, tokens are internally embedded as high-dimensional vectors during model processing. Similarly, the prompt p is processed as a sequence of token embeddings, while $\mathbf{c}$ represents the aggregated vector representation of the context. The selection uses deterministic (greedy) decoding for consistency and reliability:

$$w_t^* = \arg\max_{w} P(w \mid p, \mathbf{c}, w_{1:t-1}), \tag{9.2}$$

where w_t^* is the selected token at position t.

9.3.1.2 Constrained Argument Generation

Once function f^* is selected, the model generates arguments under schema constraints. The probability of generating the complete argument object is:

$$P(\mathbf{a}_{f^*} \mid p, f^*) = \prod_{t=1}^{T_a} P(w_t \mid p, f^*, w_{1:t-1}, \mathcal{C}), \tag{9.3}$$

where $\mathbf{a}_{f^*}$ represents the argument vector for function f^*, T_a is the number of tokens in the argument sequence, w_t denotes the discrete token (symbol) at position t in the argument sequence, and C represents the set of schema constraints that restrict the valid token vocabulary at each generation step. This formulation reflects a division of responsibilities: the language model generates tokens probabilistically based on its training (computing $P(w_t \mid p, f^*, w_{1:t-1})$), while constrained decoding enforces the constraint set C by masking invalid tokens at each decoding step, setting their probabilities to zero, and renormalizing the distribution over the remaining valid tokens, effectively restricting generation to only schema-compliant continuations.

9.3.2 Constrained Decoding Mechanisms

Although large language models are trained to generate unstructured token sequences, function calling introduces structure into the output space. The platform enforces schema compliance through constrained decoding [5].[1] This decoding process operates at inference time and, when properly implemented, ensures that the function call conforms to the expected argument structure, types, and constraints. The model proposes a function call based on its training, but the inference engine validates that the final output is structurally valid before passing it to the execution environment.

The function calling process follows a structured workflow that begins with free-form function selection and transitions to constrained argument generation (as formalized in Section 9.3.1). Once a valid function is identified, the system activates schema-based constraints for argument generation.

Constrained decoding operates through token-level masking during the generation process. As the model generates each token, the inference engine maintains a parser and validator state that tracks the current position within the JSON structure. At each decoding step, the probability distribution over the next possible tokens is modified to exclude those that would lead to invalid outputs. For example, when generating a Boolean parameter, only tokens representing `true` or `false` are permitted. Similarly, for numeric parameters, the system allows only digits, decimal points, signs, and exponent markers until a complete numeric literal is formed.

The constraint enforcement process involves multiple layers of validation. Structural constraints ensure well-formed JSON through context-free grammar rules that govern braces, brackets, commas, and proper quoting. Type constraints enforce data types such as string, number, boolean, array, and object. Value constraints validate ranges, enums, patterns, and cross-field dependencies. For instance, a schema specifying

```
{"speed_mph": {"type": "number", "minimum": 20, "maximum": 90}}
```

would allow only numeric tokens during generation and then validate that the final number falls within the specified range.

Cross-field constraints present a particular challenge, as they cannot be enforced purely through grammar rules. Conditions such as `min_speed` < `max_speed`, or lane selection being valid only for multi-lane roads, require semantic validation that occurs during or immediately after the generation process. When such constraints are violated, the system rejects the output and requests regeneration under the same constraints.

1 Examples include the OpenAI API, which applies schema-based token masking when tools are registered, and Ollama, which performs schema-constrained decoding locally for models such as LLaMA 3.2.

The fundamental principle underlying constrained decoding is the separation of responsibilities between the language model and the platform infrastructure. Language models generate tokens probabilistically based on their training but lack inherent understanding of structural requirements. The platform provides the constraint enforcement mechanisms that, when properly implemented, ensure well-formed JSON and type-correct arguments at inference time. In essence, the language model proposes content while the platform constrains the generation space.

In safety-critical applications such as connected and autonomous vehicles, constrained decoding serves as the first line of defense against malformed function calls. However, this mechanism must be complemented by additional validation layers. Post-decode validation provides additional schema verification after generation. Domain-level guards implement business-logic checks, such as jurisdiction-specific speed limits, within the tool-execution layer. Execution safety gates mediate vehicle actions, ensuring that critical operations like setting acceleration speed never take physical effect directly from the language model output without multiple validation layers. This multi-layer validation architecture enforces the fundamental principle that LLM function calls are advisory intents requiring independent safety validation and arbitration before execution. A single validation failure should prevent execution to maintain system safety.

9.4 Function Calling in Practice

This section demonstrates how to implement a complete function calling system using inference frameworks and Python-based execution. The implementation consists of three main components: (1) a function registry that maps function names to executable code, (2) an inference framework that generates structured function calls, and (3) a dynamic dispatch mechanism that executes functions based on LLM output. While the examples use local inference frameworks, the same architecture can be deployed with remote APIs or cloud-based services.

9.4.1 Function Registry

A function registry maps function names (as strings) to Python functions, creating a lookup table that connects the LLM's text-based function calls to executable code. Before registering functions, we must first define their interfaces using JSON Schema, which specifies the function name, description, and parameter structure. This schema serves as the contract between the language model and the execution environment.

9.4.1.1 Defining Function Schemas

Each function must be described using a JSON Schema that defines its interface. The schema specifies the function name, a natural language description (which helps the LLM understand when to call it), and the parameter structure:

Listing 9.1 Function schema definition

```
tool_schema = {
  "name": "get_vehicle_status",
  "description": "Return the vehicle's current speed and heading.",
  "parameters": {
    "type": "object",
    "properties": {
      "speed": { "type": "boolean" },
```

```
            "heading": { "type": "boolean" }
        },
        "required": []
    }
}
```

The schema defines the function's interface: its name (`get_vehicle_status`), a description that helps the LLM understand its purpose, and the parameter structure (two optional boolean parameters). This schema will be used by the inference framework to guide function call generation, and it also documents what the corresponding Python function should accept.

9.4.1.2 Registering Python Functions

Once the schema is defined, we implement the corresponding Python function and register it in the function registry:

Listing 9.2 Example: Registering Python Functions

```
def get_vehicle_status(speed=False, heading=False):
    return {
        "speed": "52 mph" if speed else None,
        "heading": "NE" if heading else None
    }

def set_acc_speed(speed_mph):
    return f"Successfully set ACC max speed to {speed_mph} mph"

function_registry = {
    "get_vehicle_status": get_vehicle_status,
    "set_acc_speed": set_acc_speed
}
```

The example defines two functions: `get_vehicle_status` retrieves vehicle sensor data (speed and heading), while `set_acc_speed` updates the Adaptive Cruise Control speed setting. The `function_registry` dictionary maps string names (which match the schema names) to the actual Python function objects. This mapping enables the system to look up and execute functions based on the function names produced by the language model. The Python function signatures must match the parameter structure defined in the corresponding schema.

9.4.2 Inference Framework

Inference frameworks support structured function calling through schema-compatible APIs, enabling language models to generate function calls in JSON format based on user prompts and predefined function schemas. This demonstrates that function calling is model-agnostic (as discussed in Section 9.3) and can be implemented across diverse deployment environments, including local inference frameworks, remote APIs, or cloud-based services.

9.4.2.1 Generating Structured Function Calls

The inference framework uses the function schemas (defined in Section 9.4.1) to guide the LLM in generating appropriate function calls. We invoke the framework with a user prompt and the function schema, and the framework generates a structured function call:

Listing 9.3 Generating structured function calls

```
import ollama

prompt = "What is my current speed and heading?"
response = ollama.chat(
    model="llama3.2",
    messages=[{"role": "user", "content": prompt}],
    tools=[tool_schema]
)
```

The key parameter is `tools=[tool_schema]`, which passes the function schema to the inference framework. This schema tells the LLM what functions are available, their descriptions, and parameter structures. The framework uses this information to generate a structured function call that matches the schema.

9.4.2.2 LLM Response

The inference framework returns a response object containing the generated function call. To extract the function call from the response, we use the following commands:

Listing 9.4 Extracting function call from LLM response

```
tool_call = response.message.tool_calls[0]
function_name = tool_call.function.name
arguments = tool_call.function.arguments

print(f"Function name: {function_name}")
print(f"Arguments: {arguments}")
```

For the prompt *"What is my current speed and heading?,"* the output would be:

Listing 9.5 Output from extraction commands

```
Function name: get_vehicle_status
Arguments: {"speed": true, "heading": true}
```

The `arguments` field is a JSON string that must be parsed before use. This structured output is then executed using the function registry, as shown in the next subsection.

9.4.3 Dynamic Dispatch

Dynamic dispatch is the mechanism that executes functions based on the LLM's structured output. When the LLM generates a function call, the system uses the function registry to look up the corresponding Python function and invokes it with the provided arguments. This pattern allows a clean separation between language interpretation (handled by the LLM) and system execution (handled by the dispatch mechanism). Continuing from the extraction example (Section 9.4.2), where we extracted `function_name` and `arguments` from the response, we now execute the function using the registry. The `arguments` field is a JSON string, so it must first be parsed:

Listing 9.6 Dynamic function dispatch

```
import json

# Parse the JSON string arguments
arguments_dict = json.loads(arguments)

```

```
# Retrieve the function from the registry
func = function_registry[function_name]

# Execute the function with the parsed arguments
result = func(**arguments_dict)

print(f"Result: {result}")
```

For the `get_vehicle_status` example, the output would be:

Listing 9.7 Function execution output

```
Result: {'speed': '52 mph', 'heading': 'NE'}
```

This result comes from the predefined `get_vehicle_status` function (shown in Section 9.4.1), which was executed with the arguments `speed=True` and `heading=True` extracted from the LLM response. The function returns the predefined values `"52 mph"` and `"NE"` that are hard-coded in the function definition, demonstrating how the function provides simulated vehicle data. In a real implementation, these values would come from actual vehicle sensors or subsystems.

This pattern enables the system to execute any registered function dynamically based on the LLM's output, without requiring hardcoded conditional logic for each function.

9.5 Function Schema Library and Design Principles

Function calling systems depend not only on model capability but also on the quality of the interfaces they are given. Poorly designed function schemas can lead to hallucinated arguments, incorrect function selection, or unstable execution behavior. In the context of CAVs, such failures may result in degraded system performance or unsafe actions.

9.5.1 JSON Schema Structure and Components

The effectiveness of function calling in a generative AI system depends critically on how tools are defined and described. Since the language model must both recognize the correct function to invoke and generate its arguments, the function interface serves as a contract between the LLM and the execution environment. JSON Schema [6] provides a flexible and widely adopted standard for specifying this interface.

Each function exposed to the LLM must be specified as a JSON object with the following required fields:

- **`name`** A unique string identifier for the function, such as `get_vehicle_status`.
- **`description`** A natural language phrase explaining what the function does and when it should be used.
- **`parameters`** A nested object that defines the input fields required by the function, their types, and validation rules.

The `parameters` field is itself a JSON Schema, typically with the top-level type set to `"object"`, and a `properties` dictionary that specifies each input field. Fields may be marked as `required`, and may include type annotations such as `string`, `number`, or `boolean`.

9.5.2 CAV Function Schema Patterns

This section presents additional CAV function schema examples that illustrate design patterns beyond the basic examples shown in Section 9.4.1. These schemas demonstrate how to structure functions for different CAV use cases, including system analysis, perception, and sensor access.

9.5.2.1 System Analysis and Explanation Functions

Analysis functions enable the LLM to explain vehicle behavior by accessing logs and sensor data. These functions typically have optional parameters with sensible defaults, allowing flexible querying.

Listing 9.8 Behavior Explanation Function Schema

```
{
  "name": "explain_braking",
  "description": "Analyze sensor and control logs to explain recent braking
      behavior.",
  "parameters": {
    "type": "object",
    "properties": {
      "context_window_sec": {
        "type": "number",
        "description": "Time window in seconds to analyze before the braking
            event",
        "minimum": 0.5,
        "maximum": 10.0,
        "default": 3.0
      }
    },
    "required": []
  }
}
```

This schema demonstrates the use of optional parameters with defaults. The `context_window_sec` parameter allows the LLM to specify a custom analysis window, but defaults to 3.0 seconds if omitted. The `description` field within the parameter helps the LLM understand the parameter's purpose and constraints.

9.5.2.2 Perception and Sensor Functions

Sensor access functions enable the LLM to retrieve real-time or historical sensor data. These functions require explicit parameter specification since sensor selection and timing are critical for accurate data retrieval.

Listing 9.9 Sensor Log Access Function Schema

```
{
  "name": "get_sensor_log",
  "description": "Return data from a selected sensor at a given time offset.",
  "parameters": {
    "type": "object",
    "properties": {
      "sensor": {
        "type": "string",
        "description": "Sensor identifier (e.g., 'lidar_front', 'camera_rear', '
            radar_side')",
```

```
            "enum": ["lidar_front", "lidar_rear", "camera_front", "camera_rear", "
                radar_front", "radar_side"]
          },
          "time_offset_sec": {
            "type": "number",
            "description": "Time offset in seconds from current time (negative for
                past, zero for current)",
            "minimum": -60.0,
            "maximum": 0.0
          }
        },
        "required": ["sensor", "time_offset_sec"]
      }
    }
```

This schema illustrates several design patterns: (1) using `enum` to restrict sensor names to valid options, preventing hallucinated sensor identifiers; (2) requiring both parameters since sensor selection and timing are essential; (3) providing clear parameter descriptions that guide the LLM in generating appropriate values. The `enum` constraint is particularly important for safety-critical systems, as it prevents the LLM from requesting non-existent sensors.

Caution 9.1 Schema Constraints May Not Be Enforced

Schema constraints for `set_acc_speed` (i.e., `minimum` and `maximum` on `speed_mph`) are not enforced by all model runtimes by default. The level of constraint enforcement varies across implementations. Two patterns are commonly used.

First, constrained decoding that is wired to a JSON Schema. Some cloud-based APIs support strict mode or structured outputs, which constrain decoding so that generated arguments adhere to declared types and numeric bounds at inference time for supported models. Second, some local runtimes provide schema-driven constrained decoding. For example, local inference frameworks support JSON-Schema–based structured outputs, implemented via context-free grammars or similar mechanisms, which can enforce types and limits during generation on compatible local models.

However, constraint enforcement is not universal. Some runtimes may only enforce basic type constraints (e.g., ensuring a number is generated for a numeric field) but not range constraints (e.g., `minimum` and `maximum`). Additionally, very complex schemas or unsupported keywords may be relaxed by the runtime or fail with errors. Therefore, production systems should always perform server-side validation, regardless of whether constrained decoding is available. In safety-critical paths, design schemas that are simple and testable, and complement them with explicit runtime checks before actuation.

9.5.3 Function Description Design and Prompt Alignment

The `description` field of a function is the primary semantic signal the LLM uses to determine when the function should be called. Since the LLM selects functions through semantic matching, descriptions must be written in natural, precise language that matches likely user prompts and overlaps with expected user phrasing.

Descriptions should name the specific capability exposed by the function, ideally using terminology aligned with how drivers naturally refer to the underlying feature (e.g., "lane," "speed," "braking," etc.). Avoid vague phrases such as "fetch data" or "get information"; instead, describe exactly what the function does.

Listing 9.10 Good Example

```
"description": "Return the vehicle's current speed and heading."
```

Listing 9.11 Poor Example

```
"description": "Get data from the system."
```

9.5.3.1 Prompt Alignment Principles

To achieve effective semantic alignment, consider how users phrase their queries. The description should include key terms that users are likely to use when requesting the function.

Example 9.1 Prompt Alignment Examples

The following examples demonstrate how to align function descriptions with user language:

Example 1: Object Detection If a driver asks:

```
"What objects are in front of me right now?"
```

then a well-aligned function schema might include:

```
"name": "detect_objects",
"description": "Detect objects in front of the vehicle using current sensor
    data."
```

Here, the use of terms such as "objects," "in front," and "current" directly supports semantic alignment with the user's phrasing, enabling the LLM to correctly select this function for the query.

Example 2: Speed Control For speed-related queries, align with common driver terminology:

```
User prompt: "Set my cruise control to 65"
Well-aligned description: "Set the adaptive cruise control maximum speed in
    miles per hour."
```

The description uses "adaptive cruise control" (which drivers may refer to as "cruise control") and "maximum speed," matching the user's intent.

Example 3: Behavior Explanation For explanation queries, match the user's question structure:

```
User prompt: "Why did we brake suddenly?"
Well-aligned description: "Analyze sensor and control logs to explain recent
    braking behavior."
```

The description includes "explain," "braking," and "behavior," which align with the user's question about sudden braking.

9.5.3.2 Common Description Pitfalls

Avoid descriptions that are too generic, too technical, or misaligned with user language:

- **Too Generic:** "Get vehicle information" (doesn't specify what information)
- **Too Technical:** "Query CAN bus for velocity vector data" (uses jargon drivers don't use)
- **Misaligned:** "Retrieve telemetry" (users say "speed" or "status," not "telemetry")

Instead, use domain-appropriate language that matches how drivers naturally express their needs.

9.6 Summary

Function calling (also known as tool calling in agentic AI contexts) enables large language models to move beyond static text generation by interfacing with executable tools that interact with real-world systems. This capability is fundamental to agentic AI, where models autonomously select and invoke external tools to accomplish complex tasks. In the context of CAVs, this mechanism serves as a bridge between natural language prompts and grounded system behaviors such as sensing, reasoning, and control.

The chapter established function calling as a foundational architectural component in generative AI-based CAV systems, serving three critical roles: an interface abstraction that standardizes communication between language models and vehicle subsystems, a safety boundary that enforces validation and constraints before system actuation, and a control plane that enables language models to coordinate with real-time control loops. The chapter formalized the theoretical foundation through prompt-function alignment and schema-constrained decoding, demonstrating how constrained decoding enforces schema compliance through token-level masking and parser state tracking, with multi-layer validation ensuring safety in critical applications.

The chapter demonstrated practical implementation strategies using JSON Schema for function interface definition, dynamic dispatch in Python for executing functions based on LLM output, and inference frameworks that generate structured function calls. Design principles for effective schema design were presented, including the importance of clear function descriptions that align with user language, proper use of parameter descriptions, enums for constrained choices, and schema patterns for different CAV use cases such as system analysis, perception, and sensor access.

As CAV applications increasingly demand interpretable, real-time, and context-aware behavior, function calling will continue to serve as a foundational interface that connects generative reasoning with executable system actions. However, function calling should be understood not as an endpoint, but as a building block that enables more sophisticated capabilities. When combined with retrieval-augmented generation for context grounding, agent orchestration for multi-step reasoning, and multimodal understanding for scene interpretation, function calling becomes part of a larger architecture that supports truly intelligent and adaptive CAV systems. The subsequent chapters explore how these components integrate to create comprehensive agentic systems capable of complex reasoning, planning, and coordination in dynamic vehicular environments.

References

1 Shishir G. Patil, Tianjun Zhang, Xin Wang, and Joseph E. Gonzalez. Gorilla: large language model connected with massive APIs. *arXiv preprint arXiv:2305.15334*, 2023. Foundational work on function calling (tool calling) for agentic AI systems.

2 Shunyu Yao, Jeffrey Zhao, Dian Yu, Nan Du, Izhak Shafran, Karthik Narasimhan, and Yuan Cao. React: synergizing reasoning and acting in language models. *arXiv preprint arXiv:2210.03629,* 2022. Pioneering framework for agentic AI combining reasoning and tool use.

3 Timo Schick, Jane Dwivedi-Yu, Roberto Dessi, Roberta Raileanu, Maria Lomeli, Luke Zettlemoyer, Nicola Cancedda, and Thomas Scialom. Toolformer: language models can teach themselves to use tools. *arXiv preprint arXiv:2302.04761,* 2023. Self-supervised learning approach for tool calling in agentic AI.

4 Anthropic. Model context protocol (MCP), 2024. https://modelcontextprotocol.io/. Protocol specification for connecting AI applications to external tools and data sources.

5 Saibo Geng, Martin Josifoski, Maxime Peyrard, and Robert West. Grammar-constrained decoding for structured NLP tasks without finetuning. In *Proceedings of the 2023 Conference on Empirical Methods in Natural Language Processing*, pages 10941–10957. Association for Computational Linguistics, Singapore, 2023.

6 JSON Schema Working Group. JSON schema: a media type for describing JSON documents. Technical report, Internet Engineering Task Force, 2020. https://json-schema.org/. Draft specification for describing and validating JSON data structures.

Part IV

Agentic Intelligence and Coordination

As autonomy advances, vehicles increasingly rely on agentic architectures that reason over goals, context, and long-term consequences. This part introduces generative artificial intelligence-enhanced agents as the core abstraction for autonomous behavior in connected and autonomous vehicle systems, emphasizing architectures that integrate perception, memory, reasoning, and action selection. The discussion extends from single-agent designs to multi-agent systems and orchestration frameworks that coordinate complex behaviors across interacting components. By framing agents as reasoning entities rather than rule-based controllers, this part establishes the conceptual bridge between individual intelligence and coordinated autonomy.

10

GenAI Agent Architectures

10.1 Introduction

The emergence of generative artificial intelligence (GenAI), particularly large language models (LLMs) and vision–language models (VLMs), has fundamentally transformed the capabilities of AI agents. This transformation is particularly profound in connected and autonomous vehicles (CAVs), where the complexity of real-world environments demands agents that can understand context, communicate naturally, and adapt to novel situations.

Consider a CAV approaching a busy urban intersection where pedestrians are crossing at different paces, traffic signals are about to change, and a nearby vehicle suddenly signals a lane change. A classical rule-based agent might fail in this scenario, since its policy was designed for clean, predictable conditions. Even a utility-based or learning agent might struggle because it lacks the ability to flexibly interpret unstructured inputs such as ambiguous gestures, unclear signage, or unexpected behaviors.

This chapter argues that GenAI transforms classical AI agents from rigid, rule-based systems into flexible, adaptive, and communicative entities through semantic lift operators that enable sophisticated reasoning and communication. Unlike conventional agent architectures that rely on predefined state-action mappings and symbolic representations, GenAI agents operate on semantic representations that capture meaning, context, and intent. This fundamental shift enables agents to handle unstructured inputs, generate explanations, and adapt to novel situations (capabilities that are essential for safety-critical CAV applications but fundamentally different from approaches in conventional autonomy textbooks). The transformation is not merely an incremental improvement but represents a paradigm shift in how agents perceive, reason, and act in complex real-world environments.

This shift in capability can be formalized mathematically. Classical agents operate through the policy $\pi(x, M, g)$, where x represents sensory input, M denotes memory, and g represents goals. GenAI agents extend this through semantic lift operators $\mathcal{S}(x, M, g;\ \mathcal{K})$ that transform raw inputs into grounded, interpretable representations, enabling advanced reasoning and natural communication. Here, $\mathcal{K}$ represents the external knowledge base that provides context and grounding for semantic understanding.

Figure 10.1 Visualizes this shift, demonstrating how semantic lift operators bridge the gap between classical and GenAI agent architectures.

GenAI enhances three critical cognitive substrates: memory, planning, and reflection. Memory becomes retrieval-augmented, enabling associative recall of relevant experiences. Planning extends to multi-step reasoning with natural language deliberation. Reflection enables self-critique and explanation generation, building trust and enabling continuous improvement. These

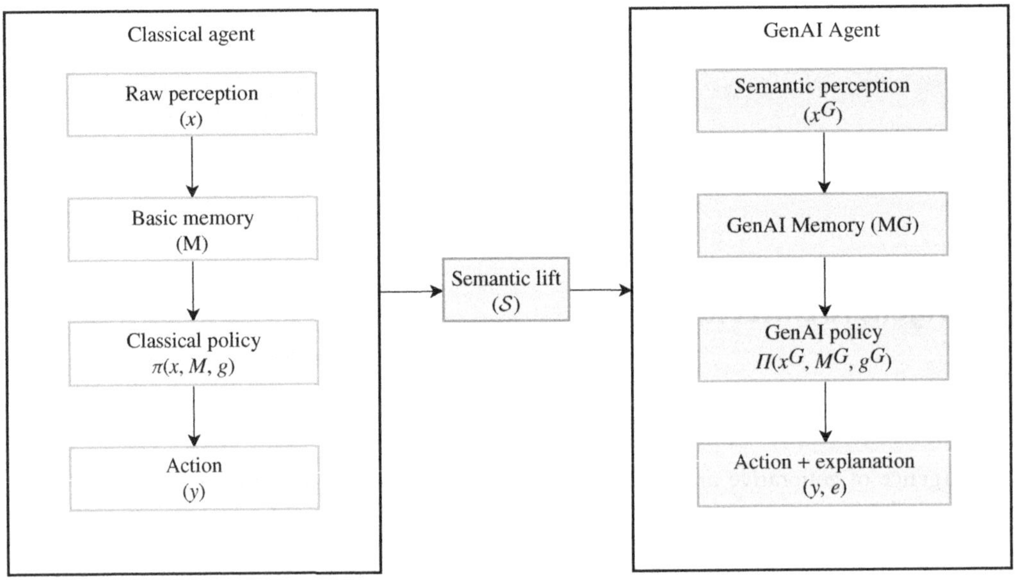

Figure 10.1 Transformation from classical to GenAI agents. Semantic lift operators $\mathcal{S}$ transform raw inputs into grounded, interpretable representations, enabling sophisticated reasoning, and communication. The GenAI agent produces both actions and explanations, building trust and enabling oversight.

enhancements are particularly valuable for CAVs, which must integrate multimodal perception, long-horizon planning, and human-centered interaction. GenAI agents can understand complex traffic situations, communicate intentions to passengers and other vehicles, and adapt to novel scenarios that classical approaches typically cannot handle.

Essential classical foundations establish the baseline, followed by exploration of how GenAI enhances each component of the agent architecture. The ReAct framework provides the implementation methodology, and concrete CAV applications demonstrate these capabilities. The enhanced individual agents provide the foundation for multi-agent systems, which will be explored in the following chapter.

10.2 Classical Agent Foundations and Architecture

10.2.1 Classical Agent Framework

A classical *agent* can be defined as any system that perceives its environment, processes information internally, and takes actions that affect the environment. This interaction is formalized through the perception–decision–action loop, with three essential components: sensors that capture inputs, reasoning modules that determine actions, and actuators that execute the chosen outputs.

Classical agents are characterized by four key properties. **Autonomy** refers to the ability to operate without continuous human supervision. **Reactivity** denotes the capacity to respond to environmental changes in real time. **Proactivity** represents the pursuit of goals that are not immediately triggered by external events. **Social ability** encompasses the competence to interact and communicate with other agents or humans.

These properties provide the foundation for intelligent behavior, but classical approaches have fundamental limitations in handling complex, unstructured, or novel situations that GenAI addresses through semantic understanding and flexible reasoning.

10.2.2 Classical Agent Types and Their Limitations

Early research in AI introduced several archetypes of agents, each representing different capabilities but sharing common limitations that GenAI addresses.

Reflex agents act directly on current percepts using simple condition–action rules. An automotive example is a basic anti-lock braking system that engages when wheel slip exceeds a threshold. While effective in narrow contexts, reflex agents lack flexibility for novel situations.

Goal-based agents extend reflexive behavior by incorporating explicit goal representations. A classical route planner that evaluates paths to reach a destination exemplifies this category. However, these agents still rely on prespecified objectives and typically cannot readily adapt to changing goals or contexts.

Utility-based agents evaluate outcomes according to preference functions, enabling trade-offs among competing objectives. A lane change decision system that balances safety margins, travel efficiency, and passenger comfort represents this paradigm. Yet these agents remain constrained by the rigidity of their utility functions.

Learning agents adapt their behavior through experience, as exemplified by reinforcement learning systems. A trajectory prediction system that learns pedestrian and vehicle behavior patterns from historical driving data represents this category. However, learning agents often lack interpretability and struggle with rapid adaptation to novel situations without extensive retraining.

10.2.3 Core Agent Architecture and Control Loop

The behavior of an agent can be understood as the interplay of several functional components. While implementations differ across domains, most architectures for intelligent agents in autonomy share a common structure. The key components form the perception–decision–action loop:

- **Perception:** Transforms raw sensor data into representations useful for decision-making. In CAVs, this includes object detection, lane recognition, and signal interpretation.
- **Memory:** Stores relevant information from past observations and experiences. Memory provides temporal continuity and contextual grounding, enabling agents to integrate the past with the present.
- **Decision or Policy Module:** selects an action given the current percepts, memory, and goals. This component integrates multiple inputs to guide behavior.
- **Planning:** generates multi-step strategies for achieving goals, often under uncertainty.
- **Action:** executes the selected behavior through actuators. In CAVs, this corresponds to throttle, brake, and steering control.

The simplest agent can be represented as a direct mapping from inputs to outputs, $y = \pi(x)$. However, to function effectively in dynamic and uncertain environments, agents require internal memory and goal representations. These enrich the loop by enabling context-sensitive decision-making and long-horizon planning. The architecture emphasizes that intelligent behavior does not arise solely from perception or from actuation, but from the integration of memory, planning, and decision-making in a continuous feedback cycle. Information flows from sensors through perception, memory, policy, and planning to actuators, with feedback loops enabling continuous adaptation. For CAVs, this translates into the ability to interpret dynamic environments, reason across temporal horizons, and act in ways that are both safe and efficient.

10.3 GenAI Agent Mechanisms

GenAI transforms classical agent capabilities through enhanced cognitive substrates. The conceptual foundations of how GenAI augments memory, planning, and reflection (the three core components that enable sophisticated agent behavior) are presented here, with mathematical formalization following in the next section. These enhancements integrate across the three cognitive substrates and work together to transform classical agent capabilities, but their specific architectural integration depends on application requirements and design choices. The following subsections detail how each component is enhanced, providing the conceptual foundation for various implementation approaches.

10.3.1 Semantic Lift

The fundamental mechanism by which GenAI enhances classical agents is through *semantic lift operators* $\mathcal{S}(x, M, g;\ \mathcal{K})$ that transform raw inputs into grounded, interpretable representations. This transformation provides three key benefits:

From Raw to Meaningful: Classical agents operate on raw sensor data, feature vectors, and symbolic representations. Semantic lift transforms these into rich, contextual representations that capture meaning, intent, and relationships. For example, instead of processing pixel values, a GenAI agent understands "pedestrian waiting to cross" or "vehicle signaling lane change."

From Rigid to Flexible: Classical agents rely on hand-crafted features and predefined rules. Semantic lift enables flexible interpretation of novel situations through natural language understanding and contextual reasoning. The agent can handle unstructured inputs like ambiguous gestures, unclear signage, or unexpected behaviors.

From Isolated to Contextual: Classical agents process inputs in isolation. Semantic lift integrates multiple sources of information: visual, textual, temporal, and spatial into unified representations that capture the full context of a situation. This enables more sophisticated reasoning about complex scenarios.

These semantic lift benefits underlie all the specific enhancements in memory, planning, and reflection described in the following subsections. The mathematical formalization of these transformations is presented in Section 10.4.

Figure 10.2 illustrates the fundamental flow of a GenAI agent, showing how semantic lift operators integrate into the agent's decision-making process.

10.3.2 Memory: Classical Storage to GenAI Retrieval

Memory provides agents with continuity across time, allowing them to move beyond purely reactive behavior. Classical memory systems store information in structured formats, but GenAI transforms memory through semantic understanding and associative recall.

Classical memory systems differentiate three forms: episodic memory stores specific experiences as structured event logs, semantic memory contains factual knowledge in predefined formats, and

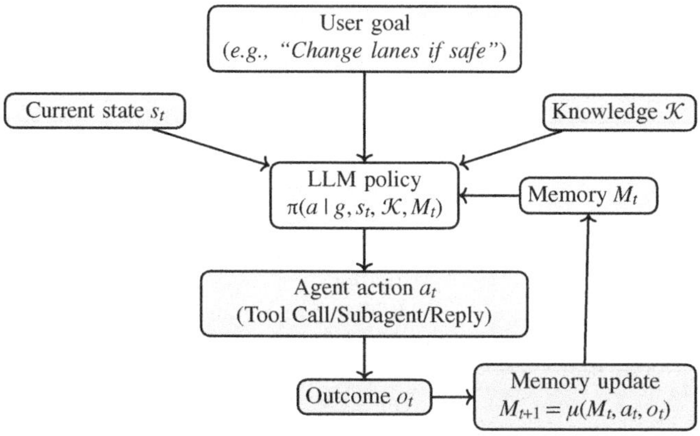

Figure 10.2 GenAI agent flow showing the interaction between goals, state, knowledge, memory, and actions. The policy π takes inputs from the current state s_t (equivalent to x in the general formulation), user goal g, knowledge base $\mathcal{K}$, and memory M_t to produce action a_t, which results in outcome o_t and updates memory to M_{t+1}. The conditional notation $\pi(a \mid g, s_t, \mathcal{K}, M_t)$ indicates that the policy can be stochastic (a probability distribution over actions) or deterministic (a function mapping to a single action).

Table 10.1 Comparison of classical and GenAI memory / planning / reasoning approaches.

Memory type	Classical approach	GenAI approach
Episodic	Structured event logs	Semantic embeddings with associative recall
Semantic	Predefined knowledge bases	Natural language descriptions and flexible querying
Procedural	Executable procedure libraries	Context-aware adaptation and learning

procedural memory stores motor routines as executable procedures [1]. This taxonomy provides a useful framework for understanding agent memory architectures. In CAVs, these correspond to sensor histories, traffic laws, and maps, and maneuver libraries, respectively.

GenAI transforms classical memory through semantic lift operators that enable RAG for memory [2], natural language storage and retrieval, context-aware recall, and semantic summarization. These enhancements allow agents to recognize patterns across different situations and apply learned strategies flexibly. Table 10.1 summarizes the key differences between classical and GenAI memory approaches.

GenAI memory enables CAVs to recall relevant past experiences, such as previous encounters with unusual road conditions, and adapt their behavior accordingly. The semantic understanding allows agents to recognize patterns across different situations and apply learned strategies flexibly.

10.3.3 Planning: Classical Search to GenAI Reasoning

Planning allows agents to consider sequences of actions rather than isolated moves. Classical planning relies on structured search and optimization, while GenAI enhances planning through natural language reasoning and flexible deliberation.

Classical planning formalizes future-oriented reasoning through deterministic search methods (depth-first search (DFS), breadth-first search (BFS), A*) for fully observable environments and probabilistic frameworks (Markov decision processes (MDPs), partially observable MDPs

Table 10.2 Comparison of classical and GenAI planning approaches.

Planning aspect	Classical approach	GenAI approach
Reasoning	Structured search algorithms	Natural language chain-of-thought reasoning
State Representation	Rigid symbolic states	Semantic representations with context
Optimization	Deterministic utility maximization	Context-aware adaptation to novel situations
Temporal Scope	Fixed horizon optimization	Multi-step reasoning with long-term consequences

(POMDPs)) for uncertain environments. These approaches seek to find optimal action sequences that maximize expected utility through structured optimization problems.

GenAI transforms classical planning through chain-of-thought reasoning [3], semantic planning, context-aware optimization, and multi-step deliberation. These enhancements enable agents to operate on semantic representations rather than rigid state spaces, adapt to changing conditions, and consider complex interdependencies. Table 10.2 contrasts classical and GenAI planning approaches.

GenAI planning enables CAVs to generate sophisticated coordination strategies, such as multi-step lane change maneuvers that consider traffic flow, weather conditions, and driver preferences. The natural language reasoning allows agents to explain their planning process and adapt strategies based on changing conditions.

10.3.4 Reflection and Reasoning: Classical Analysis to GenAI Self-critique

Reflection is the process by which agents evaluate past decisions and adjust future strategies. For example, a driver who narrowly avoids a collision may consciously adjust their following distance in the future.

Reasoning, closely linked to reflection, can take several forms: deductive reasoning applies rules to derive conclusions, inductive reasoning generalizes from observations, and abductive reasoning hypothesizes explanations for incomplete information. For CAVs, reflection may involve post-event analysis, learning from near-misses, and adapting operational safety margins.

GenAI transforms classical reflection through self-critique generation, explanation generation, adaptive learning, and context-aware analysis. These enhancements enable agents to evaluate decision quality, provide natural language explanations, refine policies based on feedback, and consider the full context of decisions. Table 10.3 compares classical and GenAI reasoning approaches.

GenAI reflection enables CAVs to learn from near-miss events, refine safety margins, and provide transparent justifications for their choices. The self-critique capabilities allow agents to continuously improve their decision-making processes and build trust with passengers through explainable behavior.

Together, these enhancements transform classical agents from rigid, rule-based systems into flexible, adaptive, and communicative entities that can handle the complexity and uncertainty inherent in real-world autonomous driving scenarios. The mathematical formalization of these conceptual enhancements is presented in the following section, providing the rigorous foundation for implementation and analysis.

Table 10.3 Comparison of classical and GenAI reasoning approaches.

Reasoning type	Classical approach	GenAI approach
Deductive	Rule-based logical inference	Context-aware rule application with explanation
Inductive	Statistical pattern recognition	Semantic pattern discovery with generalization
Abductive	Hypothesis generation from data	Natural language hypothesis formation and evaluation
Meta-cognition	Post hoc analysis	Real-time self-critique and explanation generation

10.4 Mathematical Formalization

The conceptual enhancements can be expressed rigorously using mathematical notation. The formal mathematical framework underlying GenAI agents shows how semantic lift operators transform classical agent formulations while maintaining mathematical rigor and safety guarantees.

10.4.1 General Policy Equation

At the core of agent theory lies the notion of a *policy*, denoted π, which maps perceptual inputs, internal memory, and goals into actions:

$$y = \pi(x, M, g), \tag{10.1}$$

where x represents current sensory input (also denoted s_t in temporal contexts), M the internal memory state, g the agent's goals, and y the selected action. The policy π can be deterministic (a function mapping inputs to a single action) or stochastic (a conditional probability distribution over actions, denoted $\pi(a \mid x, M, g)$). This formulation generalizes the perception–decision–action loop described in Section 10.2.

10.4.2 Utility Optimization View

A complementary perspective frames decision-making as an optimization problem. The agent selects the action that maximizes a utility function $U(\cdot)$, which encodes preferences, safety requirements, or performance metrics:

$$y = \arg\max_{a} U(g, M, x, a). \tag{10.2}$$

For planning over action sequences, the optimization extends to:

$$a^*_{1:T} = \arg\max_{a_{1:T}} U(g, M, x, a_{1:T}), \tag{10.3}$$

where the agent seeks a sequence of actions $a_{1:T}$ that maximizes expected utility.

Here, a represents a single action from the set of possible actions available to the agent, and y is the optimal action selected. The utility function $U(g, M, x, a)$ evaluates how desirable action a is given the current goal g, memory M, and state x. The relative importance of objectives (safety, efficiency, comfort) depends on the operational context and failure regime, with safety dominating in critical scenarios and efficiency/comfort taking precedence in routine conditions while maintaining minimum safety thresholds.

10.4.3 GenAI Policy Formulation

The transformation from classical to GenAI agents can be expressed mathematically by comparing their policy formulations. Classical agents output only an action y given raw inputs (x, M, g), while GenAI modules (LLMs/VLMs) are best viewed as *semantic transformers* that *lift* raw inputs into grounded, interpretable representations, and that produce not only an action but also an *explanation*.

10.4.3.1 Semantic Lift of Inputs

A semantic lift operator $\mathcal{S}$ (induced by LLM/VLM modules and external knowledge $\mathcal{K}$) maps raw inputs into grounded semantic objects:

$$\left(x^{\mathrm{G}}, M^{\mathrm{G}}, g^{\mathrm{G}}\right) = \mathcal{S}\left(x, M, g;\ \mathcal{K}\right). \tag{10.4}$$

Here, x^{G} may include scene graphs or affordances inferred by a VLM, M^{G} is retrieval-augmented memory (episodic precedents, norms), and g^{G} is a structured multi-criteria goal parsed from preferences or natural language.

10.4.3.2 Safety-shielded Hybrid Policy

Retain a classical core π for timing-critical and certifiable behavior, and add a GenAI advisor $\mathcal{A}_{\mathrm{LLM}}$ under a hard safety shield:

$$\Pi = \underbrace{\sigma_{\mathrm{safety}} \circ \pi}_{\text{classical core}} \oplus \underbrace{\mathcal{A}_{\mathrm{LLM}}}_{\text{semantic advisor}}, \tag{10.5}$$

where σ_{safety} enforces constraints (e.g., collision avoidance), $\circ$ denotes function composition (the safety shield is applied to the output of the classical policy), and $\oplus$ denotes arbitration that is designed to enforce σ_{safety} constraints.

10.4.3.3 Action-explanation Output

A GenAI-augmented agent using the hybrid policy Π emits both an action and an explanation:

$$(y, e) = \Pi\left(x^{\mathrm{G}}, M^{\mathrm{G}}, g^{\mathrm{G}}\right), \tag{10.6}$$

where y is the selected action and e is a structured or natural-language rationale suitable for humans or other agents.

10.4.3.4 Decision as Constrained Optimization with Semantics

The resulting decision extends the classical utility optimization from Equation (10.3) to operate on semantically lifted inputs. The decision can be written as a constrained optimization over a GenAI-aware utility:

$$y = \arg\max_{a} U^{\mathrm{G}}\left(a \,\middle|\, x^{\mathrm{G}}, M^{\mathrm{G}}, g^{\mathrm{G}}\right) \quad \text{s.t.} \quad \mathcal{C}(a, x) \le 0, \tag{10.7}$$

where U^{G} is the GenAI utility function that operates on semantically lifted representations $(x^{\mathrm{G}}, M^{\mathrm{G}}, g^{\mathrm{G}})$ rather than raw inputs, aggregating multiple objectives (efficiency, comfort, fairness, explainability) with context-adapted weights. The constraint $\mathcal{C}$ encodes hard safety constraints independent of GenAI, ensuring that the classical safety guarantees are maintained.

10.4.3.5 Explanations as First-class Outputs

Explanations are generated as part of the policy and stored for audit and reflection:

$$e = \text{Explain}_{\text{LLM}}\left(y,\ x^{\text{G}}, M^{\text{G}}, g^{\text{G}}\right). \tag{10.8}$$

The classical mapping $y = \pi(x, M, g)$ remains the safety-critical backbone. The semantic lift $\mathcal{S}$ changes *what* the agent reasons over (from raw to grounded representations), and the hybrid policy Π adds an LLM advisor without compromising safety. The additional output e makes explainability native, which is essential for passenger trust and certification. This semantic lift framework also provides the foundation for multi-agent coordination, where the enhanced representations $(x^{\text{G}}, M^{\text{G}}, g^{\text{G}})$ enable sophisticated communication and coordination between agents.

10.5 ReAct Framework

The theoretical enhancements described in previous sections require a practical implementation framework. The *ReAct* framework (Reason and Act) [4] provides this operational methodology, showing how to integrate GenAI capabilities into working agent systems. Unlike classical agent architectures where reasoning and acting are separated into distinct phases, ReAct integrates the two processes into a unified loop, enabling agents to deliberate, execute, and re-deliberate in rapid succession.

The key principle of ReAct is the interleaving of reasoning traces with action execution. Instead of committing to a full plan in advance, the agent produces intermediate reasoning steps, evaluates them, and acts incrementally. This allows new percepts or retrieved memories to influence subsequent reasoning in real time.

10.5.1 Formal Framework

The ReAct framework operationalizes the enhanced agent policy by decomposing it into interleaved reasoning and action steps. Formally, the ReAct process can be expressed as a composition of reasoning and acting functions:

$$y = f_{\text{act}}(f_{\text{reason}}(x, M, g), x, M, g), \tag{10.9}$$

where y is the final action selected by the agent, f_{reason} generates a structured deliberation trace, and f_{act} maps this trace into executable actions. This composition can be repeated iteratively, producing a sequence of interleaved reasoning and action steps:

$$\{(r_1, a_1), (r_2, a_2), \dots, (r_T, a_T)\}, \tag{10.10}$$

where r_t denotes a reasoning step and a_t the corresponding action at time t. This iterative sequence represents the ReAct loop in practice, illustrated in Figure 10.3, where reasoning and action alternate over multiple time steps.

10.5.2 ReAct in CAV Applications

The ReAct paradigm has particular relevance for CAVs, where the interleaving of reasoning and action enables dynamic adaptation to complex traffic scenarios. Consider a CAV navigating an

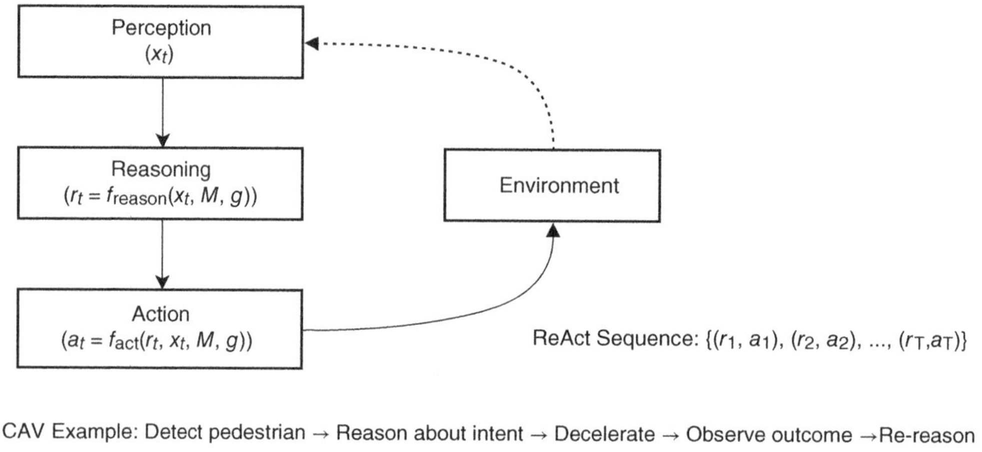

Figure 10.3 The ReAct implementation framework interleaves reasoning and action in a continuous loop, providing the operational methodology for GenAI agents. This framework enables the integration of semantic understanding, GenAI memory, and reflection capabilities into working systems.

intersection: the agent begins with **perception**, detecting pedestrians, vehicles, and signals in the environment. Through **reasoning**, the agent hypothesizes that a pedestrian may be about to cross despite the signal, based on behavioral cues and context. This reasoning leads to an **action:** proactive deceleration to ensure safety. Following the action, the agent engages in **reflection**, evaluating the outcome and adjusting its behavior accordingly—if the pedestrian does not cross, the agent adjusts acceleration to maintain traffic flow.

This cycle continues as new percepts are integrated, with reasoning and action alternating in real time. Unlike rigid controllers, a ReAct agent can dynamically refine its behavior in light of uncertainty and evolving conditions, continuously simulating possibilities, taking actions, and updating its model based on feedback. ReAct operationalizes the cognitive faculties of reasoning and reflection within an agent's action loop, providing a practical bridge between the abstract enhancements and concrete decision-making processes in CAV applications.

10.6 Individual Agent Applications in CAVs

The ReAct framework enables the implementation of GenAI agents across various CAV applications. GenAI agents address real-world challenges that classical approaches cannot handle through semantic lift operators, GenAI memory, planning, and reflection capabilities. This section presents concrete applications of individual GenAI agents in CAV systems, demonstrating how the theoretical enhancements translate into practical capabilities. Multi-agent coordination applications, including fleet management, emergency response coordination, and intersection management, are explored in detail in Chapter 11, which builds directly upon these individual agent foundations.

10.6.1 Navigation Applications

GenAI navigation agents process natural language traffic reports and construction notices, adapt routes based on weather conditions and driver preferences, recall past route experiences, and provide natural language explanations of routing decisions. This enables agents to handle unstructured information and novel situations that classical route planners cannot process. For example, when a driver requests "avoid the construction on Main Street," the agent can interpret this natural language instruction, retrieve relevant traffic information from memory, and generate an alternative route with an explanation of why the detour was chosen.

10.6.2 Hazard Detection

GenAI hazard detection agents interpret ambiguous situations such as pedestrians who appear waiting but then suddenly cross, or vehicles exhibiting unusual behavior patterns. Through semantic reasoning and memory-augmented detection, agents distinguish between actual hazards and false alarms while continuously learning from near-miss events. The GenAI memory capabilities enable agents to recall similar past situations, improving detection accuracy and reducing false positives. For instance, an agent that previously encountered a pedestrian who appeared stationary but then crossed unexpectedly can apply this experience to better assess similar situations in the future.

10.6.3 Driver Interaction

Driver interaction agents leverage natural language understanding to interpret conversational queries, generate contextual explanations of driving decisions, adapt communication to individual user preferences, and proactively share information about upcoming maneuvers. These capabilities enable effective communication with passengers of varying technical backgrounds. The enhanced reflection capabilities allow agents to generate explanations that are tailored to the user's level of technical understanding, building trust through transparent communication. For example, when a vehicle performs an unexpected lane change, the agent can explain: "I changed lanes because the vehicle ahead was slowing down, and I wanted to maintain a safe following distance."

10.7 Limitations and Challenges of GenAI Agents

While GenAI agents offer significant advantages over classical approaches, they also introduce new challenges and limitations that must be carefully addressed in safety-critical CAV applications. Understanding these limitations is essential for designing robust systems that leverage GenAI capabilities while maintaining safety and reliability.

10.7.1 Verification, Validation, and Certification

Assessing GenAI agent systems requires distinguishing three related concepts. **Verification** addresses whether the system was built correctly, ensuring implementation matches specification under expected conditions. **Validation** addresses whether the correct system was built, ensuring it meets user needs in its intended environment. **Evaluation** addresses how the system behaves under realistic and adversarial conditions, assessing performance, robustness, and safety across diverse scenarios. The flexible, non-deterministic nature of GenAI agents complicates traditional verification and certification processes. Unlike classical rule-based systems with provable correctness properties, GenAI systems exhibit probabilistic behavior that is difficult to formally verify. New certification methodologies must assess system behavior under uncertainty while maintaining safety standards, requiring probabilistic verification techniques, statistical validation methods, and continuous evaluation protocols that account for adaptive behavior. For GenAI agents, these distinctions are particularly important because the probabilistic and adaptive nature of these systems requires evaluation frameworks that go beyond traditional verification and validation approaches, necessitating continuous assessment of system behavior as agents learn and adapt.

10.7.2 Computational and Latency Constraints

GenAI models, particularly large language models, require substantial computational resources and can introduce significant latency in decision-making loops. In real-time CAV applications where millisecond response times are critical, the overhead of semantic lift operations, retrieval-augmented memory queries, and natural language generation can compromise system responsiveness. This challenge requires careful architectural design, such as asynchronous processing for non-critical paths, model quantization, and edge computing strategies that balance capability with performance.

10.7.3 Hallucination and Reliability

Despite semantic understanding capabilities, GenAI models can generate plausible but incorrect information, a phenomenon known as hallucinations. In safety-critical contexts, such errors can lead to incorrect route planning, misinterpreted traffic situations, or misleading explanations to passengers. The hybrid policy architecture with safety shields addresses this concern, but requires rigorous validation and continuous monitoring to ensure that GenAI outputs do not compromise safety-critical decisions.

10.7.4 Data Dependencies and Evaluation Drift

GenAI agents rely on training data that may not fully represent the diverse conditions encountered in real-world CAV operations. Distribution shift (when operational conditions differ from training data) can degrade performance unpredictably, such as when agents encounter weather conditions, road geometries, or traffic patterns not present in training data. Addressing this requires distribution shift detection mechanisms, online adaptation strategies, and fallback to classical systems when confidence thresholds are not met.

GenAI-enabled agents are inherently adaptive systems that learn and evolve over time. As agents operate in real-world environments, their policies evolve through online learning, memory updates, and reflection mechanisms. This adaptation creates evaluation drift: baselines established during initial deployment can become stale as agent behavior changes. For example, an agent that initially demonstrates safe behavior under specific conditions may adapt its decision-making in ways that invalidate earlier safety assessments. Addressing evaluation drift requires continuous monitoring frameworks that track agent behavior evolution, periodic re-evaluation against updated baselines, and adaptive evaluation methodologies that account for policy changes.

10.7.5 Interpretability and Debugging

While GenAI agents generate explanations, the internal reasoning processes of large language models remain opaque. When agents make unexpected decisions, debugging becomes challenging because the causal chain from input to output is not easily traceable. Engineers must rely on enhanced logging of intermediate reasoning steps, explainability tools that visualize attention patterns and decision paths, and diagnostic capabilities that can trace failures back to specific inputs or memory retrievals to understand and improve agent behavior.

10.7.6 Integration Complexity

Integrating GenAI capabilities into existing CAV architectures requires careful coordination between classical control systems and GenAI modules. The semantic lift framework and hybrid policy architecture provide a foundation, but practical implementation involves managing interfaces between deterministic and probabilistic components, data flows that transform between raw sensor data and semantic representations, and failure modes where GenAI outputs must be validated before affecting safety-critical decisions. This complexity demands sophisticated system design with clear separation of concerns, well-defined interfaces, and thorough integration testing that validates both individual components and their interactions under various failure scenarios.

These limitations do not negate the value of GenAI agents, but they highlight the importance of hybrid architectures, rigorous validation, and careful system design. The safety-shielded hybrid policy approach, where classical systems provide the safety-critical backbone and GenAI modules enhance capabilities in non-critical paths, represents a pragmatic strategy for addressing these challenges while realizing the benefits of semantic understanding and flexible reasoning.

10.8 Summary

This chapter has explored how generative AI transforms classical AI agents from rigid, rule-based systems into flexible, adaptive, and communicative entities. We began with classical agent foundations: systems that perceive, decide, and act through policies that map sensory inputs, memory, and goals to actions. While effective in structured environments, these agents struggle with the complexity and ambiguity of real-world scenarios like urban driving, where rigid state-action mappings cannot handle nuanced, context-dependent decisions.

The transformation occurs through semantic lift operators that transform raw inputs into grounded, interpretable representations. This enables agents to understand meaning, context, and generate appropriate responses. GenAI enhances three critical cognitive substrates: memory becomes retrieval-augmented with associative recall, planning extends to multi-step reasoning with natural language deliberation, and reflection enables self-critique and explanation generation.

The mathematical formalization captures this evolution through enhanced agent policies where agents not only take actions but also generate explanations. The safety-shielded hybrid policy approach combines classical safety-critical systems with GenAI enhancements, preserving mathematical rigor while adding flexibility.

The ReAct framework operationalizes these theoretical enhancements, showing how reasoning and action can be interleaved in real-time systems. By alternating between deliberation and execution, agents adapt to changing conditions while maintaining safety guarantees.

These capabilities find concrete expression in CAV applications: navigation agents that process natural language traffic reports, hazard detection systems that interpret ambiguous situations, driver interaction agents that communicate intentions clearly, and fleet coordination systems that negotiate priorities through semantic communication.

However, these enhancements introduce new challenges, including computational and latency constraints, hallucination risks, verification and certification difficulties, data dependencies and evaluation drift, interpretability challenges, and integration complexity. These limitations highlight the importance of hybrid architectures that combine classical safety-critical systems with GenAI enhancements, ensuring that the benefits of semantic understanding and flexible reasoning are realized while maintaining the reliability required for autonomous vehicle deployment.

As we move to the next chapter, we extend beyond individual agents to consider systems where multiple agents interact, coordinate, and achieve collective objectives. The coordination challenges and solutions explored in the next chapter build directly upon these individual agent foundations, introducing the challenges and opportunities of coordinated intelligence in transportation systems.

References

1 Endel Tulving. Episodic and semantic memory. In Endel Tulving and Wayne Donaldson, editors, *Organization of Memory*, pages 381–403. Academic Press, New York, 1972.

2 Patrick Lewis, Ethan Perez, Aleksandra Piktus, Fabio Petroni, Vladimir Karpukhin, Naman Goyal, Heinrich Küttler, Mike Lewis, Wen-tau Yih, Tim Rocktäschel, et al. Retrieval-augmented generation for knowledge-intensive NLP tasks. In *Proceedings of the 34th International Conference on Neural Information Processing Systems*, NIPS '20, Red Hook, NY, USA, 2020. Curran Associates Inc. ISBN 9781713829546.

3 Jason Wei, Xuezhi Wang, Dale Schuurmans, Maarten Bosma, Brian Ichter, Fei Xia, Ed Chi, Quoc Le, and Denny Zhou. Chain-of-thought prompting elicits reasoning in large language models. *Advances in Neural Information Processing Systems*, 35:24824–24837, 2022.

4 Shunyu Yao, Jeffrey Zhao, Dian Yu, Nan Du, Izhak Shafran, Karthik Narasimhan, and Yuan Cao. React: synergizing reasoning and acting in language models. *arXiv preprint arXiv:2210.03629*, 2022. Pioneering framework for agentic AI combining reasoning and tool use.

11

GenAI Multi-agent Systems

11.1 Introduction

Building on the individual agent capabilities explored in Chapter 10, this chapter examines how GenAI, particularly Large Language Models (LLMs), transforms the coordination capabilities of multi-agent systems (MAS). This transformation is particularly profound in CAVs, where the complexity of real-world traffic scenarios demands sophisticated coordination that goes beyond classical symbolic approaches.

Consider a scenario where five autonomous vehicles approach a complex intersection simultaneously: a delivery truck needs to make a left turn, an emergency vehicle approaches from the right with sirens activated, two passenger vehicles are proceeding straight, and a motorcycle is filtering through traffic. Each vehicle has distinct objectives, constraints, and time pressures. Without effective coordination, this situation could result in gridlock, safety violations, or inefficient traffic flow.

This scenario exemplifies the fundamental challenge of multi-agent systems: **how can multiple intelligent agents coordinate effectively when each possesses independent goals, constraints, and decision-making processes?** The core tension lies in balancing individual autonomy with collective coordination, where each agent must pursue its own objectives while contributing to system-wide goals.

Classical coordination approaches rely on rigid protocols and predefined rules. A vehicle might broadcast "position: (x, y), velocity: v, intent: turn_left," but this symbolic format fails to capture the rich contextual information that human drivers naturally communicate. The fundamental limitation is that classical MAS reduces complex coordination to simple message passing, losing the nuance and adaptability that make human coordination effective.

GenAI MAS addresses these limitations by providing semantic intelligence: the ability to understand meaning, context, and generate appropriate responses. Unlike classical MAS that rely on rigid message passing protocols, GenAI-equipped systems can interpret natural language, reason with unstructured data, and generate explanations of their coordination decisions.

GenAI enhances three critical coordination aspects: communication, negotiation, and collective decision-making. Communication becomes semantic and context-aware, enabling natural language understanding between agents. Negotiation extends to adaptive strategies that evolve based on context and experience. Collective decision-making enables sophisticated coordination that considers the full context of each agent's situation.

These enhancements are particularly valuable for CAVs, which must coordinate in complex traffic scenarios involving multiple vehicles, pedestrians, and infrastructure elements. GenAI MAS can understand complex traffic situations, communicate intentions between vehicles, and adapt to novel scenarios that classical approaches typically cannot handle.

11.2 Multi-agent Systems Foundations and Architecture

MAS are computational systems composed of multiple interacting intelligent agents, each capable of autonomous decision-making while operating within a shared environment [1]. Unlike single-agent systems that focus on individual optimization, MAS addresses the complexity of collective behavior, where agents must balance their own objectives with system-wide goals.

11.2.1 From Individual Agents to Multi-agent Systems

The transition from individual agents to multi-agent systems represents a fundamental shift in complexity. While individual agents operate according to the policy introduced in the previous chapter, multi-agent systems must coordinate multiple such policies through an enhanced formulation that accounts for inter-agent dependencies. Each agent in a MAS retains the cognitive substrates, including memory, planning, and reflection, discussed in the previous chapter. However, these capabilities must operate in a context where other agents' decisions directly influence each agent's optimal behavior, creating a complex web of interdependencies that classical approaches struggle to handle effectively.

The individual agent policy $\pi(x, M, g)$ provides the foundation for multi-agent coordination, where π represents the decision-making function that maps sensory input x, memory state M, and goals g into actions. The multi-agent formulation extends this policy by introducing information from other agents and enhancing all components through semantic transformation (as detailed in Section 11.3). This creates a system where each agent's decision depends not only on its own state but also on the semantic understanding of other agents' intentions, constraints, and capabilities. The result is collective behaviors that exceed the capabilities of isolated agents.

11.2.2 Core Components of Multi-agent Systems

A multi-agent system consists of several fundamental components that work together to enable collective intelligence. While implementations differ across domains, most MAS architectures share common structural elements that form the foundation for effective coordination.

Autonomous agents are entities that perceive their environment, make decisions, and take actions based on their own goals, capabilities, and decision-making processes. In CAVs, each vehicle represents an agent with distinct objectives such as reaching a destination, maintaining safety, and optimizing fuel consumption.

The **Shared environment** provides constraints, resources, and opportunities influencing agent behavior. For CAVs, this encompasses road networks, traffic conditions, weather patterns, and infrastructure elements. The environment serves as both the arena for agent interaction and the source of contextual information that shapes coordination decisions.

Communication mechanisms enable agents to exchange information, coordinate actions, and negotiate conflicts. These mechanisms can be direct, occurring between agents themselves, or indirect, mediated through environmental markers and shared state information. The effectiveness of communication directly determines the quality of coordination that can be achieved.

Coordination processes represent the core challenge of MAS, encompassing task allocation, conflict resolution, and collective decision-making. These processes transform individual agent capabilities into collective intelligence, enabling the system to achieve goals that would be impossible for individual agents acting in isolation.

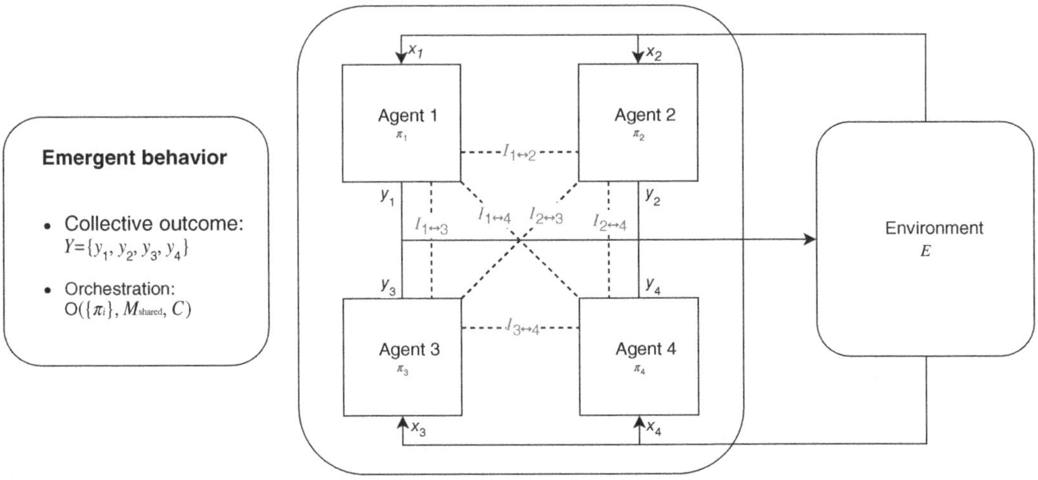

Figure 11.1 Multi-agent system architecture with agent coordination.

Figure 11.1 illustrates this architecture, showing how multiple agents interact with each other and their shared environment to produce emergent system-level behaviors.

11.2.3 Coordination Paradigms and Their Limitations

Multi-agent systems employ different coordination paradigms depending on the application requirements and system constraints [2, 3]. In the context of connected and autonomous vehicles, these paradigms manifest in different approaches to traffic management and vehicle coordination.

Centralized coordination relies on a single entity making decisions for all agents, providing optimal solutions but creating bottlenecks and single points of failure that can compromise system reliability [3]. In CAV applications, this might involve a central traffic management system that controls all vehicles at an intersection. While this approach can optimize traffic flow for small intersections, it becomes impractical as the number of vehicles increases and creates a single point of failure that could disrupt entire traffic networks.

Decentralized coordination enables agents to make decisions independently based solely on their local information, with minimal or no active coordination between agents [2]. In CAV contexts, this corresponds to vehicles making individual decisions based primarily on their own sensor data, operating essentially as independent agents. While this approach scales well and is robust to individual failures, it may lead to suboptimal collective behavior when vehicles lack information about system-wide objectives or other vehicles' intentions, potentially resulting in inefficient traffic flow or safety issues.

Distributed coordination represents a middle ground where agents actively coordinate through peer-to-peer interactions and communication, combining the benefits of decentralization with the effectiveness of coordination [3]. Unlike decentralized coordination, where agents operate independently, distributed coordination involves vehicles communicating directly with nearby vehicles to coordinate maneuvers, share intentions, and negotiate right-of-way. This approach enables vehicles to share information and coordinate actions while maintaining distributed decision-making authority (no central controller), creating systems that are both scalable and effective for real-world traffic scenarios.

The coordination challenge stems from three fundamental limitations: limited information exchange restricts agents to sharing only predefined, symbolic information; rigid decision-making forces coordination to follow fixed protocols that typically cannot adapt to novel situations; and the lack of semantic understanding prevents agents from comprehending the meaning behind actions or intentions. These limitations motivate the need for a fundamentally different approach that can handle the semantic complexity of real-world coordination scenarios.

GenAI MAS address these fundamental limitations through semantic lift operators (introduced for individual agents in Section 10.3.1 of Chapter 10) that transform each component of the coordination problem, enabling the sophisticated coordination capabilities described in Section 11.3.5.

11.3 GenAI Coordination Mechanisms

GenAI transforms classical MAS by enhancing the three core coordination components: communication, negotiation, and collective decision-making. Semantic intelligence enables agents to understand the meaning and context behind coordination messages, reason about coordination conflicts, and generate adaptive responses, enabling enhanced coordination strategies that classical approaches typically cannot support.

11.3.1 Semantic Lift for Multi-agent Coordination

The fundamental mechanism by which GenAI enhances classical MAS is through *semantic lift operators* that transform coordination inputs from rigid, symbolic representations into rich, contextual understanding. These operators act on the four core components of multi-agent coordination: local observations, inter-agent information, memory, and goals. Unlike individual agent systems, where semantic lift (Section 10.3.1) transforms sensor data into meaningful perceptions, in multi-agent systems, semantic lift transforms coordination messages and protocols into semantic understanding that enables agents to interpret intentions, negotiate conflicts, and coordinate actions based on contextual reasoning rather than rigid protocols.

Classical MAS coordination operates through rigid, symbolic message formats where agents exchange predefined data structures like "position: (x, y), velocity: v, intent: merge." These symbolic messages lack the contextual richness needed for effective coordination in complex scenarios. Semantic lift transforms these symbolic exchanges into natural language understanding, enabling agents to comprehend not just what other agents are doing, but why they are doing it and how their actions relate to the broader coordination context. For example, instead of processing symbolic coordinates and intent flags, a GenAI MAS understands "I need to merge because my exit is approaching in 500 m, and I'm running late for an important meeting." This semantic understanding enables agents to coordinate based on underlying reasons, priorities, and constraints rather than just surface-level state information.

The transformation extends beyond message interpretation to the coordination protocols themselves. Classical MAS rely on fixed coordination protocols that cannot adapt to novel situations. Semantic lift enables dynamic protocol generation where coordination strategies emerge from semantic reasoning about the specific context, agent capabilities, and coordination objectives. For example, when encountering ambiguous traffic situations where standard protocols don't apply, unclear intentions from other agents, or unexpected behaviors like sudden lane changes, the system generates appropriate coordination responses through semantic reasoning rather than following predefined templates.

Furthermore, semantic lift integrates coordination information across multiple dimensions: agent intentions, environmental conditions, historical interaction patterns, and social dynamics. This integration creates unified representations that capture the context of coordination scenarios, enabling agents to make coordination decisions that consider agent states, relationships, and environmental factors together rather than processing isolated message exchanges independently.

11.3.2 Communication: Message Passing to Semantic Dialogue

GenAI MAS enables natural language communication that captures the contextual nuances of human-like coordination dialogue, transforming how agents exchange information and coordinate their actions. Instead of exchanging symbolic coordinates and basic intent flags, GenAI agents can express complex intentions in natural language. A vehicle can communicate nuanced intentions like "I can wait if you're in a hurry" rather than just broadcasting position and velocity data. This natural language capability enables agents to convey subtle nuances, priorities, and contextual information that would be impossible to express through symbolic protocols alone.

GenAI agents can understand the context and implications of communication from other agents. They can interpret contextual cues such as urgency indicators, priority signals, and implicit constraints, understand urgency levels, and respond appropriately to complex coordination requests. This context awareness allows agents to adapt their communication style and content based on the specific situation, the other agents involved, and the overall coordination goals. The system can dynamically adjust communication strategies based on the situation: in emergency scenarios, communication becomes more direct and urgent; in routine coordination, it can be more conversational and detailed. This adaptive communication ensures that the right information is conveyed in the right way at the right time, improving coordination outcomes while managing communication bandwidth effectively.

11.3.3 Negotiation: Protocol-based to Reasoning-based

Classical MAS negotiation follows rigid protocols with predefined strategies. GenAI MAS enables negotiation through semantic reasoning and adaptive strategies that can handle complex, multi-objective coordination scenarios by understanding underlying motivations and generating contextually appropriate solutions. Instead of following fixed conflict resolution protocols, GenAI agents can reason about conflicts semantically. They can understand the underlying causes of coordination conflicts and generate creative solutions that satisfy multiple objectives. For example, when two vehicles need to merge simultaneously, the system can understand that one vehicle is in a hurry while the other has passengers who are motion-sensitive, leading to a solution that prioritizes the urgent vehicle while minimizing discomfort for the sensitive passengers.

The system can learn from past negotiation outcomes and adapt strategies accordingly. Agents can develop preferences for certain negotiation approaches based on their success rates and the specific characteristics of other agents. This learning capability improves negotiation effectiveness over time, as the system becomes better at predicting which strategies will work best in different situations and with different types of agents.

GenAI agents can balance multiple competing objectives during negotiation, considering factors like safety, efficiency, fairness, and individual preferences in a unified framework. This multi-objective approach ensures that coordination decisions consider multiple relevant factors simultaneously rather than optimizing for a single objective at the expense of others.

11.3.4 Collective Decision-making: Aggregation to Emergent Intelligence

Classical MAS collective decision-making typically involves simple aggregation of individual preferences or voting mechanisms. GenAI MAS enables collective reasoning that emerges from semantic interaction, where agents can discuss, explain, and negotiate to reach decisions that leverage the combined knowledge and reasoning capabilities of all agents. Instead of simple voting, GenAI agents can engage in semantic discussion to build consensus. They can explain their reasoning, understand others' perspectives, and work toward solutions that satisfy the collective while respecting individual constraints. This semantic discussion enables agents to understand not just what other agents want, but why they want it, leading to better compromise solutions that address underlying concerns rather than just surface-level preferences.

The system can discover novel coordination strategies through semantic interaction that no individual agent could have planned in advance. This emergent intelligence enables handling of coordination scenarios that were not anticipated during system design by allowing the collective to develop innovative solutions through collaborative reasoning. For example, in a complex traffic scenario involving multiple emergency vehicles, construction vehicles, and regular traffic, the system might discover a novel coordination pattern that optimizes the flow for all participants.

The entire system can learn from collective experiences and adapt coordination strategies over time. Successful strategies are reinforced and unsuccessful ones are modified or abandoned, improving coordination effectiveness across the entire multi-agent system. This collective learning enables the system to adapt to changing conditions and improve coordination outcomes over time.

11.3.5 Coordination Strategies

The semantic capabilities of GenAI enable several enhanced coordination strategies that are impossible with classical approaches. These strategies leverage semantic understanding and reasoning to create coordination systems that can adapt to diverse real-world scenarios with varying constraints, priorities, and agent characteristics. GenAI MAS can generate coordination plans dynamically based on context, creating strategies that adapt to specific situations. For example, in traffic intersection coordination, the system can generate custom coordination strategies for unusual scenarios like emergency vehicles, construction zones, or special events. This dynamic planning capability enables the system to handle novel situations that were not anticipated during system design.

GenAI MAS can adapt their negotiation strategies based on context, history, and the specific agents involved. The system can learn from past interactions and adjust negotiation approaches to improve coordination effectiveness over time. This adaptive capability allows the system to select and refine coordination strategies based on what works best in different situations and with different types of agents.

Unlike classical MAS that provide limited insight into coordination decisions, GenAI MAS can explain their coordination reasoning in natural language, building trust and enabling human oversight. This is particularly important in safety-critical applications like CAV coordination, where understanding system behavior is essential. The explainability capability enables human operators to understand, verify, and trust the coordination decisions made by the system.

These enhanced strategies are supported by frameworks that provide the infrastructure for implementing semantic coordination patterns such as RoundRobin and Swarm coordination [4]. The choice of coordination pattern depends on the specific application requirements: RoundRobin coordination might be appropriate for intersection management where vehicles need to take turns, while Swarm coordination could be more suitable for traffic flow optimization where vehicles adapt based on local conditions. Section 11.5 demonstrates these capabilities through concrete CAV applications.

11.4 Mathematical Formalization

The conceptual enhancements can be expressed rigorously using mathematical notation. The formal mathematical framework underlying GenAI MAS shows how semantic lift operators transform classical coordination formulations while maintaining mathematical rigor and safety guarantees.

11.4.1 The Core Coordination Problem

At the heart of multi-agent coordination lies a fundamental challenge: each agent must make decisions that depend not only on its own information but also on what other agents are doing. This interdependence creates a complex coordination problem where each agent must understand its own position, anticipate others' actions, remember past interactions, and work toward shared goals while maintaining individual objectives.

For a system of n agents, each agent i must choose an action based on its local information and the actions of others. The fundamental coordination problem is:

$$y_i = \pi_i(x_i, I_{-i}, M_i, g_i), \tag{11.1}$$

where x_i represents the agent's local observations, I_{-i} represents information from other agents, M_i denotes the agent's memory, and g_i is the agent's goal. The coordination challenge emerges from the interdependence of these policies, where the quality of overall system performance depends on how well each agent understands and responds to others.

11.4.2 Why Classical MAS Fall Short

Classical MAS intent sharing reduces complex goals to simple symbols:

$$I_{-i}^{\text{classical}} = \{p, v, b\}, \tag{11.2}$$

where p represents position, v represents velocity, and b represents basic intent. This symbolic representation fails to capture nuanced intentions such as "I can wait to let you go first." The classical policy formulation:

$$y_i = \pi_i(x_i, I_{-i}^{\text{classical}}, M_i, g_i) \tag{11.3}$$

uses rigid, symbolic representations for all components, which cannot capture the semantic richness of real-world coordination scenarios.

11.4.3 Mathematical Framework for GenAI MAS

Semantic lift operators $\mathcal{S}$ transform each component of the coordination problem from symbolic to semantic representations, applying domain knowledge $\mathcal{K}$, historical data $\mathcal{D}$, contextual information $\mathcal{C}$, and reasoning capabilities $\mathcal{R}$:

$$I_{-i}^{\text{G}} = \mathcal{S}(I_{-i}^{\text{classical}}; \mathcal{K}) \quad \text{(enhanced intent understanding)} \tag{11.4}$$

$$M_i^{\text{G}} = \mathcal{S}(M_i; \mathcal{D}) \quad \text{(rich memory systems)} \tag{11.5}$$

$$g_i^{\text{G}} = \mathcal{S}(g_i; \mathcal{C}) \quad \text{(context-aware goals)} \tag{11.6}$$

$$\pi_i^{\text{G}} = \mathcal{S}(\pi_i; \mathcal{R}) \quad \text{(intelligent reasoning)} \tag{11.7}$$

The enhanced policy becomes:

$$y_i = \pi_i^{\mathrm{G}}(x_i, I_{-i}^{\mathrm{G}}, M_i^{\mathrm{G}}, g_i^{\mathrm{G}}). \tag{11.8}$$

This framework provides the foundation for formalizing the three core coordination enhancements: communication, negotiation, and collective decision-making. Each enhancement applies specialized semantic lift operators to transform classical coordination mechanisms into semantic-aware processes.

11.4.4 Communication

Building on the core framework, the transformation from symbolic message passing to semantic dialogue is formalized as:

$$C_i^{\mathrm{classical}} = \{p_i, v_i, b_i\} \quad \rightarrow \quad C_i^{\mathrm{G}} = \mathcal{S}_{\mathrm{comm}}(C_i^{\mathrm{classical}}; \mathcal{L}, \mathcal{K}), \tag{11.9}$$

where $\mathcal{L}$ represents language understanding capabilities and $\mathcal{K}$ represents domain knowledge. The enhanced response generation is:

$$R_i^{\mathrm{G}} = \mathcal{G}_c(C_i^{\mathrm{G}}, x_c, h), \tag{11.10}$$

where $\mathcal{G}_c$ generates contextually appropriate responses based on semantic understanding, x_c represents the current context, and h represents the conversation history.

11.4.5 Negotiation

While enhanced communication enables semantic message exchange, enhanced negotiation formalizes how agents resolve conflicts and optimize decisions. Enhanced negotiation strategies are formalized through adaptive utility functions. The classical utility function $U_i(a_i, x_i, I_{-i})$ encodes agent i's preferences for action a_i given observations x_i and information from other agents I_{-i}. The enhanced utility function is:

$$U_i^{\mathrm{G}}(a_i, x_i, I_{-i}^{\mathrm{G}}) = \mathcal{S}_{\mathrm{neg}}(U_i; \mathcal{H}, \mathcal{R}), \tag{11.11}$$

where $\mathcal{H}$ represents historical interaction data and $\mathcal{R}$ represents reasoning capabilities. The multi-objective optimization is:

$$a_i^* = \arg\max_{a_i} \sum_{j=1}^{m} w_j \cdot U_i^{\mathrm{G},j}(a_i, x_i, I_{-i}^{\mathrm{G}}), \tag{11.12}$$

where a_i represents the action chosen by agent i, a_i^* is the optimal action, and w_j are adaptive weights for different objectives (safety, efficiency, fairness, etc.).

11.4.6 Collective Decision-making

Beyond individual agent negotiation, enhanced collective decision-making formalizes how groups of agents reach consensus and discover emergent coordination strategies. Emergent coordination strategies are formalized through collective reasoning:

$$\pi_{\mathrm{collective}}^{\mathrm{G}} = \mathcal{S}_{\mathrm{collective}}(\{\pi_i^{\mathrm{G}}\}_{i=1}^{n}; \mathcal{E}, \mathcal{C}), \tag{11.13}$$

where $\mathcal{E}$ represents emergent intelligence, and $\mathcal{C}$ represents consensus-building capabilities. The consensus-building process is:

$$c_t = \mathcal{F}_c(\{p_i^t\}_{i=1}^n, x_c, h), \tag{11.14}$$

where $\mathcal{F}_c$ is the consensus-building function that synthesizes proposals from all agents, p_i^t represents the proposal from agent i at time t, c_t is the consensus reached at time t, x_c is the current context, and h is the history of interactions. The collective learning process is:

$$\pi_c^{\mathrm{G}}(t+1) = \mathcal{L}_c(\pi_c^{\mathrm{G}}(t), o, f), \tag{11.15}$$

where $\mathcal{L}_c$ is the collective learning function that updates the collective policy based on outcomes o and feedback f from the coordination experience.

11.4.7 Coordination Strategies

The enhanced coordination capabilities described above enable dynamic strategy generation that adapts to specific contexts. Generated planning strategies can be formalized as:

$$P_i^{\mathrm{G}} = \mathcal{S}_{\mathrm{plan}}(\pi_i^{\mathrm{G}}, x_i, I_{-i}^{\mathrm{G}}; \mathcal{G}, \mathcal{A}), \tag{11.16}$$

where $\mathcal{G}$ represents planning generation capabilities and $\mathcal{A}$ represents adaptation mechanisms. This formulation shows how the semantic lift framework enables coordination strategies that emerge dynamically from context, completing the mathematical formalization of how GenAI transforms classical MAS coordination.

11.5 GenAI Applications in CAVs

The enhanced coordination capabilities described in Section 11.3, including semantic communication, adaptive negotiation, and collective decision-making, find concrete application in CAVs, where vehicles must cooperate as agents through various V2X communication modalities. This section demonstrates how these theoretical enhancements address real-world coordination challenges in traffic scenarios.

11.5.1 GenAI Traffic Intersection Coordination

Traffic intersection coordination represents one of the most challenging scenarios for CAV MAS, encompassing classical intersections, highway on-ramps, and roundabouts. The enhanced communication capabilities from Section 11.3 enable vehicles to express intentions, constraints, and preferences in natural language, while adaptive negotiation resolves conflicts and collective decision-making optimizes traffic flow. This combination enables effective coordination even in complex scenarios with multiple vehicles having competing priorities.

At conventional traffic intersections, GenAI coordination can replace rigid traffic light timing with dynamic, context-aware coordination. Vehicles can communicate their destinations, urgency levels, and constraints, enabling the system to optimize traffic flow based on real-time conditions rather than fixed schedules.

On-ramp merging represents a particularly challenging coordination scenario where vehicles must integrate into high-speed traffic flow. GenAI MAS can coordinate merging vehicles with highway traffic through semantic understanding of speed differentials, available gaps, and merging preferences. For example, a vehicle can communicate "I need to merge, but I have a passenger who gets motion sick, so I need to merge smoothly," enabling the system to coordinate an optimal merging sequence that considers individual vehicle constraints and priorities.

Roundabout coordination requires continuous negotiation between vehicles entering, circulating, and exiting the roundabout. GenAI MAS can handle the complex priority rules and continuous flow coordination through semantic understanding of vehicle intentions and roundabout dynamics. The system can coordinate vehicles to maintain smooth flow while respecting roundabout rules and optimizing for overall traffic efficiency.

11.5.2 Fleet Management and Emergency Response

Beyond intersection coordination, GenAI MAS enables coordination across broader CAV applications. In fleet management and logistics, the enhanced communication and collective decision-making capabilities enable multiple delivery vehicles to coordinate routes, schedules, and resource allocation. When a delivery vehicle encounters unexpected traffic congestion, semantic communication allows it to explain the situation to other vehicles, and collective decision-making coordinates route adjustments across the fleet to minimize overall delivery delays.

Emergency response coordination demonstrates how adaptive negotiation and collective decision-making handle time-critical scenarios. Multiple emergency vehicles must coordinate their response to ensure effective and timely assistance. The enhanced negotiation capabilities enable vehicles to resolve priority conflicts and optimize approach routes, while collective decision-making coordinates resource allocation and response strategies based on real-time conditions and the overall emergency situation.

11.6 Implementation Considerations and Practical Challenges

The transition from classical to GenAI MAS introduces fundamental implementation challenges that stem directly from the theoretical enhancements. While individual agent challenges (discussed in Chapter 10), such as computational latency and verification, apply here, multi-agent coordination introduces unique challenges related to inter-agent communication, consensus building, and collective behavior that must be addressed for practical deployment.

11.6.1 Computational Complexity and Scalability

The semantic lift operators that transform classical MAS into GenAI systems introduce computational overhead that scales with the number of coordinating agents. The operators $\mathcal{S}(I_{-i}^{\text{classical}}; \mathcal{K})$, $\mathcal{S}(M_i; \mathcal{D})$, and $\mathcal{S}(g_i; \mathcal{C})$ require computational resources for natural language understanding and generation, and this processing must occur for each agent in the system. The enhanced policy $\pi_i^{\text{G}}(x_i, I_{-i}^{\text{G}}, M_i^{\text{G}}, g_i^{\text{G}})$ must compute coordination decisions within strict time limits for safety-critical applications like CAVs, and coordination requires synchronization across multiple agents.

As the number of agents increases, the computational cost of semantic coordination grows nonlinearly because each agent must process information from all other agents. In a system with n agents, each agent processes $O(n)$ messages, leading to $O(n^2)$ total processing requirements. This computational overhead represents a fundamental trade-off: the enhanced coordination capabilities enabled by semantic intelligence come at the cost of increased computational complexity that scales with the number of agents. The system must scale to support dense urban environments with hundreds of vehicles coordinating simultaneously, requiring efficient algorithms and distributed processing architectures that can maintain coordination quality as the number of agents increases.

11.6.2 Communication Overhead

The enhanced communication capabilities we formalized mathematically introduce communication overhead compared to classical symbolic approaches. This challenge is unique to multi-agent coordination and stems directly from the transformation from simple message passing to semantic dialogue. Natural language communication $C_i^{\mathrm{G}} = \mathcal{S}_{\mathrm{comm}}(C_i^{\mathrm{classical}}; \mathcal{L}, \mathcal{K})$ requires more bandwidth than symbolic messages $\{p_i, v_i, b_i\}$, and in a system with n agents, this bandwidth requirement scales with the number of agent pairs communicating.

The adaptive communication protocols that adjust based on context and history require continuous information exchange, increasing network load as each agent maintains communication channels with multiple other agents simultaneously. The enhanced negotiation strategies that balance multiple objectives require richer information exchange to support semantic reasoning, further increasing communication volume compared to simple symbolic coordination. In distributed coordination scenarios, network partitions can isolate groups of agents, preventing coordination and requiring protocols that can handle partial connectivity and detect network failures.

This communication overhead illustrates the practical implications of the theoretical enhancements: while semantic communication enables more sophisticated coordination, it requires optimization to maintain efficiency in bandwidth-constrained environments. The multi-agent nature amplifies this challenge because communication scales with the number of agents and their interactions.

11.6.3 Safety and Verification

The semantic nature of GenAI coordination introduces new challenges for safety verification and validation that are unique to multi-agent systems. The enhanced utility functions $U_i^{\mathrm{G}}(a_i, x_i, I_{-i}^{\mathrm{G}}) = \mathcal{S}_{\mathrm{neg}}(U_i; \mathcal{H}, \mathcal{R})$ must be verified to ensure they produce safe coordination decisions across all possible contexts and agent combinations. The combinatorial nature of multi-agent interactions makes exhaustive verification challenging, as the number of possible agent interaction patterns grows exponentially with the number of agents.

The collective decision-making process $\pi_c^{\mathrm{G}} = \mathcal{S}_c(\{\pi_i^{\mathrm{G}}\}_{i=1}^n; \mathcal{E}, \mathcal{C})$ can produce emergent behaviors that are difficult to predict and validate. Unlike single-agent systems, where behavior is more predictable, multi-agent systems can exhibit emergent coordination patterns that were not explicitly designed. Multi-agent coordination requires consensus building, but semantic reasoning can lead to coordination deadlocks where agents cannot reach agreement. Verification must ensure that consensus mechanisms always terminate and produce safe outcomes even when agents have conflicting preferences.

When individual agents fail or behave unexpectedly, the coordination system must detect these failures and adapt without compromising safety. Verification must account for partial system failures and ensure graceful degradation. When semantic coordination fails, the system must gracefully fall back to classical coordination methods while maintaining safety guarantees, requiring seamless transition protocols that preserve safety during mode switching.

These safety considerations highlight the need for verification frameworks that can validate the behavior of semantic coordination systems while maintaining the flexibility and adaptability that make them valuable. The multi-agent nature adds complexity because safety must be verified not just for individual agents but for the collective system behavior, including all possible agent interaction patterns.

11.6.4 Integration Challenges

The deployment of GenAI MAS requires integration with existing classical systems and infrastructure, with additional complexity from multi-agent coordination requirements. GenAI agents must communicate with classical agents using both semantic and symbolic protocols, requiring dual communication capabilities. In multi-agent systems, this means managing heterogeneous agent populations where some agents use semantic communication while others use symbolic protocols.

Existing traffic management systems must be adapted to handle both classical symbolic data and enhanced semantic information. Multi-agent coordination adds the challenge of managing coordination protocols that span both classical and enhanced agents. The transition from classical to GenAI coordination must be gradual, requiring systems that can operate in mixed environments. Multi-agent systems must handle coordination when only a subset of agents have enhanced capabilities, ensuring that enhanced agents can coordinate effectively with classical agents while maintaining system-wide safety.

These integration challenges demonstrate how theoretical enhancements must be implemented to ensure practical deployment while maintaining system reliability and effectiveness in heterogeneous multi-agent environments where classical and enhanced agents must coordinate together.

11.6.5 Privacy and Security

Privacy and security represent critical considerations for GenAI MAS in CAV applications, with unique challenges arising from multi-agent information sharing. Semantic communication enables richer information exchange, but this creates privacy risks when agents share detailed intentions, constraints, and preferences. Privacy-preserving coordination mechanisms must balance the need for semantic communication with the protection of sensitive personal information, ensuring that coordination can occur without exposing individual vehicle location, destination, or passenger details to other coordinating agents.

Unlike single-agent systems, where privacy concerns are primarily about external access, multi-agent coordination requires privacy protection between agents themselves. Vehicles must coordinate effectively while protecting their own location, destination, and passenger information from other coordinating agents. The system must ensure secure communication between vehicles and infrastructure to prevent unauthorized access or manipulation. This requires encryption and authentication mechanisms for V2V and vehicle-to-infrastructure (V2I) communications, such as the Security Credential Management System (SCMS) [5], which provides a security framework for CAV communications, including certificate management, privacy protection, and secure message authentication.

Multi-agent systems introduce new attack vectors where malicious agents could disrupt coordination, inject false information, or compromise consensus mechanisms. Security protocols must detect and mitigate these coordination-specific threats, such as by validating agent identities, verifying message authenticity, and detecting anomalous coordination behavior. These privacy and security considerations are fundamental to multi-agent coordination, where information sharing is essential for effective coordination but must be managed to protect individual agent privacy and system security.

11.7 Summary

This chapter addressed the fundamental challenge of multi-agent coordination: how can multiple intelligent agents coordinate effectively when each possesses independent goals, constraints, and decision-making processes? The core tension lies in balancing individual autonomy with collective coordination.

Classical MAS approaches coordination through rigid, symbolic representations that reduce complex coordination to simple message passing. GenAI transforms this limitation through semantic lift operators that enhance each component of coordination: communication becomes natural language dialogue, negotiation becomes adaptive reasoning, and collective decision-making becomes emergent intelligence through semantic interaction.

The mathematical framework formalizes this transformation, showing how semantic lift operators transform classical coordination formulations into GenAI policies that operate on semantic representations. This enables coordination strategies that emerge dynamically from context rather than following predefined templates.

In CAV applications, these capabilities enable intersection coordination that adapts to real-time conditions, fleet management that optimizes routes across multiple vehicles, and emergency response that coordinates multiple responders. The enhanced communication, negotiation, and collective decision-making capabilities find concrete application in these scenarios.

However, these enhancements introduce multi-agent-specific challenges: computational complexity that scales non-linearly with the number of agents, communication overhead from semantic message exchange, safety verification for emergent collective behaviors, integration with heterogeneous agent populations, and privacy protection between coordinating agents. These implementation considerations demonstrate that semantic coordination requires a system design that balances coordination effectiveness with practical constraints in real-world deployment.

References

1 Michael Wooldridge. *An Introduction to MultiAgent Systems*. John Wiley & Sons, Chichester, UK, 2nd edition, 2009. ISBN 978-0-470-51946-2. Foundational textbook on multi-agent systems covering agent architectures, communication, coordination, and collective intelligence.

2 Peter Stone and Manuela Veloso. Multiagent systems: a survey from a machine learning perspective. *Autonomous Robots*, 8(3):345–383, 2000. https://doi.org/10.1023/A:1008942012299. Comprehensive survey of multi-agent systems focusing on coordination paradigms, learning, and distributed problem solving.

3 Gerhard Weiss. Multiagent systems: a modern approach to distributed artificial intelligence. In Gerhard Weiss, editor, *Multiagent Systems: A Modern Approach to Distributed Artificial Intelligence*, MIT Press, Cambridge, MA, 1999. Overview of coordination paradigms including centralized, decentralized, and distributed approaches in multi-agent systems.

4 Qingyun Wu, Gagan Bansal, Jwala Zhang, Yiran Wu, Shaokun Zhang, Erkang Zhu, Li Li, Jiale Liu, Ahmed Hassan Awadallah, Ryen White, et al. AutoGen: enabling Next-Gen LLM applications via multi-agent conversation. *arXiv preprint arXiv:2308.08155*, 2023. Microsoft AutoGen framework for multi-agent conversation systems.

5 Benedikt Brecht, Denis Therriault, Andre Weimerskirch, William Whyte, Virendra Kumar, Thorsten Hehn, and Roy Goudy. A security credential management system for V2X communications. *arXiv preprint arXiv:1802.05323*, 2018. https://doi.org/10.48550/arXiv.1802.05323.

12

Orchestration Frameworks for Agentic AI

12.1 Introduction

Consider a connected and autonomous vehicle (CAV) navigating through a complex urban intersection during rush hour. The vehicle must simultaneously process real-time sensor data, coordinate with traffic signals, communicate with nearby vehicles, and make split-second decisions about lane changes, speed adjustments, and emergency responses. This scenario demonstrates the complex orchestration challenges that agentic artificial intelligence (AI) systems face in real-world applications.

The sophisticated capabilities of agentic AI systems, including natural language understanding, autonomous reasoning, multi-step planning, and proactive behavior, require equally sophisticated orchestration mechanisms to coordinate their operation. The practical frameworks and methodologies that enable their implementation and deployment in real-world CAV applications build directly on the agentic AI principles and capabilities established in the previous chapter. This chapter focuses on **single-system orchestration frameworks** (LangGraph and AutoGen) that operate within shared memory, coordinating agents and workflows within a single CAV system. For **distributed agentic systems** that require communication protocols across multiple CAV systems, see Chapter 16, which covers the integration of model context protocol (MCP) and agent-to-agent (A2A) for multi-system coordination.

Orchestration frameworks serve as the computational infrastructure that transforms theoretical concepts into operational systems. These frameworks provide the mechanisms for structuring multi-step reasoning workflows, managing state transitions and memory updates, coordinating tool invocations (external system interactions) and application programming interface (API) calls, enabling multi-agent collaboration, and implementing robust error handling and recovery mechanisms. Without effective orchestration, the advanced capabilities of agentic AI systems would remain theoretical constructs rather than practical solutions.

The challenge of orchestrating agentic AI systems is particularly acute in CAV environments, where safety, reliability, and real-time performance are paramount. These systems must coordinate complex reasoning processes while maintaining deterministic behavior (predictable, repeatable outcomes), handle uncertainty and partial information, and adapt to rapidly changing conditions. Orchestration frameworks address these challenges through different architectural approaches and design philosophies.

Two dominant paradigms have emerged for orchestrating agentic AI systems: graph-based orchestration and dialogue-based orchestration. Each paradigm offers distinct advantages and trade-offs, making them suitable for different types of applications and use cases. Understanding these paradigms and their implementation frameworks is essential for designing effective agentic AI systems for CAVs.

The chapter covers the theoretical foundations, practical implementations, and comparative analysis of these orchestration paradigms. Graph-based orchestration using LangGraph provides deterministic, verifiable workflows suitable for safety-critical applications. Dialogue-based orchestration using AutoGen offers flexibility and adaptability for complex, dynamic scenarios. A comprehensive comparison and selection guidance helps choose the appropriate framework for specific CAV applications.

12.2 Graph-based Orchestration with LangGraph

Graph-based orchestration represents a structured, deterministic approach to coordinating agentic AI systems. This paradigm models workflows as directed graphs where nodes represent reasoning steps, tool invocations, or decision points, and edges define conditional transitions between these components. LangGraph, developed by LangChain, provides a comprehensive framework for implementing graph-based orchestration in agentic AI systems [1].

12.2.1 Mathematical Foundation

The mathematical foundation of graph-based orchestration builds upon fundamental graph theory concepts and extends them to model dynamic, stateful workflows, establishing the formal basis for understanding workflow execution.

12.2.1.1 Graph Theory Fundamentals

A graph-based orchestration workflow is formally represented as a directed graph $G = (V, E)$, where:

- $V = \{v_1, v_2, \dots, v_n\}$ is the set of nodes representing computational components
- $E \subseteq V \times V$ is the set of directed edges representing possible transitions
- Each edge $(u, v) \in E$ indicates that execution can transition from node u to node v

The graph structure provides the foundation for workflow representation, where nodes encapsulate computational logic and edges define the flow of execution.

12.2.1.2 State Space and Dynamics

The system maintains a *shared state* that flows through all nodes in the workflow. This shared state at time t is represented by a state vector S_t that captures all relevant information:

$$S_t = \{M, C, V_c, R\}, \tag{12.1}$$

where M represents memory containing historical information and learned patterns, C represents context including current environmental conditions and user inputs, V_c represents control variables that manage the flow of execution, and R represents results that store intermediate outputs and outcomes. The shared state ensures that all nodes have access to the same information, enabling consistent decision-making and maintaining workflow context throughout execution.

The evolution of the shared state follows a discrete-time dynamical system:

$$S_{t+1} = f_{n_t}(S_t, I_t), \tag{12.2}$$

where f_{n_t} represents the computation performed by the active node at time t, and I_t represents additional input parameters.

12.2.1.3 Node Functions and Routing

Each node in the graph performs a specific computation on the shared state. Node functions can implement various types of operations: **reasoning operations** for logical inference, decision-making, and problem-solving; **tool invocations** for external API calls, database queries, and system interactions; **memory operations** for reading from and writing to the shared state; and **control operations** for flow control, branching, and termination decisions.

The selection of which node to execute at each time step is determined by a routing function that examines the current state and graph topology:

$$n_t = r(S_t, G), \tag{12.3}$$

where r is the routing function that maps the current state S_t and graph structure G to the next node n_t to execute. The routing function implements the graph's edge logic, selecting the next node based on the current state and the graph's structure, and can implement various types of logic, including conditional branching, probabilistic transitions, and deterministic sequences.

12.2.1.4 Execution Model

The execution of a graph-based workflow follows a systematic process. At each time step t:

1. The system evaluates the current shared state S_t
2. The routing function $r(S_t, G)$ determines the next node n_{t+1} to execute
3. The selected node performs its computation on the shared state: $S_{t+1} = f_{n_{t+1}}(S_t, I_t)$
4. The shared state is updated and the process continues

This process continues until a termination condition is met or the workflow completes successfully. The termination condition can be represented as a predicate $\tau(S_t)$ that evaluates to true when the workflow should stop.

12.2.2 Implementation Example

LangGraph is a Python library developed by LangChain that provides a comprehensive framework for building stateful, multi-step applications with LLMs [1]. It implements the graph-based orchestration paradigm by allowing developers to define workflows as directed graphs where nodes represent computational steps and edges define conditional transitions between them.

LangGraph workflows consist of three fundamental components: nodes (deterministic functions operating on shared state), edges (conditional transitions determining execution flow), and state (a shared object holding all information that flows through all nodes). These components create structured, deterministic workflows where execution flow is explicitly defined and verifiable. To illustrate the practical implementation of graph-based orchestration, consider a lane-change workflow implemented using LangGraph. This workflow demonstrates how complex, safety-critical operations can be structured as deterministic, verifiable graphs. The lane-change workflow includes several key components: safety checking, execution, retry logic, and abort conditions. Each component is implemented as a node in the graph, with conditional transitions that determine the flow of execution based on the results of each step. Figure 12.1 illustrates the visual representation of the lane-change workflow, showing how the graph structure enables retries and fallback through memory-based transitions. The workflow starts at the safety check node and follows conditional paths based on the safety assessment and retry count.

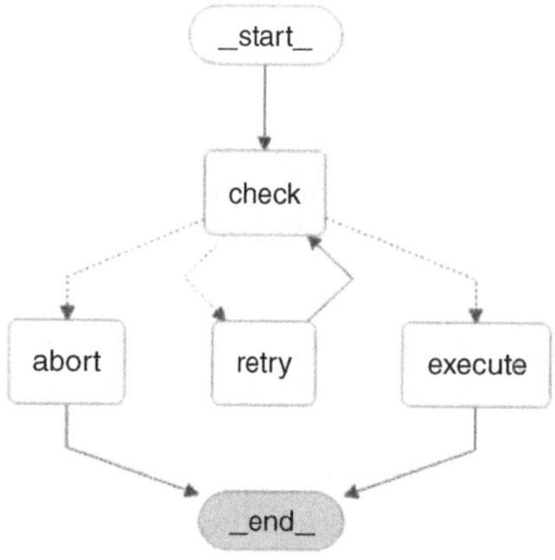

Figure 12.1 Lane change workflow graph showing the retry logic with safety checks, execution, and abort conditions.

12.2.3 State Structure

The foundation of any LangGraph workflow is the shared state object that flows through all nodes. State is explicitly defined and updated at every step, with each node reading from and writing to a common state object. This enables debuggable, memory-aware workflows. Mathematically, this corresponds to the shared state S_t from Equation (12.1), which flows through all nodes and gets updated according to Equation (12.2). The following code defines the state structure for the lane-change workflow:

Listing 12.1 State structure definition.

```
class LaneChangeState:
    def __init__(self):
        # Memory (M): Historical information and learned patterns
        self.attempts = 0  # Track retry attempts

        # Context (C): Current environmental conditions
        self.is_safe = False  # Current safety assessment

        # Control Variables ($V_c$): Manage flow of execution
        self.execution_status = None  # Current execution state

        # Results (R): Intermediate outputs and outcomes
        self.safety_check_result = None  # Latest safety check result
```

This code defines the `LaneChangeState` class that represents the shared state S_t from Equation (12.1). The `__init__` method initializes four categories of state components, each corresponding to the mathematical structure: Memory (M) stores `attempts` to track retry history; Context (C) stores `is_safe` to represent current safety assessment; control variable (V_c) store `execution_status` to manage execution flow; and Results (R) stores `safety_check_result` to hold intermediate outputs. The state is a globally shared state, not per-node local memory. Each node updates the state through its function (implementing f_{n_t} from Equation (12.2)), and routing uses the updated state to select the next node (implementing $r(S_t, G)$ from Equation (12.3)). Design state objects to clearly separate different types of information (memory, context, control, results) to maintain clean workflow logic and enable effective debugging.

12.2.4 Nodes as Workflow Steps

Each *node* represents a step in the process. Nodes are deterministic functions that operate on the shared state, performing specific operations like safety checks, actions, or decision-making. Mathematically, each node implements the function f_{n_t} from Equation (12.2), where n_t is the node identifier. The following code demonstrates how to define nodes in LangGraph:

Listing 12.2 Defining nodes in LangGraph.

```
graph = StateGraph(LaneChangeState)

# Define workflow nodes using RunnableLambda (wraps functions for LangGraph)
graph.add_node("check", RunnableLambda(check_safety))
graph.add_node("execute", RunnableLambda(execute_change))
graph.add_node("retry", RunnableLambda(retry_wait))
graph.add_node("abort", RunnableLambda(abort_action))
```

This code defines a LangGraph workflow with four nodes. The first line creates a `StateGraph` using the `LaneChangeState` class as the state structure. Nodes are added to the graph using the `add_node` method, which takes a node name (e.g., `"check"`) and a function wrapped in `RunnableLambda` to make it compatible with LangGraph's execution model. The four nodes added are: `check` (f_{check}) assesses lane safety and updates Context and Results, `execute` ($f_{execute}$) initiates the lane change and updates Results and Control Variables, `retry` (f_{retry}) pauses and retries the safety check while updating Memory and Control Variables, and `abort` (f_{abort}) exits due to unsafe conditions or repeated failures, updating Results and Control Variables.

12.2.5 Edges as Outcome-based Transitions

Edges determine how the workflow progresses based on the current state. They encode the control logic that routes execution between nodes based on outcomes and conditions. Mathematically, edges implement the routing function $r(S_t, G)$ from Equation (12.3), mapping the current state and graph structure to the next node. The following code demonstrates how to define conditional edges:

Listing 12.3 Defining conditional edges in LangGraph.

```
graph.set_entry_point("check")

# Define conditional transitions based on routing function
graph.add_conditional_edges("check", should_retry, {
    "execute": "execute",
    "retry": "retry",
    "abort": "abort"
})

graph.add_edge("retry", "check")      # Loop back to recheck
graph.add_edge("execute", END)        # Terminal success (END from langgraph.
    constants)
graph.add_edge("abort", END)          # Terminal failure (END from langgraph.
    constants)
```

This code defines the graph structure and conditional transitions. The first line sets the entry point to the `check` node, where execution begins. The conditional edges from the `check` node use the `should_retry` function (defined below) to examine the current state and return one of three strings (`"execute,"` `"retry,"` or `"abort"`), which map to the corresponding target node. The simple edges define deterministic transitions: `retry` loops back to `check` for re-evaluation, while `execute` and `abort` both transition to `END`, terminating the workflow. The `END` constant is imported from `langgraph.constants` and represents the terminal state. Together with the node definitions above, this code implements the complete graph structure shown in Figure 12.1.

The routing function `should_retry(state)` implements the routing function $r(S_t, G)$ from Equation (12.3) to enable looping, fallback, and adaptive control. The routing logic is defined by a user-provided function that examines the current state and determines the next step:

Listing 12.4 Routing function for conditional transitions.

```
def should_retry(state):
    """Route: execute if safe, retry if unsafe and < 3 attempts, else abort"""
    if state.is_safe:
        return "execute"
    elif state.attempts < 3:
        return "retry"
    else:
        return "abort"
```

This function implements the routing logic by examining the shared state. The first condition checks if `state.is_safe` is `True`, returning `"execute"` to proceed with the lane change. The second condition checks if the number of retry attempts is less than 3, returning `"retry"` to loop back and recheck safety. The final condition handles the case where safety checks have failed multiple times, returning `"abort"` to terminate the workflow. This routing mechanism brings reactive intelligence into structured agent workflows, following the systematic execution process defined in the mathematical foundation.

12.2.6 Advantages of Graph-based Orchestration

Graph-based orchestration offers several advantages for safety-critical CAV applications. The structured, deterministic nature ensures predictable and verifiable execution, essential for safety-critical scenarios where behavior must be deterministic and testable, as detailed in the mathematical foundation and implementation examples above. The graph structure supports scalability and modularity, allowing workflows to be decomposed into smaller, manageable components that can be developed, tested, and optimized independently, enabling complex workflows to be constructed by combining simpler sub-workflows.

12.2.7 Use Cases in CAV Systems

Graph-based orchestration is well-suited for safety-critical, predictable scenarios in CAVs. Beyond the safety-critical maneuvers demonstrated in the implementation example above, diagnostic and maintenance workflows benefit from the explicit structure of graph-based orchestration. These workflows can be designed to systematically check system components, identify issues, and implement appropriate responses, with the deterministic nature of the approach ensuring consistent diagnostic procedures. Regulatory compliance scenarios require verifiable workflows that can be certified for compliance with safety standards and regulations, and graph-based orchestration provides the explicit, analyzable structure needed for these certification processes.

12.3 Dialogue-based Orchestration with AutoGen

12.3.1 Conceptual Foundation

While graph-based orchestration excels at deterministic, safety-critical workflows, many real-world scenarios require a different approach. Complex decision-making situations often involve uncertainty, multiple stakeholders, and dynamic conditions that cannot be fully anticipated during system design. Classical workflow approaches struggle with these challenges because they require all intended execution paths to be predefined, limiting the system's ability to adapt to novel situations.

Dialogue-based orchestration addresses these limitations by modeling workflows as conversations among agents, where the flow of execution emerges from the dialogue rather than being predetermined by a graph structure. This paradigm recognizes that many complex problems are best solved through collaborative reasoning, where multiple experts contribute their perspectives and adapt their approach based on the evolving situation. Complex workflows can emerge naturally from interaction patterns among agents with distinct roles and capabilities, allowing workflows to develop dynamically based on agents' responses to each other and their assessment of the current situation. This paradigm is particularly well-suited for scenarios where the required workflow cannot be predetermined due to complexity or uncertainty. It excels when flexibility and adaptability are essential for handling unexpected situations, when natural language interaction is desired for human-in-the-loop scenarios, and when multiple agents need to collaborate on complex problems.

12.3.2 AutoGen Framework

AutoGen, developed by Microsoft, provides a comprehensive framework for implementing dialogue-based orchestration, enabling developers to define agents and coordinate their interactions through natural language conversations [2]. The framework addresses the core challenges of dialogue-based orchestration by providing structured approaches to agent definition, conversation management, and workflow execution.

12.3.2.1 Agent Definitions

Each agent in AutoGen is defined with specific roles, capabilities, and communication patterns. To illustrate this concept, consider a lane change scenario where we need two agents: one to evaluate safety conditions and another to execute the maneuver. The following code demonstrates how to create these agents:

Listing 12.5 Defining specialized agents in AutoGen.

```
from autogen_agentchat.agents import AssistantAgent, RoundRobinGroupChat,
    TextMentionTermination

# Define agents with distinct roles and capabilities
safety_agent = AssistantAgent(
    name="SafetyAgent",
    system_message="Check lane safety using tool. Report clearly.",
    tools=[check_lane_safety],
    model_client=model_client,  # LLM client for agent reasoning
)

execution_agent = AssistantAgent(
    name="ExecutionAgent",
    system_message="If it's safe, perform the lane change. End with 'COMPLETE'.",
    model_client=model_client,
)
```

This code creates two agents using AutoGen's `AssistantAgent` class. The `SafetyAgent` is defined with four key parameters: `name` identifies the agent, `system_message` defines its role and behavior (checking lane safety and reporting clearly), `tools` provides access to the `check_lane_safety` function for evaluating safety conditions, and `model_client` enables the agent's reasoning capabilities. The `ExecutionAgent` uses a different `system_message` that instructs it to perform the lane change if safe and end responses with "COMPLETE" to signal task completion. Both agents share the same `model_client`, which provides the LLM-based reasoning engine.

Each agent is defined by three key components that map directly to the AutoGen parameters:

Role: Defined by the `system_message` parameter (e.g., "Check lane safety using the tool. Report clearly.")
Capabilities: Defined by the `tools` parameter (e.g., `[check_lane_safety]`)
Decision Policy: Implemented through the `model_client` parameter, which enables the agent's reasoning and decision-making

AutoGen provides multiple conversation management strategies that determine how agents take turns, select the next speaker, and coordinate their activities.

12.3.3 Conversation Management

AutoGen supports multiple execution strategies for agent coordination, each implementing different forms of agent selection policies. The conversation flow is managed through a group chat system that coordinates agent interactions, managing turn-taking, message flow, and termination conditions according to the chosen strategy:

- **`RoundRobinGroupChat`:** Cycles through agents in a fixed order. After agent u_t finishes, the next agent u_{t+1} is chosen by cycling through the N available agents, independent of message content.
- **`SelectorGroupChat` (LLM-based):** Uses an LLM selector to examine the conversation history and choose the next agent. After agent u_t finishes, the LLM selector examines the history and chooses the next speaker.
- **`SelectorGroupChat` with a custom selector function:** A user-provided function (passed as selector_func), rather than the LLM, examines the conversation history and returns the next speaker—for example, routing to the safety agent when safety concerns are detected, otherwise to the execution agent.
- **`Swarm`:** Enables direct agent-to-agent communication, enabling emergent, decentralized behavior where agents delegate directly to each other.

The choice of conversation strategy significantly impacts the workflow's predictability and flexibility. `RoundRobinGroupChat` provides predictable turn-taking, while `Swarm` enables more emergent, adaptive behavior. The following code demonstrates how to set up conversation management using the RoundRobin strategy:

Listing 12.6 Setting up conversation management in AutoGen using RoundRobin strategy.

```
from autogen_agentchat.teams import RoundRobinGroupChat
from autogen_agentchat.conditions import TextMentionTermination

# Create a group chat
group_chat = RoundRobinGroupChat(
    [safety_agent, execution_agent],
```

```
    termination_condition=TextMentionTermination("COMPLETE"),
    max_turns=6,
)
```

This code creates a `RoundRobinGroupChat` that implements the conversation dynamics described above. The group chat receives the list of participating `agents` as its first argument; these agents cycle through turns in sequence. Its `termination_condition` uses `TextMentionTermination("COMPLETE")` to end the conversation as soon as any agent's `message` contains the word "COMPLETE", and `max_turns=6` caps the conversation at six turns to prevent infinite loops. The starting prompt is not supplied to the group chat itself; it is passed to the run call shown next. The `RoundRobinGroupChat` coordinates the conversation by allowing each agent to take turns in sequence, ensuring predictable turn-taking while maintaining the emergent workflow pattern.

The conversation system operates through a two-step process: agent selection (deciding which agent speaks next) and message generation (what that agent says). Different strategies focus on different approaches to agent selection, while message generation is straightforward, where the selected agent generates a response based on the conversation history. The key insight is that strategy determines WHO speaks, while the agent determines WHAT they say.

Advanced Note 12.1 Mathematical Foundation of Agent Selection Strategies

The agent selection process can be mathematically modeled using the following notation: H_t represents the conversation history at time t (sequence of all messages exchanged), u_t represents the agent that speaks at time t (e.g., safety_agent, execution_agent), and $\mathcal{U}$ represents the set of all available agents in the conversation.

Agent Selection Strategies:

For the `RoundRobinGroupChat` strategy, the selection rule is:

$$u_{t+1} = u_{((t)\ mod\ N)+1}, \tag{12.4}$$

where u_t represents the agent speaking at time t, u_{t+1} represents the next agent to speak, and N represents the total number of agents.

For the `SelectorGroupChat` strategy:

$$u_{t+1} = \arg\max_{u \in \mathcal{U}} \pi_{\text{select}}(u \mid H_t), \tag{12.5}$$

where $\mathcal{U}$ is the set of available agents and π_{select} is the selection probability based on conversation history H_t. In practice, when using LLMs, this is typically implemented using sampling rather than exact maximization:

$$u_{t+1} \sim \pi_{\text{select}}(u \mid H_t). \tag{12.6}$$

For a custom selector function (SelectorGroupChat with selector_func):

$$u_{t+1} = f_{\text{select}}(H_t, S_t), \tag{12.7}$$

where f_{select} is a user-provided selector function that determines the next agent based on conversation history H_t and current system state S_t.

For the `Swarm` strategy:

$$u_{t+1} = \text{SwarmSelector}(H_t, \mathcal{U}), \tag{12.8}$$

where SwarmSelector is a function that selects the next agent based on swarm intelligence principles.

(Continued)

Advanced Note 12.1 (Continued)

This mathematical foundation provides insight into how different strategies operate: deterministic strategies like `RoundRobinGroupChat` follow fixed patterns, while probabilistic strategies like `SelectorGroupChat` adapt based on conversation context, and a custom selector function or `Swarm` enables domain-specific routing logic.

Once the conversation management system is configured, executing the workflow is straightforward regardless of the chosen strategy. The following code initiates the conversation:

Listing 12.7 Initiating workflow execution.

```
# Run the conversation
result = await group_chat.run(task="Can I change lanes now?")
```

The await `group_chat.run(task=...)` call initiates the entire conversation workflow, triggering the multi-agent dialogue where agents take turns, evaluate conditions, make decisions, and eventually complete the task. The call is asynchronous, and the starting prompt ("Can I change lanes now?") is supplied here as the task. The system automatically handles conversation flow, agent selection, and termination based on the configured strategy. The behavior that emerges from the conversation depends on the chosen strategy: with `RoundRobinGroupChat`, the workflow follows a predetermined sequence, while with `SelectorGroupChat` or `Swarm`, the execution path develops dynamically based on agents' responses to each other and their assessment of the current situation, enabling the system to adapt to changing conditions and unexpected situations. Agents alternate turns until "COMPLETE" is mentioned, with tool calls, memory, and reasoning handled via the conversation loop. The termination condition ensures that conversations complete appropriately and do not run indefinitely.

12.3.4 Advantages of Dialogue-based Orchestration

Dialogue-based orchestration offers unique advantages for complex, dynamic scenarios in CAV applications, as detailed in the conceptual foundation and implementation examples above. The conversational model enables adaptive problem-solving and collaborative reasoning, allowing agents to contribute their expertise and dynamically adjust their approach based on real-time conditions. The incremental nature of dialogue-based decision-making allows agents to build upon each other's contributions, refining and improving solutions through iterative discussion. This enables more nuanced decision-making for complex, multi-step problems where the optimal solution emerges through collaborative exploration rather than predetermined logic.

12.3.5 Use Cases in CAV Systems

Dialogue-based orchestration is well-suited for complex, dynamic scenarios in CAVs. Emergency response coordination represents a critical application where dialogue-based orchestration excels, enabling agents to rapidly coordinate their response without predetermined protocols, adapting to the specific nature of the emergency. Multi-modal decision-making scenarios require integration of diverse information sources including sensors, maps, traffic data, and weather conditions. These scenarios benefit from having different agents specialize in different data types while collaborating to reach optimal decisions. The implementation of dialogue-based orchestration in CAV systems requires careful consideration of several factors: agent roles must be clearly defined, conversation management strategies should be chosen based on application requirements, and the system must include robust termination conditions, timeout mechanisms, performance monitoring, and fallback mechanisms to ensure reliability in safety-critical applications.

12.4 Framework Comparison and Selection

Selecting the appropriate orchestration framework is a critical decision that significantly impacts the success of CAV applications. Both LangGraph and AutoGen offer powerful capabilities, but they excel in fundamentally different scenarios. The fundamental distinction between these frameworks lies in their approach to workflow execution and control flow. LangGraph employs a deterministic, pre-defined workflow structure where all intended execution paths are specified in advance, making it ideal for safety-critical applications where behavior must be deterministic and testable. AutoGen employs an emergent, adaptive workflow where execution paths develop dynamically through agent conversations, providing flexibility and adaptability at the cost of some predictability, making it excel in scenarios requiring collaborative reasoning and adaptive problem-solving. Table 12.1 provides a detailed comparison of key aspects across both orchestration paradigms:

12.4.1 Framework Selection Principles

The choice between graph-based and dialogue-based orchestration fundamentally depends on the nature of the problem domain and specific application requirements, as summarized in Table 12.1. Graph-based approaches excel when deterministic behavior is essential for safety-critical scenarios, when execution paths can be fully specified in advance, and when verifiability and predictability

Table 12.1 Comprehensive comparison of orchestration frameworks for CAV applications.

Aspect	LangGraph (graph-based)	AutoGen (dialogue-based)
Execution model	Directed graph with nodes and edges	Conversation among specialized agents
Control flow	Deterministic, pre-defined transitions	Emergent, adaptive dialogue
State management	Explicit shared state object	Distributed across agent memories
Coordination	Single orchestrator with routing logic	Multi-agent collaboration
Predictability	High - all paths predefined	Medium - emergent behavior
Verification	Excellent—analyzable paths	Good - conversation audit trail
Performance	Typically, fast—optimized execution for typical workloads	Variable—depends on conversation length
Development	Moderate—graph design required	High—agent coordination needed
Debugging	Excellent—state inspection	Good—conversation trace
Best Use Cases	Safety-critical, deterministic tasks	Complex reasoning, human interaction
CAV Applications	Lane changes, emergency braking	Driver assistance, route planning

are paramount. Dialogue-based approaches are more suitable when the problem domain involves uncertainty, when workflows must adapt dynamically to changing conditions, and when collaborative reasoning and human interaction are required. The selection process involves analyzing the relationship between framework capabilities and application requirements. Key considerations include the predictability needs, verification requirements, performance constraints, development complexity, and the balance between structure and flexibility that best serves the specific CAV application.

12.4.2 Integration Strategies

Complex CAV systems often require multiple orchestration approaches to address different aspects of the system. The integration of graph-based and dialogue-based orchestration can leverage the strengths of both paradigms. For example, a layered architecture might use graph-based orchestration for safety-critical functions while employing dialogue-based orchestration for user interaction and adaptive decision-making.

These orchestration frameworks can also be integrated with distributed communication protocols. LangGraph and AutoGen can serve as the orchestration layer within each CAV system, while protocols such as MCP and A2A handle communication between systems. In such architectures, the orchestration framework coordinates agents and workflows locally, while MCP/A2A manages distributed coordination across multiple CAV systems, as detailed in Chapter 16.

The choice of integration strategy depends on the system architecture and the specific requirements of different subsystems. Understanding these integration patterns is crucial for designing effective CAV systems that can handle both deterministic safety requirements and adaptive decision-making needs.

12.5 Summary

This chapter examined orchestration frameworks for agentic AI systems in CAV applications, focusing on two dominant paradigms: graph-based orchestration using LangGraph and dialogue-based orchestration using AutoGen. These frameworks provide the computational infrastructure needed to transform theoretical agentic AI concepts into operational systems.

Graph-based orchestration models workflows as directed graphs with nodes representing computational components and edges representing transitions. LangGraph implements this paradigm, providing deterministic, verifiable workflows suitable for safety-critical applications through explicit state management and conditional transitions. Graph-based orchestration excels in safety-critical, predictable scenarios such as diagnostic workflows and regulatory compliance applications where deterministic behavior and verifiability are essential.

Dialogue-based orchestration models workflows as conversations among agents, where workflows emerge dynamically from agent interactions rather than being predetermined. AutoGen implements this paradigm, enabling developers to define agents with specific roles, capabilities, and decision policies, and coordinate their interactions through conversation management strategies. Dialogue-based orchestration excels in complex, dynamic scenarios such as emergency response coordination and multi-modal decision-making, where workflows must adapt to changing conditions and uncertainty.

Framework selection depends on application requirements, as detailed in the comprehensive comparison table. LangGraph excels when deterministic behavior is essential for safety-critical scenarios and when verifiability and predictability are paramount. AutoGen is more suitable when the problem domain involves uncertainty and when workflows must adapt dynamically to changing conditions. Hybrid approaches combine both paradigms for optimal results, and these orchestration frameworks can be integrated with distributed communication protocols such as MCP and A2A to enable comprehensive CAV systems that coordinate both locally and across multiple vehicles.

References

1 LangChain. LangGraph: build stateful, multi-actor applications with LLMs, 2024. https://github.com/langchain-ai/langgraph.

2 Qingyun Wu, Gagan Bansal, Jwala Zhang, Yiran Wu, Shaokun Zhang, Erkang Zhu, Li Li, Jiale Liu, Ahmed Hassan Awadallah, Ryen White, et al. AutoGen: enabling next-gen LLM applications via multi-agent conversation. *arXiv preprint arXiv:2308.08155,* 2023. Microsoft AutoGen framework for multi-agent conversation systems.

Part V

Communication Protocols and Collective Intelligence

Connected autonomy enables intelligence to extend beyond individual vehicles through cooperation with other vehicles, infrastructure, and cloud services. This part focuses on the communication protocols and architectural frameworks that support collective intelligence in connected and autonomous vehicle ecosystems. Rather than treating communication as a purely networking concern, the chapters frame protocols as semantic interfaces between autonomous agents. Topics include protocol design principles, structured context exchange through the model context protocol, agent-to-agent communication, and integrated architectures that enable scalable, interpretable coordination across distributed systems.

13

Communication Protocols for Agentic AI

13.1 Introduction

The integration of generative artificial intelligence (GenAI) systems into connected and autonomous vehicles (CAVs) presents unique challenges that extend beyond rule-based vehicular communication requirements. As large language models (LLMs) and other GenAI systems become central to vehicular decision-making, the effectiveness of these systems depends critically on their ability to interact with external services and coordinate with other intelligent agents. While the previous chapter addressed orchestration frameworks for coordinating agents within a single CAV system, this chapter focuses on communication protocols that enable coordination across distributed CAV systems and external services.

Consider a representative scenario where a CAV approaches an urban intersection during heavy rain, with its onboard sensors partially occluded by a large truck. The vehicle's perception system cannot directly observe a pedestrian hidden behind the obstruction, yet the vehicle must make safe navigation decisions. The agent must integrate information from multiple sources, including querying external weather services for precipitation data, coordinating with nearby vehicles that may have clearer sensor views, and accessing real-time traffic management systems for intersection status. This scenario illustrates the fundamental requirement for both external service integration and peer-to-peer coordination that characterizes CAV systems.

The challenge of enabling such connectivity has led to the development of numerous communication protocols and standards. However, not all protocols are equally suited for the unique requirements of CAV systems, which demand low-latency performance suitable for safety-critical operations, safety guarantees, and seamless integration with both deterministic services and intelligent peer agents. The selection of appropriate protocols directly impacts system reliability, performance, and safety in vehicular environments.

To clarify the scope of communication mechanisms considered in this chapter, Figure 13.1 presents a high-level communication stack for agentic AI in connected vehicles, organized into three distinct layers that serve different purposes in the overall communication architecture.

The **Semantic Interaction Layer** (top layer) encompasses agent communication protocols such as the model context protocol (MCP) [1] and agent-to-agent (A2A) [2]. These protocols operate at the highest level of abstraction, defining semantic intent, coordination logic, and the meaning of interactions between agents and services. They specify *what* agents want to communicate and *how* they coordinate, but do not prescribe the underlying message transport mechanisms.

The **Messaging Layer** (middle layer) provides delivery semantics and message routing mechanisms. Protocols at this layer, including Hypertext Transfer Protocol (HTTP), Message Queuing Telemetry Transport (MQTT), Google Remote Procedure Call (gRPC), and WebSocket, handle the

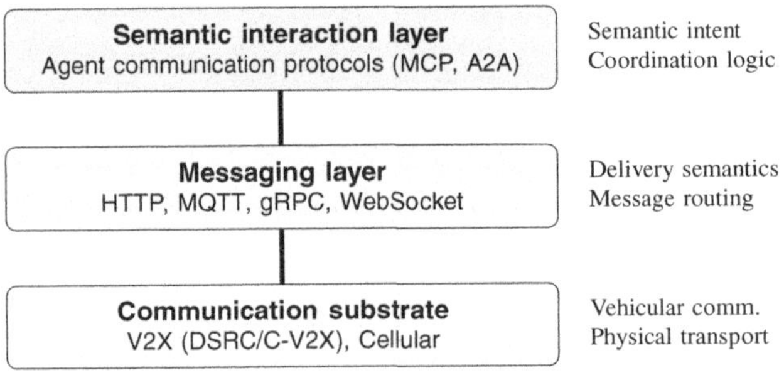

Figure 13.1 Big-picture communication stack for agentic AI in connected vehicles. Agent communication protocols operate at the semantic interaction layer, abstracting over messaging mechanisms such as HTTP and MQTT, and underlying connected-vehicle communication substrates such as vehicle-to-everything (V2X).

mechanics of message transmission, including connection management, reliability guarantees, and routing decisions. These protocols specify *how* messages are delivered but remain agnostic to the semantic content of those messages.

The **Communication Substrate** (bottom layer) represents the physical and vehicular-specific communication technologies that provide the actual transport mechanism. This includes V2X technologies such as dedicated short-range communications (DSRC) and C-V2X,[1] as well as cellular networks. These technologies handle the low-level details of wireless transmission, signal propagation, and vehicular network protocols.

The layered architecture shown in Figure 13.1 illustrates that agent communication protocols abstract over messaging mechanisms, which in turn utilize the underlying communication substrate. This abstraction enables agent protocols to focus on semantic coordination without being tied to specific transport technologies, while messaging mechanisms can leverage different substrates depending on availability and requirements. For example, an agent protocol might use HTTP over cellular networks in one scenario and MQTT over V2X in another, without changing the semantic interaction logic.

13.2 Protocol Classification Framework

The design of intelligent agents in CAVs increasingly relies on structured protocols to support dynamic, context-sensitive behavior. As these agents reason over natural language, sensor data, and external services, their effectiveness is shaped not only by internal models such as LLMs, but also by their ability to communicate with both tools and other agents.

Scholarship [3] highlights an important evolution from single-agent tool invocation to multi-agent coordination. This shift entails a deeper coupling between agents' internal state and external interactions. Protocols must support not only tool binding, but also task sharing, skill advertising, and artifact exchange. Such capabilities are foundational for distributed cognition [4] in CAVs.

This chapter adopts a two-dimensional classification framework based on interaction orientation and protocol scope to organize the diverse landscape of agent communication protocols operating at the semantic interaction layer (as introduced in Figure 13.1). The context-oriented branch

1 The 5.9 GHz ITS band originally comprised 75 MHz. In 2020, the U.S. Federal Communications Commission reallocated 45 MHz to unlicensed use and retained 30 MHz for ITS operations using C-V2X, rather than DSRC; DSRC is therefore now a legacy direct-V2X technology.

encompasses protocols for accessing external services and tools, while the peer-oriented branch covers protocols for agent-to-agent collaboration and coordination. This hierarchical classification enables systematic analysis of protocol capabilities and their suitability for different CAV applications. It is important to emphasize that this taxonomy applies exclusively to protocols operating at the semantic interaction layer and does not include underlying communication technologies such as V2X, which operate at the communication substrate layer. As discussed in Section 13.1, V2X provides standardized message formats (e.g., basic safety messages [BSM], traveler information messages [TIM]), security services [5, 6], and low-latency broadcast mechanisms at the vehicular communication substrate level. Agent protocols may consume and reason over information delivered via V2X, but they operate at a higher level of abstraction, defining semantic intent rather than physical transport mechanisms. Figure 13.2 presents the complete taxonomy structure, showing how protocols are organized across these two primary dimensions.

Figure 13.2 organizes protocols along two primary dimensions. The first dimension, **interaction orientation**, divides protocols into two branches: *Context-oriented* (shown in the upper branch) and *peer-oriented* (shown in the lower branch). The second dimension, **protocol scope**, further subdivides each orientation branch into *General-Purpose* and *Domain-Specific* categories. The figure shows specific protocol instances at the leaves of this hierarchical structure.

13.2.1 Interaction Orientation

The *context-oriented* branch (upper path in Figure 13.2) encompasses protocols that enable agents to access external services, data sources, and structured tools. These protocols typically follow request-response models, offering type-safe interfaces to deterministic services (services that produce predictable outputs for given inputs). Context-oriented protocols may support either stateless interactions (e.g., REST APIs, GraphQL, agents.json) or stateful interactions (e.g., MCP), depending on their design. Protocol types in this category include REST APIs, GraphQL, and function calling frameworks. The *peer-oriented* branch (lower path in Figure 13.2) encompasses protocols that support collaboration between autonomous agents through stateful task exchange, asynchronous messaging, and decentralized task lifecycles. Protocol types in this category include multi-agent communication protocols, task delegation frameworks, and collaborative planning systems.

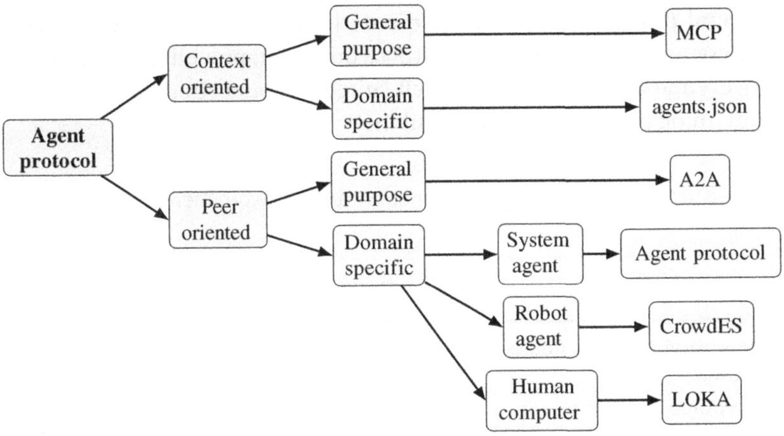

Figure 13.2 Overview of agent communication protocols within the two-dimensional taxonomy framework. The diagram illustrates how protocols can be classified based on interaction orientation (context-oriented vs. peer-oriented) and scope (general-purpose vs. domain-specific).

13.2.2 Protocol Scope

The *general-purpose* category includes protocols designed to operate across domains and tasks. Protocol types in this category include function calling frameworks and JSON-RPC, which provide generic interfaces that can be adapted to various application domains. Specific protocol instances in this category include the MCP [1], which provides a standardized framework for exposing tools, data, and prompts to AI applications across diverse domains, supporting stateful interactions and dynamic capability discovery. The *domain-specific* category includes protocols tailored to particular application domains and task models, often incorporating specialized semantics, constraints, and interaction patterns. Protocol types in this category include protocols designed for robotics coordination, human–computer interaction, or system-level agent integration. Specific protocol instances include `agents.json`, an open specification built on the OpenAPI standard that describes contracts for API and agent interactions, optimized for large language models and enforcing stateless interactions for API-focused use cases. Other specific protocol instances include A2A (peer-oriented, general-purpose) and various peer-oriented domain-specific protocols. The peer-oriented, domain-specific category further subdivides into specialized protocols for robot-agent coordination (e.g., CrowdES), human-computer interaction (e.g., LOKA), and system-agent integration (e.g., Agent Protocol).

The interaction characteristics for each protocol category reveal distinct capabilities and use cases. Context-oriented protocols focus on tool integration and resource access, while peer-oriented protocols emphasize delegation, artifact exchange, and negotiation. General-purpose protocols provide broad applicability across domains, whereas domain-specific protocols offer specialized capabilities for particular application areas. Table 13.1 summarizes these interaction characteristics, providing a concise reference for protocol selection decisions.

Applying this taxonomy to CAV systems reveals that vehicular environments require protocols that can handle both deterministic service access and intelligent peer coordination. The unique requirements of vehicular environments, including low-latency constraints suitable for safety-critical operations and heterogeneous system integration, favor protocols that offer several key capabilities: type safety and reliability for deterministic service access, semantic richness (the ability to express complex relationships and meanings) for complex peer-to-peer coordination and task delegation, extensibility (the ability to accommodate new features and platforms) to accommodate diverse vehicle platforms and infrastructure, and performance to meet low-latency decision-making requirements typical of CAV applications.

Based on this analysis, several protocols emerge as particularly well-suited for CAV applications. MCP [1] represents the context-oriented, general-purpose quadrant, providing standardized interfaces for primitive discovery and invocation. MCP is well-suited for connecting agents to deterministic services, while maintaining type safety and extensibility.

Table 13.1 Interaction characteristics of agent protocol categories within the two-dimensional taxonomy framework.

Protocol scope	Context-oriented	Peer-oriented
General-purpose	Stateless, service access, function invocation	Multi-agent, task-based, collaborative planning
Domain-specific	Tool/resource integration, sensor data access	Delegation, artifact exchange, peer negotiation

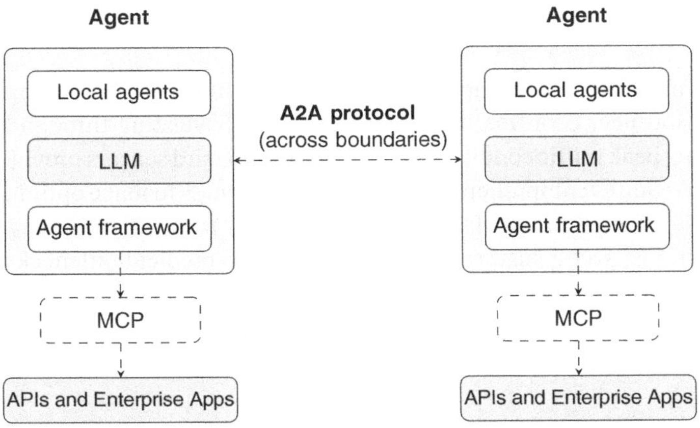

Figure 13.3 Visual overview of MCP vs. A2A architecture showing how MCP connects agents to external tools and services, while A2A enables direct agent-to-agent communication.

For peer-oriented communication, A2A [2] represents the peer-oriented, general-purpose quadrant, enabling sophisticated task-based communication between autonomous agents.

These protocols are not mutually exclusive; rather, they complement each other to provide comprehensive connectivity, as illustrated in Figure 13.3. MCP handles the vertical integration with tools and services, while A2A enables horizontal coordination among peer agents.

13.3 Protocol Evolution and Development

Having established the taxonomy framework in Section 13.2, we now examine the historical development of agent protocols to provide crucial context for selecting appropriate protocols in CAV applications. The evolution of these protocols reflects broader trends in computing, from early message-passing systems to AI-native architectures.

The development of agent protocols can be traced through three distinct phases that reflect the evolution of computing paradigms and AI capabilities.

Early Phase: Initial protocols focused primarily on simple message passing between agents, with limited support for complex interactions. Protocols such as Foundation for Intelligent Physical Agents—Agent Communication Language (FIPA-ACL) provided basic communication primitives but lacked the sophistication required for AI agents [7]. This period was characterized by centralized architectures and relatively simple interaction patterns.

Integration Phase: The rise of cloud computing and microservices architecture led to the development of REST-based protocols and service-oriented architectures. This period saw the emergence of protocols designed for tool integration and external service access, laying the groundwork for context-oriented communication. The shift toward distributed systems and API-first design principles influenced protocol development significantly.

AI-native Phase: The integration of large language models and generative AI systems has driven the development of protocols specifically designed for AI agent communication [3]. This phase has seen the emergence of protocols that support natural language processing, semantic understanding, and dynamic tool discovery. The focus has shifted toward protocols that can handle the unique requirements of LLM-based systems, including context management, function calling, and multi-modal interactions.

13.4 Protocol Examples and CAV Applications

The theoretical taxonomy becomes meaningful when applied to real-world vehicular scenarios that require both external data access and peer coordination. Consider a CAV navigating through a complex highway interchange during peak traffic conditions. The vehicle's onboard sensors provide local perception data, but the agent requires comprehensive situational awareness to make optimal routing decisions. Through context-oriented protocols, the agent can query traffic management systems for current congestion patterns, access historical traffic flow data to predict bottlenecks, and retrieve real-time incident reports from transportation authorities. Simultaneously, the vehicle must coordinate with nearby vehicles through peer-oriented protocols to negotiate lane changes, share sensor data for improved perception, and collectively optimize traffic flow.

13.4.1 Context-oriented Protocol Applications

Context-oriented protocols enable CAV agents to access external services and tools, providing the foundation for intelligent decision-making based on real-time data. Weather services allow agents to query precipitation intensity and adjust driving behavior for rain, snow, or fog conditions. Traffic information from mapping services helps optimize routing decisions, while infrastructure status updates provide information about bridge conditions, road closures, and construction zones from transportation authorities.

Function calling frameworks represent a crucial bridge between GenAI systems and vehicular protocols [8]. When an LLM processes natural language instructions such as "find the safest route home considering the weather," it must translate this into specific function calls such as `get_traffic_conditions()`, `calculate_optimal_route()`, or `assess_weather_impact()`. MCP [1] provides a standardized interface for this translation, allowing the LLM to discover and invoke appropriate functions without needing to understand the underlying protocol details. Custom function libraries provide vehicle-specific functions for sensor data processing, actuator control, and system monitoring.

13.4.2 Peer-oriented Protocol Applications

Peer-oriented protocols facilitate coordination between vehicles that may have different sensor capabilities or clearer views of the environment. These protocols allow vehicles to share sensor data, negotiate right-of-way, and collectively optimize traffic flow. Key applications include the following:

Platoon Formation: Vehicles coordinate to form energy-efficient groups that share speed and spacing information. The lead vehicle communicates its intentions to following vehicles, while each vehicle maintains its own safety margins and can break away when necessary. This coordination requires protocols that support low-latency communication suitable for safety-critical operations and graceful degradation when individual vehicles need to exit the formation.

Intersection Management: Multiple vehicles negotiate right-of-way through direct communication rather than relying solely on traditional traffic lights. A vehicle approaching an intersection can broadcast its arrival time and intended path, allowing other vehicles to calculate safe crossing sequences. This distributed approach can be more efficient than centralized traffic control, especially when vehicles have different priorities or capabilities.

Emergency Response: When an ambulance needs to clear a path through traffic, vehicles must quickly coordinate to create safe passage routes. This requires protocols that can handle high-priority message routing, real-time position updates, and rapid consensus formation among multiple vehicles.

13.4.3 GenAI Integration Requirements

The integration of GenAI systems with these protocols introduces unique requirements that rule-based vehicular systems do not face. Natural language processing capabilities mean that protocols must support conversational interfaces where humans can interact with vehicles using natural language. A passenger might ask "Why are we taking this route?" and the system must be able to explain its reasoning by accessing relevant data through context-oriented protocols and coordinating with other vehicles through peer-oriented protocols.

Context management becomes particularly important in GenAI applications, where conversations and interactions may span extended periods. A vehicle might engage in a multi-turn conversation with a passenger about route planning, requiring the system to maintain context about previous decisions, current traffic conditions, and future plans. Protocols must support this context persistence while also handling the dynamic nature of vehicular environments where conditions can change rapidly.

Dynamic tool discovery represents another key requirement for GenAI systems in CAV applications. As vehicles encounter new situations or enter new environments, they may need to access previously unknown services or capabilities. A vehicle entering a construction zone might need to discover new protocols for communicating with construction equipment or accessing temporary traffic management systems. The ability to dynamically discover and integrate new tools is essential for robust operation in diverse environments.

CAV systems can employ hybrid approaches that combine both context-oriented and peer-oriented protocols, with the choice between protocols depending on the specific requirements of each interaction. External data access favors context-oriented protocols, while low-latency coordination favors peer-oriented protocols. The integration of multiple protocols and intelligent protocol switching based on current conditions and requirements is explored in detail in Chapter 16.

13.5 Summary

This chapter established a comprehensive taxonomy of agent communication protocols operating at the semantic interaction layer, organized along two primary dimensions: interaction orientation (context-oriented vs. peer-oriented) and protocol scope (general-purpose vs. domain-specific). The taxonomy distinguishes agent protocols from underlying messaging mechanisms (HTTP, MQTT, gRPC) and communication substrates (V2X), clarifying that agent protocols define semantic intent and coordination logic while abstracting over lower-layer transport mechanisms.

The historical evolution of agent protocols reveals three distinct phases: early message-passing protocols, integration-phase REST-based architectures, and AI-native protocols designed for LLM-based systems. The taxonomy reveals that CAV systems require protocols that can handle both deterministic service access and intelligent peer coordination. Two protocols emerge as particularly well-suited: MCP for context-oriented tool integration and A2A for peer-oriented collaboration. These protocols complement each other, with MCP handling vertical integration with tools and services, while A2A enables horizontal coordination among peer agents.

Practical examples demonstrated how context-oriented protocols enable external service integration, while peer-oriented protocols facilitate vehicle coordination for platoon formation, intersection management, and emergency response. The integration of GenAI systems introduces additional requirements for natural language processing, context management, and dynamic tool discovery.

Chapters 14 and 15 build on this conceptual framework by presenting concrete instantiations of context-oriented and peer-oriented agent communication protocols, respectively.

References

1 Anthropic. Model context protocol (MCP), 2024. https://modelcontextprotocol.io/. Protocol specification for connecting AI applications to external tools and data sources.

2 Google DeepMind. Agent-to-agent protocol (A2A), 2024. https://github.com/google-deepmind/a2a-protocol. Protocol for inter-agent communication and collaboration in multi-agent systems.

3 Yingxuan Yang, Huacan Chai, Yuanyi Song, Siyuan Qi, Muning Wen, Ning Li, Junwei Liao, Haoyi Hu, Jianghao Lin, Gaowei Chang, et al. A survey of AI agent protocols. *arXiv preprint arXiv:2504.16736*, 2024.

4 Edwin Hutchins. *Cognition in the Wild*. MIT Press, Cambridge, MA, 1995. Foundational work on distributed cognition theory.

5 Benedikt Brecht, Thomas Nowey, Christian Mletzko, Max Mühlhäuser, and Christina Pöpper. A security credential management system for V2X communications. *IEEE Transactions on Intelligent Transportation Systems*, 19(12):3850–3871, 2018.

6 U.S. Department of Transportation. Connected vehicle deployment technical assistance: security credential management system (SCMS) technical primer. Technical report, Federal Highway Administration, 2019. Technical primer on SCMS architecture and implementation.

7 Foundation for Intelligent Physical Agents. FIPA ACL message structure specification. Technical report, FIPA, 2002. https://jmvidal.cse.sc.edu/lib/fipa_acl_messag_struc_specif.html. Standard specification for agent communication language.

8 Shishir G. Patil, Tianjun Zhang, Xin Wang, and Joseph E. Gonzalez. Gorilla: large language model connected with massive APIs. *arXiv preprint arXiv:2305.15334*, 2023. Foundational work on function calling (tool calling) for agentic AI systems.

14

Model Context Protocol for Agentic Systems

14.1 Introduction

While the previous chapter (Chapter 13) established a taxonomy of agent communication protocols and identified MCP as a context-oriented, general-purpose protocol, this chapter provides a comprehensive examination of MCP's architecture, implementation, and application in CAV systems. In the emerging landscape of CAVs, intelligent agents are increasingly tasked with processing complex, multimodal information and responding to dynamic, uncertain environments. As generative AI systems, particularly LLMs, are integrated into these agents, their utility depends not only on reasoning ability but also on their capacity to interact with external tools and services.

Consider a critical scenario: a vehicle approaching an urban intersection obscured by a large truck. The onboard sensors fail to detect a pedestrian hidden behind the obstruction. The vehicle agent must reason about traffic light signals, seek assistance from nearby vehicles with clearer views, and assess risks associated with occluded regions. Such behavior is not achievable by isolated models or standalone perception systems. Instead, it requires seamless connectivity to external services and tools that can provide timely context for safety-critical operations.

The challenge of enabling such connectivity has led to the development of numerous protocols and standards. However, not all protocols are equally suited for the unique requirements of CAV systems, which demand low-latency performance, safety guarantees, and seamless integration with deterministic services.

This chapter addresses this challenge by focusing on MCP, which represents the most promising approach for context-oriented tool integration in CAV applications. As established in Chapter 13, MCP represents the *Context-Oriented, General-Purpose* quadrant of the agent protocol taxonomy, providing standardized interfaces for primitive discovery and invocation. MCP operationalizes the notion of distributed cognition, enabling bidirectional context flow between AI models and external systems. To understand how MCP addresses these challenges, we must first examine the fundamental context problem that CAV systems face.

14.2 Context Provision in CAV Systems

The complexity of vehicular environments reveals fundamental challenges that make context provision critical for safe and efficient operation. **Context provision** refers to the process of supplying AI models with relevant, timely information and capabilities from external sources to enhance their reasoning and decision-making abilities.

14.2.1 The Context Problem

The fundamental challenge in AI systems is providing models with the right context at the right time to enable effective reasoning and decision-making. While large language models excel at reasoning within their training knowledge, they face significant limitations when dealing with real-world scenarios that require access to external information, real-time data, or dynamic system states. Models are trained on historical data and cannot access current information such as live traffic conditions, weather updates, or system status without external interfaces. They need access to changing system conditions, user preferences, and environmental factors to make contextually appropriate decisions. Additionally, they must be able to execute actions through interfaces to external systems and services.

Several approaches have emerged to address this context problem. RAG has proven effective for enhancing model responses with retrieved knowledge from static knowledge bases, enabling models to access relevant information from documents, databases, and knowledge graphs [1]. Function calling (also known as tool calling) has enabled models to execute specific functions and tools, allowing them to interact with external APIs and perform actions beyond text generation [2]. However, these approaches primarily address specific aspects of the context problem: RAG focuses on knowledge retrieval from static sources, while function calling typically involves predefined, hardcoded integrations with specific tools and APIs.

In CAV applications, the context problem becomes particularly acute. CAV systems require interaction with live sensors, APIs, and control systems that provide real-time data and can execute actions, capabilities that go beyond RAG's knowledge retrieval focus. Unlike RAG's unidirectional information flow, CAV systems need bidirectional communication where AI models can both receive context from external systems and send commands or requests back to those systems. CAV applications require dynamic discovery of available tools and capabilities at runtime, rather than relying on predefined, hardcoded function integrations. CAV systems must coordinate context from multiple heterogeneous sources (sensors, APIs, databases, user interfaces) simultaneously, requiring a more sophisticated integration approach than RAG's single knowledge base model. The diverse nature of CAV system components requires a standardized protocol for integration, ensuring interoperability across different vendors and system types.

These challenges motivate the need for a more comprehensive approach to context provision that extends beyond knowledge retrieval to encompass dynamic system integration and tool execution.

14.2.2 MCP as a Solution

MCP emerges as a comprehensive solution that addresses these challenges through a standardized approach to context integration. MCP represents a significant advancement in addressing the remaining challenges of dynamic system integration in AI systems. Introduced by Anthropic in 2024 as part of the Claude Desktop ecosystem, MCP has evolved into an open protocol specification that provides a standardized approach to bidirectional context provision and tool execution for large language models and AI agents [3].

MCP complements existing approaches by providing a standardized framework for dynamic tool discovery, real-time system interaction, and bidirectional communication between AI models and external services. MCP addresses the gaps in existing approaches by enabling runtime discovery of capabilities, standardized integration protocols, and comprehensive context coordination. This makes MCP particularly well-suited for CAV applications where models must dynamically discover and interact with diverse sensors, control systems, and real-time data sources.

The evolution of context provision approaches can be summarized through three distinct paradigms. RAG addresses the question "What do I know?" by providing static knowledge retrieval from documents and databases. Function calling answers "What can I do?" through predefined tools and API integrations. MCP extends this progression by addressing "What's available and how do I use it?" through dynamic discovery and standardized integration. This progression reflects the increasing sophistication of AI systems, from simple knowledge retrieval to predefined tool execution, and finally to dynamic capability discovery and standardized system integration.

MCP solves the context problem through a systematic approach that provides three distinct context types: static context through resources, dynamic context through tools, and guided context through prompts. The protocol implements a four-step discovery and selection mechanism (discovery, selection, invocation, integration) that enables AI models to dynamically find and use the most appropriate context sources without requiring pre-programmed knowledge of specific data sources or tools. This makes MCP particularly well-suited for CAV applications where AI models must integrate diverse, real-time information sources to make safety-critical decisions. The protocol enables standardized integration across diverse hardware platforms, sensor configurations, and communication protocols.

Example 14.1 MCP as a Universal Integration Protocol

Before MCP, integrating AI systems with external services required building custom connectors for each combination of applications and services, resulting in an $M \times N$ integration problem where M applications needed to connect to N services. MCP solves this challenge by providing a universal "USB-C port" for AI integrations. Just as USB-C standardized device connectivity across manufacturers and device types, MCP standardizes how AI applications discover and interact with external tools, resources, and services. This unified approach eliminates the need for custom integration code, enabling AI systems to dynamically discover and use capabilities from any MCP-compatible server through a single, standardized protocol interface.

The MCP specification does not define official standard URIs (uniform resource identifiers), tools, or primitives specifically for V2X or CAV systems. The examples presented in this chapter are hypothetical illustrations of how MCP could be applied to CAV scenarios, demonstrating the protocol's potential rather than documenting existing standards.

14.3 MCP Architecture

The effectiveness of MCP depends on its architectural design, which enables seamless context flow between AI applications and external systems. MCP employs a client-server architecture designed to facilitate bidirectional context flow between AI applications and external systems. The architecture consists of three core components that work together to enable seamless context provision:

Host: The LLM application that initiates connections and orchestrates the overall AI experience. The host is the user-facing application (like Claude Desktop) that determines when and how the model should interact with external context sources. The host contains one or more clients that manage the actual connections.

Client: Connectors embedded within the host application that establish and manage transport connections with MCP servers. Clients handle the low-level communication, message routing,

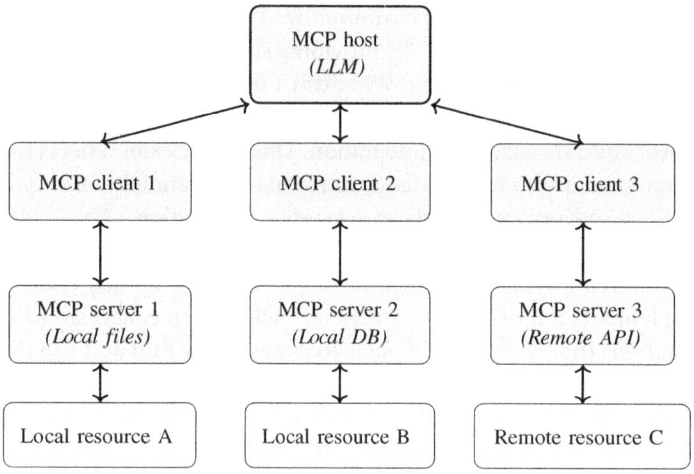

Figure 14.1 Conceptual architecture of an MCP-enabled AI agent system showing the relationship between host, clients, and servers.

and protocol implementation. They discover available context sources, route context to the AI model, and manage the bidirectional flow of information between host and servers.

Server: Services that provide specialized context and capabilities, exposing resources, tools, and prompts through standardized MCP primitives. Servers can be local processes or remote services that operate independently, each offering specific domain expertise or data access capabilities.

The interaction between these three components creates a flexible architecture that enables dynamic context discovery and integration. Figure 14.1 illustrates how these architectural components work together to facilitate context flow from external sources to AI models.

14.3.1 Architecture Benefits and Design Principles

The MCP architecture provides several key benefits that make it well-suited for CAV applications and other distributed AI systems. **Modularity** enables independent evolution of transport and data layers, allowing CAV systems to upgrade communication protocols without changing primitive definitions. **Scalability** supports both local and distributed deployments, from embedded devices to cloud-based services, enabling vehicles to run local MCP servers for immediate sensor data while connecting to cloud servers for traffic information. **Interoperability** through standardized primitive interfaces ensures seamless integration across different manufacturers and vendors, critical for CAV systems where vehicles must communicate with diverse infrastructure. **Flexibility** through bidirectional primitives enables rich interactions where CAV systems can both receive context from external sources and provide their own capabilities to other systems. **Security** principles support fine-grained access control, data isolation, and secure communication patterns essential for safety-critical applications.

These architectural benefits enable the context flow mechanisms described below.

14.3.2 How Context Flows Through the Architecture

The interaction between these components follows a systematic process that enables AI models to access external information and capabilities. The MCP architecture enables context to flow seamlessly from external sources to AI models through a well-defined four-step process:

1. **Context Discovery:** The client discovers available context sources (resources, tools, prompts) from connected servers through the data layer's primitive schemas and discovery mechanisms, with the transport layer handling the actual message exchange.
2. **Context Selection:** The host evaluates available context sources based on current needs and user intent, reasoning about primitive capabilities and selecting the most appropriate ones.
3. **Context Execution:** The client invokes selected context sources to gather relevant information, with the transport layer managing message delivery and the data layer defining the structure and semantics of the retrieved context.
4. **Context Integration:** The host incorporates retrieved context into the AI model's reasoning process, relying on the structured data formats defined by the data layer.

This systematic flow ensures that AI models receive the most relevant and up-to-date context for their decision-making processes. The process is implemented through two architectural layers: the transport layer handles the communication mechanics, while the data layer manages the content and structure of the exchanged information. Figure 14.2 visualizes this four-step process, showing how context flows from discovery through integration, with feedback loops enabling continuous adaptation.

14.3.3 Architectural Layers

The MCP architecture implements the context flow through two distinct layers that manage different aspects of the communication. The **transport layer** manages the low-level communication protocols and message delivery, ensuring reliable connection management. The **data layer** defines the structure and semantics of context information, including schemas for primitive types (resources, tools, prompts, sampling, roots, elicitation), their discovery, invocation, and integration. This modular design allows for dynamic capability extension, distributed reasoning, and fault isolation across both local and remote MCP servers.

14.3.4 Transport Layer

The transport layer establishes the communication foundation for MCP, defining how messages flow between MCP clients and servers. This layer handles low-level communication protocols, message serialization, and connection management using JSON-RPC 2.0 (JavaScript Object Notation Remote Procedure Call version 2.0) as the standardized message format.

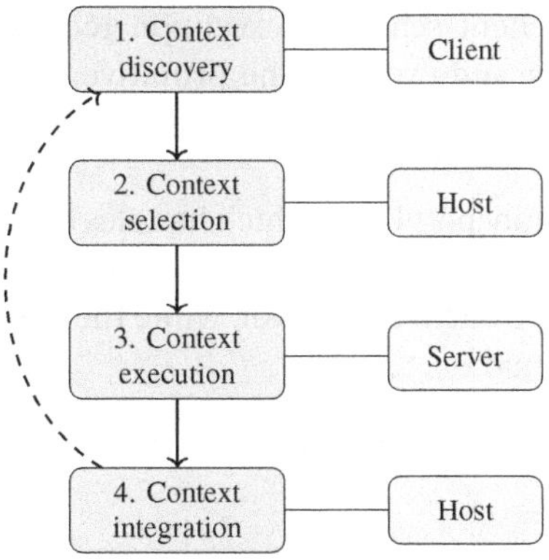

Figure 14.2 Four-step context flow process showing the sequence of operations and components involved in each step.

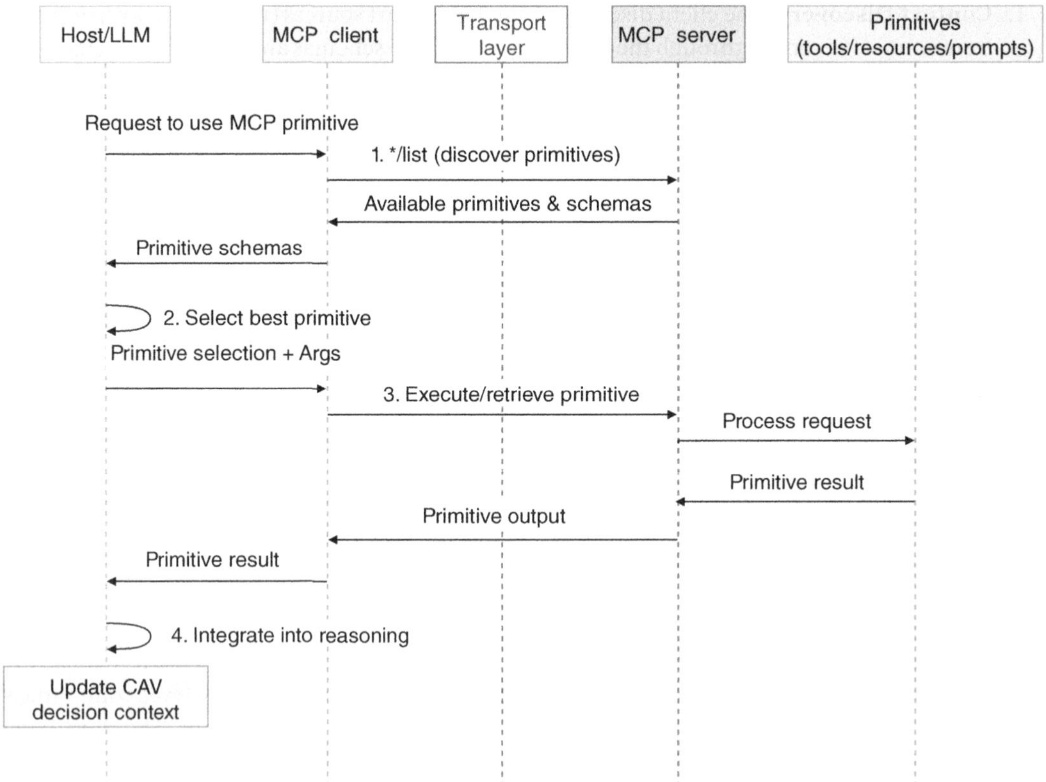

Figure 14.3 MCP primitive lifecycle flow showing the complete process from discovery to integration.

The MCP primitive lifecycle flow is illustrated in Figure 14.3, which demonstrates the complete process from discovery to integration. This sequence diagram shows how the Host/LLM, MCP client, transport layer, and MCP server interact to enable dynamic primitive discovery and execution.

The sequence diagram illustrates the detailed interaction flow: the Host/LLM requests access to MCP primitives through its MCP client, which sends discovery requests (e.g., `tools/list`) to the MCP Server. The server responds with available primitives and their schemas. The Host/LLM performs internal reasoning to select the most appropriate primitive and sends the selected primitive with required arguments to the MCP client. The client sends an execution request (e.g., `tools/call`) to the server, which processes the request and executes the primitive. The results are returned through the client to the Host/LLM, which integrates them into its reasoning process.

This lifecycle demonstrates the clear separation between the transport layer, which handles communication mechanics, and the data layer, which manages primitive semantics and execution. The transport layer facilitates all message exchange between the MCP client and server, while the data layer defines the structure and meaning of the exchanged information.

14.3.5 Transport Mechanisms

MCP uses JSON-RPC to encode messages and defines two standard transport mechanisms for client-server communication:

STDIO Transport: Uses standard input/output streams for direct process-to-process communication within the same runtime environment. The client launches the MCP server as a subprocess, and they communicate through stdin/stdout streams. This transport is ideal for embedded systems, local services, and scenarios requiring minimal overhead and security isolation.

Streamable HTTP Transport: Enables the server to operate as an independent process that can handle multiple client connections across network boundaries. This transport uses HTTP requests and server-sent events (SSE) to support both basic functionality and advanced features like streaming responses and server-initiated notifications. This transport enables distributed deployments and supports multiple concurrent client connections.

MCP is designed to be transport-agnostic, meaning that clients and servers can implement custom transport mechanisms to suit their specific needs. The protocol can be implemented over any communication channel that supports bidirectional message exchange, as long as the JSON-RPC message format and lifecycle requirements are preserved.

14.3.6 Data Layer

The data layer defines the actual content and structure of information exchanged through the MCP transport layer. This layer specifies the primitive types, their schemas, and the semantic meaning of the data that flows between MCP clients and servers. While the transport layer handles *how* messages are delivered, the data layer defines *what* those messages contain and how they should be interpreted.

The data layer is built around the concept of **primitives**, well-defined, typed interfaces that represent specific capabilities or data sources. These primitives provide a standardized way for AI agents to discover, understand, and interact with external systems, regardless of the underlying implementation details. The primitive system enables dynamic capability discovery, where agents can explore available functionality at runtime and adapt their behavior accordingly.

MCP defines multiple primitive types that serve distinct functional roles in agent-context interaction. These primitives are organized into two main categories: server features that provide context to AI models (tools, resources, prompts) and client features that provide servers with access to client environment and capabilities (roots, sampling, elicitation). The following sections provide detailed descriptions, characteristics, and practical examples of these primitives.

14.4 Server Features

MCP servers provide context to AI models through three distinct primitive types: tools (executable functions), resources (static data), and prompts (structured templates). While all three serve important roles, **tools** represent the most critical primitive for CAV applications, as they enable AI models to access real-time information and perform actions that create new context. This section focuses primarily on tools, with brief descriptions of resources and prompts. For comprehensive coverage of function calling and tool use concepts, see Chapter 9.

14.4.1 Tools

Tools are executable functions that generate context through actions, queries, or computations. They bridge the gap between AI reasoning and external systems, enabling models to access real-time information and perform operations beyond their training knowledge. In CAV systems, tools enable critical capabilities: querying live traffic conditions, retrieving weather updates, performing route calculations, executing vehicle diagnostics, and reporting incidents to infrastructure systems.

The power of tools lies in their ability to be dynamically discovered and autonomously invoked by AI models. When a vehicle agent needs to check traffic conditions, it doesn't require pre-programmed knowledge of specific APIs or endpoints. Instead, the MCP client discovers available tools from connected servers, the LLM evaluates which tool is most appropriate for the current query, and the system executes the tool with parameters extracted from the natural language request. This dynamic discovery and execution model enables CAV systems to adapt to new services and capabilities without code changes.

Consider a practical scenario: a vehicle approaching an intersection needs to determine if it's safe to proceed. The agent might discover and invoke multiple tools in sequence: first querying traffic light status, then checking pedestrian detection data, and finally accessing weather conditions. Each tool call provides context that informs the next decision, creating a chain of reasoning grounded in real-world data.

The following example demonstrates how a tool primitive for accident reporting in CAV systems might be structured:

Listing 14.1 Tool Primitive Schema for Accident Reporting

```
{
  "name": "reportAccident",  // Tool identifier for MCP calls
  "description": "Report traffic accident to infrastructure",  // Human-readable
      description
  "inputSchema": {  // JSON Schema defining required parameters
    "properties": {
      "location": {"type": "string", "description": "GPS coordinates"},  //
          Required string parameter
      "severity": {"type": "string", "enum": ["minor", "moderate", "severe", "
          fatal"]}  // Enum with predefined values
    } } }
```

Tool primitives are defined by their name (unique identifier), description (used by LLMs for selection), and inputSchema (JSON schema specifying required parameters and validation rules). When an LLM encounters a user query like "Report a severe accident at coordinates 37.7749, −122.4194," and it can match this to the `reportAccident` tool, extract the location and severity parameters, and execute the tool call through the MCP protocol. The mathematical framework for this discovery and selection process is formalized in Section 14.6, and the practical implementation details are demonstrated in Section 14.7.

14.4.2 Resources

Resources provide static, preexisting data that serves as background information or reference material for AI reasoning. They are read-only, with each resource identified by a unique URI for precise access. Resources contain well-defined data that can be parsed and understood by AI models, and can include metadata about modification times and versions. In vehicular systems, resources might include road maps, traffic regulations, vehicle specifications, or historical traffic patterns that provide essential background context for navigation and decision-making.

The following example shows how a resource primitive might expose road hazard information for CAV systems:

Listing 14.2 Resource Primitive Schema for Road Hazards Data

```
{
  "uri": "file:///road_segments/segment_123/hazards",  // Unique identifier for the resource
  "name": "road-hazards",  // Short name for reference
  "title": "Road Hazards Data",  // Human-readable title
  "description": "Road segment data: hazards, conditions, incidents",  // Detailed description
  "mimeType": "application/json"  // Expected data format
}
```

Resources are accessed by URI rather than being invoked with parameters. When an AI model needs background information about a specific road segment, it can request the resource directly using its URI, enabling efficient access to static reference data without the overhead of function calls.

14.4.3 Prompts

Prompts are structured templates that direct how context should be used, providing guidance for AI behavior and response patterns. They help ensure consistent and appropriate use of available context. Prompts are typically invoked by user choice or system configuration, provide structured formats for consistent interactions, direct how AI models should use available context, and ensure consistent communication patterns across systems. In vehicular systems, prompts might include emergency notification templates, status update formats, or user interaction patterns that ensure consistent and appropriate communication in safety-critical scenarios.

The following example demonstrates how a prompt primitive for EV route planning scenarios might be structured:

Listing 14.3 Prompt Primitive Schema for EV Route Planning

```
{
  "name": "ev-route-planning",  // Prompt template identifier
  "title": "EV Route Planning Protocol",  // Display title
  "description": "Plan route with EV charging stations and nearby facilities",  // Purpose description
  "arguments": [  // Required parameters for the prompt
    { "name": "destination", "type": "string", "description": "Final destination address" },
    { "name": "current_charge", "type": "number", "description": "Current battery level (%)" },
    { "name": "charging_preference", "type": "string", "enum": ["fast_charging", "destination_charging", "opportunity_charging"] }
  ] }
```

Prompts provide reusable interaction templates that ensure consistent communication patterns. When a user initiates an EV route planning session, the system can invoke the appropriate prompt template with the required parameters, ensuring that the AI model follows a standardized interaction pattern that has been validated for safety and effectiveness.

14.5 Client Features

The bidirectional nature of MCP becomes apparent through client features, which enable servers to **request** specific capabilities from the client's environment. MCP clients offer three key primitives that servers can request to access specific capabilities from the client environment. These primitives solve the fundamental problem that servers cannot directly access client resources, LLM capabilities, or user interactions without explicit client support.

14.5.1 Roots

The roots primitive defines the filesystem boundaries and access scope that servers can utilize. This provides controlled access to local file systems while maintaining security boundaries and preventing unauthorized access to sensitive data. Servers can access specific directories or file types as defined by the client's root configuration, enabling secure file operations and data access. Roots establish the initial context for server interactions by defining what parts of the client's filesystem are accessible. In CAV systems, roots might define access to vehicle configuration files, sensor data logs, or user preference directories while restricting access to critical system files, ensuring that external servers cannot compromise vehicle safety systems.

Roots Effectiveness: The effectiveness of the roots primitive depends significantly on the deployment scenario. In **local systems** using STDIO transport, roots provide real security boundaries where servers can directly access files within the defined scope. However, in **remote systems** using HTTP transport, roots serve primarily as guidance mechanisms since servers cannot access client files directly and must rely on data transfer through the MCP protocol. The MCP specification treats roots as guidance rather than strict enforcement, meaning their effectiveness depends on server implementation and the transport mechanism used.

The following example demonstrates how a roots primitive could define filesystem boundaries for CAV systems, showing how clients might specify accessible directories and their purposes:

Listing 14.4 Roots Primitive for CAV Filesystem Boundaries

```
{
  "uri": "file:///vehicle/config",
  "name": "Vehicle Configuration"
}
```

Example 14.2 CAV Directory Structure Example

The roots primitive can define access to:

- `file:///vehicle/config` for vehicle settings and parameters
- `file:///vehicle/sensors` for sensor calibration and logs
- `file:///vehicle/maps` for local map data and routing

14.5.2 Sampling

The sampling primitive allows MCP servers to request LLM completions from the client. This enables servers to leverage the client's language model capabilities for tasks such as text generation, analysis, or reasoning. Servers might use sampling to generate summaries of retrieved data, analyze user queries, or provide natural language responses to complex requests. In vehicular systems, sampling could be used to generate natural language explanations of sensor data, create user-friendly status reports, or provide contextual analysis of traffic conditions.

The following example demonstrates how a sampling primitive could enable AI-assisted analysis of sensor data in CAV systems, showing how servers might request LLM completions through the client:

Listing 14.5 Sampling Primitive for AI-assisted Sensor Analysis

```
{
  "method": "sampling/create",  // MCP method for requesting LLM completions
  "params": {  // Parameters for the sampling request
    "messages": [{  // Conversation messages for the LLM
      "role": "user",  // Message role (user, assistant, system)
      "content": "Analyze these sensor readings and identify potential hazards:
          \n[Camera, LiDAR, Radar data]\nVehicle speed: 65 mph, weather: rainy"
          // Detailed prompt with context
    }],
    "systemPrompt": "You are a CAV safety expert analyzing sensor data for
        hazards", // System-level instructions
    "maxTokens": 1000  // Maximum response length
  } }
```

14.5.3 Elicitation

The elicitation primitive enables servers to request user input or confirmation for specific actions. This provides a mechanism for human-in-the-loop decision-making, user consent, and interactive workflows where servers need additional information from users. Servers can request user confirmation for sensitive operations, gather additional information needed for task completion, or provide interactive user experiences. Elicitation facilitates servers in gathering specific information from clients when needed. In vehicular systems, elicitation might be used to request driver confirmation for route changes, gather user preferences for comfort settings, or provide interactive safety alerts that require user acknowledgment.

The following example demonstrates how an elicitation primitive could enable user input for EV charging preferences in CAV systems, showing how servers might request structured user input through the client:

Listing 14.6 Elicitation Primitive for EV Charging Preferences

```
{
  "method": "elicitation/create",
  "params": {
    "message": "Select EV charging preferences for your trip:",
    "requestedSchema": {
      "type": "object",
      "properties": {
        "chargingStrategy": {
          "type": "string",
          "title": "Charging Strategy",
          "description": "Preferred charging approach",
          "enum": ["minimize_stops", "premium_stations", "budget_friendly"],
          "enumNames": ["Minimize Stops", "Premium Stations", "Budget Friendly"]
        } },
      "required": ["chargingStrategy"]
  } } }
```

14.6 Mathematical Foundations

The MCP primitive discovery and execution process can be formalized mathematically to provide a rigorous theoretical foundation. This section develops the mathematical framework for understanding how primitives are discovered, selected, and integrated, with primary focus on tools as the most critical primitive for CAV applications. For comprehensive coverage of function calling concepts, architectures, and safety considerations, see Chapter 9.

The tool discovery and selection process follows a systematic mathematical model. The context provision process begins with a set of context requests $R = \{r_1, r_2, \dots, r_m\}$, where each request r_j specifies the required context type and quality parameters. The system maintains a set of available tools $S = \{s_1, s_2, \dots, s_n\}$, where each tool s_i provides context of type $t_i = \text{tool}$ as defined in Section 14.4. The selection function $f : R \times S \rightarrow S^*$ maps context requests to selected tools, where $S^* \subseteq S$ represents the tools chosen based on relevance and appropriateness. The integration function $g : S^* \times R \rightarrow C$ then combines selected tool results to produce integrated context C for the AI model.

To formalize the tool discovery process, let $P = \{p_1, p_2, \dots, p_n\}$ denote the complete set of known tool primitives, where each tool $p_i \in P$ is associated with a schema $S_i = (I_i, O_i, D_i, T_i)$. Here, I_i represents the typed set of input arguments (JSON schema), O_i represents the expected output format, D_i represents the natural language description, and $T_i = \text{tool}$ for all tools. At runtime, the MCP client queries the server via listing calls (e.g., `tools/list`) to discover available tools:

$$P_{\text{available}} = \{p_i \in P \mid \text{server\_supports}(p_i) \wedge T_i = \text{tool}\}, \tag{14.1}$$

where $\wedge$ denotes the logical AND operator, ensuring that only primitives both supported by the server and of type tool are included in the available set.

Given a natural language query q and current agent context c, the host LLM evaluates the relevance of each available tool. The selection process aims to find the tool p^* that maximizes the conditional probability of appropriateness:

$$p^* = \arg \max_{p_i \in P_{\text{available}}} \text{Prob}(p_i \mid q, c, S_i). \tag{14.2}$$

Once a tool p^* is selected, its parameters are extracted from the query based on the tool's input schema:

$$\text{args} = \text{extract\_args}(q, S_{p^*}). \tag{14.3}$$

The tool is then invoked by the MCP client:

$$\text{result} = f_{p^*}(\text{args}). \tag{14.4}$$

The result is integrated into the agent's reasoning context:

$$c_{\text{new}} = c \oplus \text{result}, \tag{14.5}$$

where $\oplus$ denotes context integration. This framework formalizes the tool discovery, selection, and execution process that enables dynamic capability access in CAV systems.

Advanced Note 14.1 Mathematical Framework for Resources and Prompts

The mathematical framework extends to other server primitives. For resources, the discovery process identifies resources by URI rather than through function calls:

$$R_{\text{available}} = \{r_i \in R \mid \text{server\_supports}(r_i) \wedge T_i = \text{resource}\}. \tag{14.6}$$

Resource access follows a direct retrieval pattern: given a resource URI u extracted from the query, the resource is retrieved as $\text{resource} = \text{read\_resource}(u)$.

For prompts, the discovery process identifies available prompt templates:

$$\text{Prompt}_{\text{available}} = \{p_i \in P \mid \text{server\_supports}(p_i) \wedge T_i = \text{prompt}\}. \tag{14.7}$$

Prompt invocation follows a template-based pattern where the prompt name n is extracted from the query, and the prompt template is retrieved with arguments: $\text{prompt} = \text{get\_prompt}(n, \text{args})$.

The mathematical framework for resources and prompts follows similar patterns to tools but with type-specific access mechanisms that reflect their distinct roles in context provision.

14.7 Practical Implementation

The mathematical framework described above translates directly into working code. This section demonstrates how to build MCP servers and clients using the FastMCP framework, showing how the mathematical models map to practical implementations. For comprehensive coverage of function calling architectures, safety validation, and tool design patterns, see Chapter 9.

14.7.1 Server Implementation

Building an MCP server begins with creating a server instance and registering primitives. The FastMCP framework provides decorators for registering tools, resources, and prompts. The following example demonstrates server setup and tool registration, showing how the mathematical framework maps to code:

Listing 14.7 Tool Registration Using FastMCP Decorator

```
from mcp.server.fastmcp import FastMCP
from typing import Dict, Any
from datetime import datetime

class MinimalMCPServer:
    def __init__(self):
        self.mcp = FastMCP(name="Minimal-MCP-Server")
        self._register_tools()

    def _register_tools(self):
        @self.mcp.tool()
        def report_road_alert(segment_id: str, user_type: str, reason: str) ->
            Dict[str, Any]:
```

```
            """Report a road user (e.g., cyclist, pedestrian) that requires
                driver/system attention."""
            return {
                "status": "ok",
                "message": f"{user_type} reported in {segment_id} - reason: {
                    reason}",
                "timestamp": datetime.now().isoformat()
            }

    def run(self):
        self.mcp.run(transport="stdio")
```

The server implementation demonstrates several key concepts from the mathematical framework. The `@self.mcp.tool()` decorator automatically registers the function as an MCP tool primitive, making it discoverable through the `tools/list` method. The function signature with type hints (e.g., `segment_id: str`) generates the required JSON schema for the tool's input parameters, which is used for validation and LLM understanding. The docstring provides the tool description that LLMs use for selection, implementing the selection process described in Equation (14.2). When a client queries `tools/list`, this tool becomes part of the discovered set $P_{\text{available}}$ (Equation 14.1).

The `run()` method starts the MCP server using STDIO transport, which enables communication through standard input and output streams. This transport method is ideal for local communication where the server runs as a subprocess of the client. The server listens for incoming MCP protocol messages, processes tool discovery requests, and executes tool calls from clients.

14.7.2 Client Implementation

On the client side, the workflow follows the mathematical model described in Section 14.6. The client discovers tools, converts them to the LLM's expected format, and processes queries:

Listing 14.8 Client-side Tool Discovery and Execution

```
import json
import ollama

# Discover available tools (implements Equation~\ref{eq:available_primitives})
response = await self.session.list_tools()
tools = response.tools

# Convert to LLM format
ollama_tools = [{"name": t.name, "description": t.description, "parameters": t.
    inputSchema}
                for t in tools]

# Process query - LLM performs selection (Equation~\ref{eq:primitive_selection})
response = ollama.chat(model="qwen2.5", messages=[{"role": "user", "content":
    user_query}],
                       tools=ollama_tools)

# Execute tool calls (implements Equations~\ref{eq:parameter_extraction} and~\ref
    {eq:primitive_invocation})
if response.get('message', {}).get('tool_calls'):
    for tool_call in response['message']['tool_calls']:
        result = await self.session.call_tool(
            name=tool_call['function']['name'],
```

```
            arguments=json.loads(tool_call['function']['arguments'])
        )
        # Integrate result (Equation~\ref{eq:context_integration})
        if result.content and len(result.content) > 0:
            print(f"Tool result: {result.content[0].text}")
        else:
            print(f"Tool executed successfully")
```

The client implementation demonstrates the complete MCP lifecycle. The `list_tools()` method sends a request to the MCP server to retrieve all available tools, implementing the discovery process from Equation (14.1). The server responds with a list of tool definitions, each containing metadata about the tool's name, description, and input schema. Since different LLM providers may expect different tool formats, the client converts the MCP tool format into the specific format required by the LLM (in this case, Ollama format).

The query processing constructs a conversation with the LLM, which performs the selection process from Equation (14.2). If the response contains tool calls, the client executes them via the MCP session, implementing the parameter extraction (Equation 14.3) and invocation (Equation 14.4) processes. The tool results are then integrated into the reasoning context (Equation 14.5), completing the full cycle from discovery through integration.

14.8 MCP in CAV Systems

The integration of MCP's server and client features enables sophisticated context management in CAV applications, but deployment in safety-critical vehicular systems requires careful consideration of security, limitations, and practical challenges.

14.8.1 CAV Application Scenarios

This context management is realized through two distinct deployment models: on-vehicle MCP servers that provide local, low-latency access to vehicle-specific resources, and infrastructure/cloud MCP servers that provide federated, region-scale information for planning and coordination.

On-vehicle MCP servers run inside the vehicle and provide fast, private access to local tools, configurations, sensor data, and received V2X information. These servers enable immediate access to sensor readings from cameras and LiDAR, locally received V2X context from nearby vehicles and roadside units, vehicle configuration and diagnostics, ADAS tools for vehicle control such as speed adjustment and ACC toggling, and emergency response prompts. This local deployment ensures minimal latency for safety-critical operations while maintaining data privacy within the vehicle.

Infrastructure/Cloud MCP servers operate in infrastructure nodes or mobility cloud platforms, exposing federated, region-scale information useful for planning, coordination, and risk reasoning. These servers provide access to road hazard and surface data for specific road segments, traffic and weather APIs for route planning, infrastructure perception data from roadside units, aggregated V2X context from vehicle fleets, and shared map and navigation resources, including intersection signal timing. This remote deployment enables vehicles to access shared, dynamic context that extends beyond their immediate sensor range.

A vehicle approaching an urban intersection during rush hour can use MCP tools from both local and remote servers to retrieve live traffic conditions, alternative route suggestions, and estimated delays from multiple sources, including traffic management cameras, incident reporting systems, and congestion data services. When encountering sudden weather changes, the system can access weather radar data, road surface temperature sensors, and nearby vehicle reports to determine safe driving speeds and recommend appropriate driving modes.

Vehicle system context integration enables real-time awareness of internal state. When a vehicle's perception system detects a pedestrian stepping into a crosswalk, MCP tools can combine camera data with LiDAR point clouds to create a 3D understanding of the pedestrian's trajectory, informing emergency braking decisions. Electric vehicles can use MCP to monitor battery health and range estimates, automatically searching for available charging stations and adjusting routes to include charging stops when needed.

CAV decision-making typically relies on **current** and **upcoming** road segments, requiring spatial and temporal context awareness. Road segment-aware context integration enables vehicles to access context specific to their current location and anticipated path. For example, a vehicle can access current road segment data for locally detected or V2X-received road users such as bicycles, upcoming road segment data for anticipated objects based on infrastructure or cloud data, V2X-shared road user information from nearby vehicles, or infrastructure-provided crosswalk detection from roadside units. This segment-scoped resource access improves spatial focus and action relevance, enabling vehicles to make contextually appropriate decisions based on their immediate and near-future spatial context.

Context coordination in CAV systems requires retrieving multiple context sources within milliseconds for safety-critical decisions. A vehicle approaching a school zone must simultaneously access traffic signal status, pedestrian detection data, and school zone regulations. During severe weather, when primary data sources become unavailable, the MCP system must immediately switch to backup sources.

14.8.2 Security and Trust Considerations

MCP incorporates security principles specific to its architecture that address the unique challenges of distributed AI systems in CAV applications. The protocol's primitive-based design enables fine-grained access control through roots (filesystem boundaries), sampling (LLM capability access), and elicitation (user input requests). Each primitive type has specific security implications: roots limit filesystem access to defined directories, sampling requests can be subject to resource limits and content filtering, and elicitation requires explicit user consent for sensitive operations. The bidirectional nature of MCP requires careful consideration of both server-to-client and client-to-server trust boundaries, particularly in CAV applications where safety-critical operations depend on validated context sources. Security requires validating all context information against multiple trusted sources to prevent malicious actors from sending false data that could compromise vehicle safety.

14.8.3 Limitations and Deployment Challenges

MCP faces specific challenges in CAV deployments that stem from the protocol's design and the unique requirements of vehicular systems. The protocol's limitations include **static primitive schemas** that cannot change mid-session, requiring reconnection to discover new capabilities. **No built-in authentication** means security must be handled at the transport layer, adding implementation complexity. **Absence of quality of service (QoS) guarantees** means latency and message ordering are left to implementation, making it difficult to ensure deterministic performance for safety-critical operations requiring sub-100 ms response times. **Limited error semantics** restricted to JSON-RPC error codes may not provide sufficient granularity for complex failure scenarios in CAV systems.

The discovery, selection, and integration of context sources introduces latency that may be unacceptable for safety-critical operations requiring sub-100 ms response times. The JSON-RPC request-response pattern adds communication overhead, and the dynamic nature of context discovery can introduce complexity that is difficult to manage and debug in real-time systems. CAV systems must also address schema versioning as context sources evolve, handle partial failures and network timeouts gracefully, and integrate with existing automotive protocols and legacy systems. The protocol's effectiveness depends on the reliability of context sources, and network failures or server downtime can directly impact vehicle safety operations, requiring robust fallback mechanisms and redundancy strategies. Scalability presents additional challenges, as the system must efficiently distribute context updates to millions of vehicles during peak rush hour while maintaining sub-100 ms response times.

14.9 Summary

MCP addresses the fundamental challenge of providing AI systems with timely, relevant context from external sources. This chapter has demonstrated how MCP extends beyond rule-based approaches like RAG and function calling to provide a comprehensive framework for dynamic context discovery, selection, and integration.

The chapter began by establishing the context problem in CAV systems, where vehicles require access to live traffic conditions, weather updates, road hazards, and infrastructure status to make safe navigation decisions. Existing approaches like RAG and function calling address specific aspects but cannot fully meet CAV requirements for dynamic discovery, bidirectional communication, and multi-source coordination.

MCP's architecture centers on a host-client-server model that enables bidirectional context flow through standardized primitives. Server features (resources, tools, prompts) provide context to AI models, while client features (roots, sampling, elicitation) enable servers to access client capabilities. The protocol's mathematical foundations formalize the context provision process, providing models for primitive representation, discovery algorithms, and context integration that enable systematic analysis and optimization.

The practical implementation of MCP primitives demonstrates how the mathematical framework translates to working systems, with concrete examples showing tool registration, client-server communication, and LLM integration. Security and trust mechanisms address the critical requirements for safe operation in distributed AI systems, particularly in safety-critical applications. However, MCP faces inherent limitations, including latency from discovery and selection processes, scalability challenges, and reliability dependencies on context sources.

MCP's application in CAV systems illustrates its potential for real-world deployment. The protocol enables sophisticated context integration for environmental awareness, vehicle system monitoring, and real-time decision-making, while addressing the unique challenges of context coordination in safety-critical environments. As MCP matures, it provides a promising foundation for building the next generation of context-aware AI systems that can operate effectively in dynamic, uncertain environments.

References

1 Patrick Lewis, Ethan Perez, Aleksandra Piktus, Fabio Petroni, Vladimir Karpukhin, Naman Goyal, Heinrich Küttler, Mike Lewis, Wen-tau Yih, Tim Rocktäschel, et al. Retrieval-augmented generation for knowledge-intensive NLP tasks. In *Proceedings of the 34th International Conference on Neural Information Processing Systems*, NIPS '20, Red Hook, NY, USA, 2020. Curran Associates Inc. ISBN 9781713829546.

2 Shishir G. Patil, Tianjun Zhang, Xin Wang, and Joseph E. Gonzalez. Gorilla: large language model connected with massive APIs. *arXiv preprint arXiv:2305.15334*, 2023. Foundational work on function calling (tool calling) for agentic AI systems.

3 Anthropic. Model context protocol (MCP), 2024. https://modelcontextprotocol.io/. Protocol specification for connecting AI applications to external tools and data sources.

15

Agent-to-agent Protocols for Collaborative Intelligence

15.1 Introduction

In CAVs, intelligent agents are required to collaborate and coordinate in complex, dynamic environments. As generative AI systems, particularly LLMs, are integrated into vehicular agents, their effectiveness depends not only on individual reasoning capabilities but also on their ability to communicate and collaborate with other intelligent agents with low-latency performance suitable for safety-critical operations.

However, enabling such agent-to-agent collaboration presents significant technical challenges that have led to the development of numerous communication protocols and standards. Not all protocols are equally suited for the unique requirements of CAV systems, which demand low-latency performance suitable for safety-critical operations, safety guarantees, and seamless integration with autonomous reasoning capabilities.

To address these limitations, this chapter focuses on A2A [1], which represents the most promising approach for peer-oriented communication in CAV applications. As established in Chapter 13, A2A represents the *Peer-Oriented, General-Purpose* quadrant of the agent protocol taxonomy. A2A operationalizes the notion of distributed intelligence: agents can delegate reasoning tasks to peers, share contextual information, and coordinate complex decision-making processes.

Figure 15.1 illustrates the A2A protocol's client-server architecture that enables agent collaboration across distributed environments. The architecture demonstrates how local agent clients can interact with multiple remote agent servers through standardized A2A communication protocols. This design supports flexible deployment scenarios where local agents can leverage remote cloud-based services, enabling rich multi-agent ecosystems that can adapt to different computational and networking constraints. A2A also supports local server deployments for development, testing, and edge computing scenarios.

15.2 A2A Architecture

A2A provides a standardized framework for peer-to-peer communication and task delegation among AI agents. A2A enables dynamic agent discovery, low-latency collaboration suitable for safety-critical operations, and bidirectional communication through a client-server architecture that supports flexible deployment scenarios.

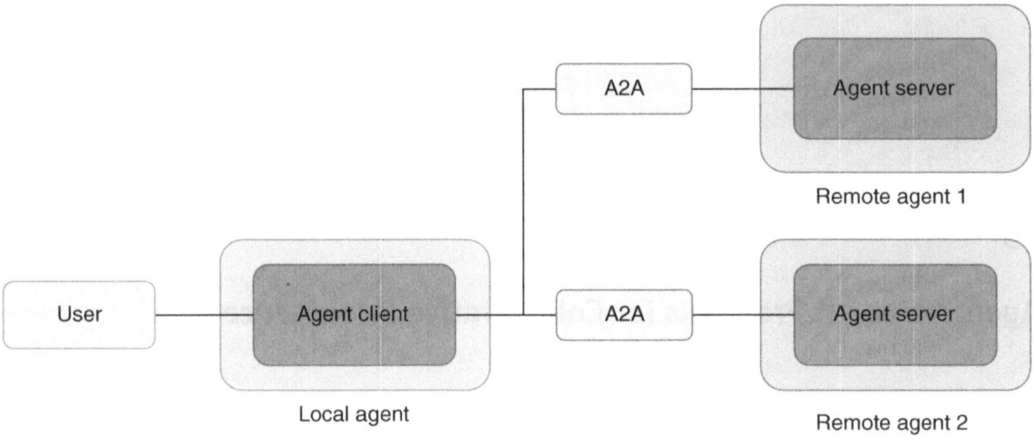

Figure 15.1 A2A protocol architecture showing client-server relationships and communication patterns.

15.2.1 Core Architecture Components

A2A operates through three fundamental components that work together to enable agent collaboration:

Discovery Mechanism: Agents can dynamically find each other through standardized metadata exchange, eliminating the need for preconfigured relationships.
Task Delegation System: Agents can delegate reasoning tasks to peers through structured requests, enabling distributed problem-solving.
Communication Framework: A standardized message architecture supports rich, multimodal communication, including text, structured data, and file attachments.

A2A's client-server architecture enables seamless integration of local and remote capabilities, allowing systems to leverage on-device processing, other vehicle agents, roadside units, and cloud-based services as needed.

15.2.2 Discovery

Discovery is the process of finding agents with the required capabilities. A2A supports multiple discovery mechanisms: public agents can be found through well-known URIs using `GET https://{domain}/.well-known/agent-card.json`, enterprise scenarios may use curated registries with centralized catalogs and search capabilities, and private environments can use hard-coded URLs, configuration files, or environment variables.

Figure 15.2 illustrates the complete A2A communication sequence, showing both the discovery phase and the subsequent task execution phase. The sequence demonstrates how agents first discover each other's capabilities through agent card exchange, then proceed to task-based communication and response handling.

15.2.2.1 Agent Cards—What Agents Expose

Agent cards are self-describing metadata documents that each agent publishes to advertise its capabilities and skills. They contain structured information about the agent's identity, available skills,

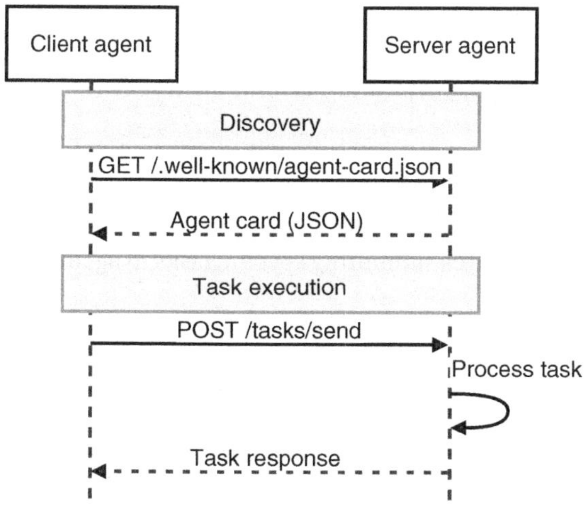

Figure 15.2 A2A communication sequence flow showing agent discovery, task submission, processing, and response generation.

input/output modes, and example queries, enabling other agents to programmatically discover and evaluate potential collaborators.

Here is a concrete example of an agent card for a vehicle visual question-answering agent:

Listing 15.1 Example Agent Card for Vehicle VQA Agent

```
{
  "name": "VehicleVQAAgent",
  "description": "Visual Question Answering for vehicle perception",
  "url": "https://vehicleagent.com/ABC123456",
  "version": "1.0.0",
  "capabilities": {
    "streaming": false,
    "pushNotifications": false
  },
  "authentication": {
    "schemes": ["Bearer", "OAuth2"]
  },
  "skills": [
    {
      "id": "identify_objects",
      "name": "Object Identification",
      "description": "Identify objects in camera images from vehicle sensors",
      "inputModes": ["text", "image"],
      "outputModes": ["text", "object_list"],
      "examples": ["What objects are behind the truck?", "Identify the traffic
          sign ahead"]
    } ] }
```

This agent card example demonstrates key A2A concepts. The `name` and `description` fields provide basic identification, while the `URL` serves as the unique endpoint for communication. The `capabilities` section specifies technical features like streaming support and push notifications. The `authentication` field declares supported security schemes (bearer tokens and OAuth2).

Most importantly, the `skills` array defines the agent's specific abilities. In this example, the "identify_objects" skill can process both text queries and image inputs, returning either text descriptions or structured object lists, with example queries like "What objects are behind the truck?"

15.2.3 Task Execution

Once agents have discovered each other through the discovery phase (as shown in Figure 15.2), they can proceed to task execution. Task execution involves submitting specific tasks to agents and receiving structured responses.

The task execution process follows these steps:

Task Submission: The client agent submits a specific task using a `POST` request to the `/tasks/send` endpoint, specifying the task requirements and any necessary parameters.

Task Processing: The server agent processes the task internally, performing the requested analysis or computation using its specialized capabilities.

Response Generation: Upon completion, the server agent returns a structured response containing the task results, status information, and any generated artifacts.

This task execution phase enables agents to leverage each other's specialized capabilities for distributed problem-solving.

15.2.3.1 Task Request and Response Examples

Here are concrete examples of how task execution works in practice:

Task Request Example:

The following code demonstrates a typical A2A task request structure:

Listing 15.2 Example A2A Task Request

```
{
  "id": "task-uuid-123",
  "message": {
    "role": "user",
    "parts": [{
      "text": "{\"skill\": \"identify_objects\",
      \"params\": {\"query\": \"What is behind the truck?\"}}"
    }] } }
```

The `id` field provides a unique identifier for tracking the task throughout its lifecycle. The `message` object contains the actual communication, with `role` set to "user" indicating this is a request from the client. The `parts` array contains the structured content: a JSON string specifying the skill to invoke ("identify_objects") and the parameters (the query "What is behind the truck?"). This task request is created by the LLM in the agent client, which uses the skills identified during the discovery phase (from the agent card) along with the user query to construct the appropriate task structure.

Task Response Example:

The following code demonstrates the corresponding task response:

Listing 15.3 Example A2A Task Response

```
{
  "id": "task-uuid-123",
  "status": {"state": "completed"},
  "messages": [
```

```
    {
      "role": "user",
      "parts": [{"text": "{\"skill\": \"identify_objects\", \"params\":
        {\"query\": \"What is behind the truck?\"}}"}]
    },
    {
      "role": "agent",
      "parts": [{"text": "A pedestrian is crossing behind the truck"}]
    } ] }
```

The response maintains the same `id` for correlation with the original request. The `status` field indicates the task completion state ("completed"). The `messages` array contains the complete conversation history: first the original user request, then the agent's response with `role` set to "agent" and the answer in the `parts` array.

In this example, the client requests the "identify_objects" skill from the VehicleVQAAgent, asking "What is behind the truck?"—a common scenario where one vehicle needs another's perception capabilities due to visual occlusion. The agent responds that "A pedestrian is crossing behind the truck," demonstrating how A2A handles the complete interaction cycle with status tracking and bidirectional message structure.

15.2.3.2 Message Structures–What to Send

A2A uses a hierarchical message architecture where tasks serve as containers for complete interactions, messages represent individual communication units within a task, and parts form the fundamental content units. Messages are role-based, with "user" messages representing requests and "agent" messages containing responses. This structure enables rich, multimodal communication, including text, structured data, and file attachments.

15.2.3.3 Task Lifecycle–How to Handle Responses

Tasks maintain state throughout their lifecycle, supporting asynchronous execution and status tracking. Each task progresses through six official states: "Created" (submitted but not started), "Working" (actively processing), "NeedsInput" (additional information required), "Completed" (successfully finished), "Failed" (error encountered), and "Canceled" (terminated). This enables agents to monitor long-running operations and handle complex collaborative scenarios effectively.

15.2.4 Communication Patterns and Implementation Considerations

A2A's design philosophy centers on flexible communication patterns that accommodate different operational requirements. The protocol supports three fundamental communication modes, each suited to different use cases in CAV systems.

Synchronous Communication: For tasks requiring immediate responses, A2A supports synchronous request-response patterns where the client agent submits a task and waits for completion before proceeding. This pattern is suitable for quick queries where a response is needed for subsequent decision-making, such as requesting traffic light status or verifying agent availability.

Streaming Communication: For scenarios requiring real-time updates, A2A supports streaming communication using server-sent events (SSE), enabling low-latency updates suitable for safety-critical operations. This pattern allows agents to receive progressive updates as tasks are processed, rather than waiting for complete results. For example, a vehicle might receive streaming updates about an intersection's traffic conditions as they evolve.

Asynchronous Communication: For long-running operations that may take significant time to complete, A2A supports asynchronous patterns where tasks are submitted and tracked via unique identifiers. The client can poll for status updates or receive notifications when tasks are completed. This pattern is essential for complex reasoning tasks that may require extended processing time.

The protocol's implementation builds on standard web technologies, including JSON-RPC 2.0 over HTTP(S), ensuring compatibility with existing infrastructure while providing the flexibility needed for diverse deployment scenarios. The architecture supports multiple transport mechanisms, from standard HTTP connections for cloud-based services to WebSocket connections for low-latency, bidirectional communication suitable for safety-critical operations between agents.

This design enables agents to adapt their communication patterns based on task requirements, network conditions, and operational constraints, making A2A suitable for both time-sensitive safety-critical operations and complex collaborative reasoning tasks that may require extended processing.

15.3 Mathematical Foundations for Agent Collaboration

To understand how A2A enables effective agent collaboration, it is helpful to formalize the underlying decision-making processes. This section presents mathematical frameworks that capture how agents discover each other, select appropriate collaborators, and coordinate their reasoning processes. These formalizations help clarify the design principles behind A2A and provide a foundation for understanding how agents make intelligent decisions about task delegation and collaborative reasoning.

15.3.1 Agent Discovery and Selection

The agent discovery and selection process involves finding available agents within a network scope and evaluating them based on multiple criteria. Agents publish their capabilities through agent cards (as described earlier in this chapter). The discovery process searches for agents that are both available and accessible. Once agents are discovered, the selection process evaluates them based on how relevant their skills are to the query q, their current availability, their overall capability, and the expected latency of communication. The agent a^* with the highest utility score across these dimensions is selected for the task. The utility function balances these factors: utility$(q, a_i) = \alpha \cdot \text{relevance} + \beta \cdot \text{availability} + \gamma \cdot \text{capability} - \delta \cdot \text{latency}$, where the weighting parameters α, β, γ, and δ control the relative importance of each criterion.

Advanced Note 15.1 Agent Discovery and Selection Mathematics

To formalize the agent discovery and selection process, we define the key components. Let $A = \{a_1, a_2, \dots, a_n\}$ represent the set of available agents in the system, where n is the total number of agents. Each agent $a_i \in A$ is characterized by its capabilities, represented as a set of skills $S_i = \{s_{i1}, s_{i2}, \dots, s_{ik}\}$, where k is the number of skills for agent i. The symbol q represents a query or task to be performed by the system.

Each agent's capabilities are formally represented through an agent card:

$$C_j = (\text{name}_j, \text{url}_j, \text{skills}_j, \text{capabilities}_j) \tag{15.1}$$

where skills_j represents the specific abilities the agent can perform, and capabilities_j represents the general capabilities such as streaming support, authentication requirements, and transport preferences (see the agent card discussion earlier in this chapter for concrete examples).

The agent discovery process can be formalized as:

$$A_{\text{discovered}} = \text{discover\_agents}(q, \text{network\_scope}) \tag{15.2}$$

$$= \{a_i \in A : \text{available}(a_i) \wedge \text{accessible}(a_i, \text{network\_scope})\}. \tag{15.3}$$

The agent selection process can then be formalized as:

$$a^* = \arg \max_{a_i \in A_{\text{discovered}}} \text{utility}(q, a_i). \tag{15.4}$$

$$\text{utility}(q, a_i) = \alpha \cdot \text{relevance}(q, S_i) + \beta \cdot \text{availability}(a_i) + \gamma \cdot \text{capability}(a_i) - \delta \cdot \text{latency}(a_i) \tag{15.5}$$

where q represents the query or task to be performed, α, β, γ, and δ are weighting parameters, and the utility function balances relevance, availability, capability, and latency considerations.

Advanced Note 15.2 Task Delegation Mathematics

The task delegation process can be modeled as a sequence of operations where the client agent selects an appropriate skill and creates a task request:

$$s^* = \arg \max_{s \in S_{a^*}} \text{confidence}(q, s). \tag{15.6}$$

$$\text{task} = \text{create\_task}(q, a^*, s^*), \tag{15.7}$$

$$\text{response} = \text{execute\_task}(\text{task}), \tag{15.8}$$

where s^* represents the optimal skill selection within the chosen agent a^* (determined by the client based on the agent card), S_{a^*} is the skill set of agent a^*, and the task request sent to the remote agent includes both the query q and the selected skill s^* as part of the message content. The response includes the task execution results from the remote agent.

15.3.2 Complete Processing Flow

The complete A2A processing flow transforms a local query q into a collaborative solution through a sequence of operations. This flow demonstrates how the mathematical formulations above are realized in practice:

Processing Steps (Local → Remote → Local):

1. **Local Query q:** The process begins when a local agent receives a query or task that requires capabilities beyond its own.
2. **Agent Discovery:** The local agent searches for available agents within the network scope, resulting in a discovered set $A_{\text{discovered}}$.

3. **Agent Selection:** From the discovered agents, the local agent selects agent a^* whose skills are most relevant to the query.
4. **Skill Selection:** Within the selected agent, the most appropriate skill s^* is chosen based on confidence that it can handle the query.
5. **Task Creation and Submission:** The client agent creates a task request containing the query q and selected skill s^*, then sends it to the remote agent a^*. The task request is created by the LLM in the agent client, which uses the skills identified during the discovery phase along with the user query.
6. **Remote Processing:** The remote agent processes the task using the specified skill and generates a response response = (status, messages, artifacts) containing status information, reasoning messages, and any generated artifacts.
7. **Response Integration:** The local agent integrates the remote agent's response into its own context and reasoning process, producing the final result.

The response consists of the task state (completed, failed, working), agent reasoning and output content in the messages, and any generated files, data, or structured results as artifacts. This flow illustrates how A2A enables distributed problem-solving by allowing agents to leverage each other's specialized capabilities.

Advanced Note 15.3 Complete Processing Flow Mathematics

The complete A2A processing flow can be formalized as a sequence of mathematical operations:

$$q \quad \text{(local query)}, \tag{15.9}$$

$$A_{\text{discovered}} = \text{discover\_agents}(q, \text{network\_scope}), \tag{15.10}$$

$$a^* = \arg \max_{a_j \in A_{\text{discovered}}} \text{relevance}(q, S_j), \tag{15.11}$$

$$s^* = \arg \max_{s \in S_{a^*}} \text{confidence}(q, s), \tag{15.12}$$

$$\text{task} = \text{create\_task}(q, a^*, s^*), \tag{15.13}$$

$$\text{response} = (\text{status}, \text{messages}, \text{artifacts}), \tag{15.14}$$

$$\text{result} = \text{integrate}(\text{response}, \text{local\_context}), \tag{15.15}$$

where q represents the query from the local agent, $A_{\text{discovered}}$ is the set of available agents found through discovery, a^* is the selected agent with optimal relevance, s^* is the selected skill within that agent, and response = (status, messages, artifacts) consists of the task state, agent reasoning content, and any generated artifacts.

15.3.3 Multi-agent Coordination

When multiple agents need to collaborate on complex tasks, A2A enables sophisticated coordination patterns. These mathematical models help understand how agents share information, resolve conflicts, and reach consensus when working together.

Advanced Note 15.4 Multi-agent Coordination Mathematics

For collaborative tasks involving multiple agents $A_{\text{collab}} = \{a_1, a_2, \dots, a_k\} \subseteq A$, the coordination process can be modeled as:

$$\text{result} = \text{fusion}(\{f_{a_i}(q_i, \text{context}_i)\}_{i=1}^{k}), \tag{15.16}$$

$$\text{context}_i = \text{integrate}(\{\text{response}_j\}_{j \neq i}, \text{local\_context}_i), \tag{15.17}$$

where $\text{fusion}(\cdot)$ represents the aggregation operation across multiple agent outputs, and each agent a_i operates on its assigned sub-task q_i with integrated context from other agents in the collaboration set.

Advanced Note 15.5 Consensus and Conflict Resolution Mathematics

When multiple agents provide conflicting responses, the consensus mechanism can be formalized as:

$$\text{consensus} = \frac{\sum_{i=1}^{k} \text{weight}_i \cdot \text{response}_i}{\sum_{i=1}^{k} \text{weight}_i}, \tag{15.18}$$

$$\text{weight}_i = \text{confidence}(\text{response}_i) \cdot \text{reliability}(a_i), \tag{15.19}$$

where k is the number of agents in the collaboration set, and the weighted vote combines responses based on confidence and agent reliability.

These applications demonstrate how the A2A protocol can be used to build sophisticated collaborative systems in CAV applications, enabling agents to work together effectively through standardized communication and coordination mechanisms.

15.4 Security and Trust

A2A security is built on a multilayered framework that addresses authentication, authorization, data integrity, and trust management across distributed agent networks. The framework must ensure that agents can securely discover, authenticate, and collaborate with each other while maintaining the privacy and integrity of sensitive information.

15.4.1 Authentication Mechanisms

A2A supports multiple authentication mechanisms to ensure secure agent communication. Bearer tokens provide standard OAuth2 (Open Authorization version 2) authentication for cloud services, enabling secure access to cloud-based agent services. API keys offer vehicle-specific credentials for service access, enabling secure communication between vehicles and infrastructure services. Certificate-based authentication can leverage existing V2X certificate infrastructure (such as IEEE 1609.2 and SCMS) for vehicle-to-vehicle (V2V) communication, ensuring authentic vehicle identity in direct V2V scenarios.

15.4.2 Trust Considerations

The A2A trust framework addresses the fundamental challenge of establishing trust between autonomous agents in dynamic, distributed environments. The trust framework includes agent reputation systems that track reliability and performance of agents over time, enabling informed decisions about which agents to trust for specific tasks. Verification mechanisms validate agent identity and capabilities through cryptographic proofs and capability demonstrations. Access control systems limit what agents can request from each other based on trust levels, capabilities, and security policies.

15.4.3 CAV-specific Security

CAV applications introduce unique security challenges and implementation considerations that require specialized approaches. Vehicle identity verification ensures authentic vehicle communication through cryptographic identity verification and certificate management. Data privacy protection safeguards sensitive vehicle and passenger information through encryption, access controls, and data minimization. Safety validation verifies that coordinated decisions are safe through formal verification, simulation testing, and runtime monitoring.

The implementation of A2A security in CAV systems requires careful consideration of the trade-offs between security, performance, and usability, balancing strong authentication and authorization with the low-latency requirements suitable for safety-critical operations in autonomous driving systems. For V2V communication scenarios, existing V2X security infrastructures such as the security credential management system (SCMS) [2, 3] provide proven certificate-based authentication mechanisms that could be leveraged for A2A agent identity verification.

15.5 A2A in CAV Systems

This section explores how A2A protocol is applied in real-world CAV systems, demonstrating the practical value of agent-to-agent collaboration in autonomous driving scenarios.

15.5.1 CAV Agent Types

CAV systems employ various types of agents that benefit from A2A communication protocols, each serving distinct roles in the autonomous driving ecosystem:

Vehicle Agents: On-board agents responsible for core autonomous driving functions, including perception (camera, LiDAR, radar processing), planning (route planning, trajectory optimization), and control (vehicle dynamics, steering, acceleration). These agents form the foundation of autonomous operation, handling the primary sensing, reasoning, and actuation tasks.

Collaborative Agents: Facilitate vehicle-to-vehicle (V2V) communication and coordination through intent sharing, negotiation processes for lane changes and merging operations, and cooperative driving scenarios. These agents enable vehicles to coordinate their actions proactively, transforming individual autonomous vehicles into a collaborative system.

Service Agents: Provide vehicle-to-network (V2N) connectivity and access to infrastructure services, including traffic management systems, weather services, and infrastructure services for parking, charging, and emergency response. These agents extend the vehicle's capabilities by connecting to cloud-based services and infrastructure systems.

15.5.2 Real-world CAV Applications

The practical applications of A2A in CAV systems demonstrate the protocol's versatility in addressing real-world challenges faced by autonomous vehicles. These applications span multiple domains of vehicular intelligence, from basic communication to complex collaborative reasoning.

Intent Sharing and Negotiation: Vehicle intent sharing represents one of the most fundamental applications, enabling proactive communication between vehicles about planned maneuvers, route changes, and emergency situations. This capability transforms reactive driving scenarios into proactive coordination, where vehicles can anticipate and respond to each other's actions before they occur. For example, a vehicle planning to change lanes can broadcast its intent to nearby vehicles, allowing them to adjust their trajectories accordingly and maintain safe following distances. When multiple vehicles have conflicting or competing intentions, intent sharing naturally extends into negotiation processes where vehicles must resolve conflicts and reach mutually acceptable solutions. This progression from simple intent communication to complex negotiation represents a key strength of A2A, enabling vehicles to handle increasingly sophisticated coordination scenarios as the complexity of their interactions grows.

Collaborative Perception: Collaborative perception addresses one of the most significant challenges in autonomous driving: sensor occlusion. Individual vehicles often have limited visibility due to obstacles, weather conditions, or blind spots created by other vehicles. Through A2A communication, vehicles can share perception data and delegate sensing tasks to peers with better viewpoints. A vehicle approaching a construction zone might request signage interpretation from a vehicle that has already passed through the area, or a vehicle with a clear view of an intersection can provide traffic light status to vehicles with obstructed views.

Intersection Coordination: Intersection coordination represents a complex application where multiple vehicles must negotiate priority and coordinate their movements to optimize traffic flow while maintaining safety. Rule-based traffic control relies on fixed signals and human decision-making, but A2A enables dynamic, low-latency coordination suitable for safety-critical operations where vehicles can negotiate right-of-way based on current traffic conditions, emergency situations, and efficiency considerations. This application demonstrates how A2A transforms static traffic management into adaptive, intelligent coordination systems.

15.5.3 Deployment Architectures

A2A can be deployed in different architectural patterns depending on the communication requirements and infrastructure availability. Understanding these deployment options helps practitioners choose the appropriate architecture for their specific CAV applications.

Vehicle-to-vehicle (V2V) Architecture: Direct V2V communication enables vehicles to coordinate their actions without intermediate infrastructure, providing low-latency communication suitable for safety-critical operations. This architecture is essential for applications requiring immediate coordination, such as emergency braking coordination or intersection negotiation, where delays introduced by cloud infrastructure could compromise safety.

Vehicle-to-network (V2N) Architecture: While V2V communication is essential for direct vehicle coordination, many CAV applications benefit from cloud-mediated communication through a V2N architecture. This approach extends A2A's peer-to-peer model to include cloud-based agent services, creating a hybrid architecture that combines the benefits of direct vehicle communication with the computational power and scalability of cloud infrastructure.

The V2N communication flow follows a three-stage process: vehicles communicate with cloud agent services, which then relay information to other vehicles. This architecture leverages standard web APIs, including HTTPS (hypertext transfer protocol secure), REST (representational state transfer), and JSON protocols, ensuring compatibility with existing web infrastructure. The communication relies on cellular and 5G networks, which provide reliable, high-bandwidth connectivity that is often more stable than direct V2X communication. The cloud infrastructure provides scalable, managed services that can handle large numbers of vehicles simultaneously.

This V2N architecture offers several advantages for A2A implementations. The use of standard protocols such as HTTP, REST, and JSON ensures compatibility with well-established web technologies and existing A2A implementations. Cloud services provide scalability that can handle many vehicles simultaneously, addressing the computational and storage requirements of large-scale CAV deployments. Cellular networks offer more reliable connectivity compared to V2X communication, particularly in areas with limited V2X infrastructure. The architecture also benefits from standard web security practices, including SSL/TLS encryption, authentication mechanisms, and access control systems. Most importantly, this approach maintains compatibility with existing A2A implementations while extending their capabilities through cloud-based services, enabling vehicles to access cloud-based agent services for complex reasoning tasks while maintaining the benefits of A2A's standardized communication protocol.

15.6 Limitations and Challenges

While A2A provides a powerful framework for agent-to-agent communication, it is important to understand the limitations and challenges that practitioners should consider when implementing A2A-based systems, particularly in CAV applications.

15.6.1 A2A Protocol Limitations

As an emerging technology, A2A presents several limitations that affect its practical deployment. The protocol's relative novelty means there is limited production experience and real-world validation of its effectiveness in complex, safety-critical scenarios. This lack of maturity creates uncertainty about the protocol's performance under various edge cases and failure modes.

The discovery mechanism, while elegant in design, introduces overhead through agent card exchange processes that may impact system responsiveness in time-critical situations. Each agent discovery requires network requests to retrieve agent cards, which can create additional latency and network traffic, particularly in scenarios with many available agents.

Network latency represents another significant concern, as the protocol's reliance on HTTP-based communication may introduce delays that could impact low-latency decision-making requirements suitable for safety-critical operations. While HTTP provides good interoperability and debugging capabilities, it may not be optimal for applications requiring millisecond-level response times.

15.6.2 CAV-specific Challenges

CAV applications introduce additional challenges that are specific to the automotive domain. Low-latency constraints suitable for safety-critical operations may conflict with A2A's HTTP-based architecture, as autonomous driving requires millisecond-level response times.

Bandwidth constraints in V2X channels have limited capacity, which can restrict the amount of data that can be exchanged between vehicles and infrastructure. This limitation becomes particularly problematic when agents need to share large amounts of perception data or complex reasoning artifacts.

Inter-vendor consistency presents another challenge, as different implementations of A2A may vary in their specific behaviors, error handling, and performance characteristics. This inconsistency can make it difficult to ensure reliable operation across heterogeneous systems from different manufacturers.

Safety validation becomes critical when multiple agents collaborate, as there is a need to ensure that merged outputs from multiple agents are safe and do not introduce conflicts or dangerous situations. The complexity of validating multi-agent decisions increases exponentially with the number of collaborating agents. Integration complexity arises from the need to integrate A2A with existing CAV systems and protocols, which may require significant adaptation and testing.

15.6.3 Implementation Considerations

These limitations suggest that A2A should be carefully evaluated for specific use cases, with particular attention to timing requirements, network conditions, and safety validation needs. While A2A provides a solid foundation for agent collaboration, successful deployment will require careful consideration of these constraints and potentially hybrid approaches that combine A2A with other communication mechanisms for time-critical operations.

15.7 Summary

This chapter examined A2A as a framework for peer-oriented communication among autonomous agents in CAV systems. A2A represents a paradigm shift from rule-based service protocols by recognizing each peer as a taskful agent capable of autonomous reasoning and long-running operations.

The protocol's core architecture operates through three fundamental components: discovery mechanisms that enable agents to dynamically find each other through agent card exchange, task delegation systems that allow agents to delegate reasoning tasks to peers, and communication frameworks that support rich, multimodal communication. The protocol's core abstractions—agent cards, tasks, artifacts, messages, and status—provide a semantic framework for agent interaction beyond simple function calling. A2A's client-server architecture supports flexible deployment scenarios, enabling agents to leverage both local and remote capabilities through standardized communication patterns, including synchronous, streaming, and asynchronous modes.

The mathematical foundations provide formal definitions for agent discovery, selection, and task delegation processes. These frameworks model how agents evaluate potential collaborators using utility functions that balance relevance, availability, capability, and latency considerations. The complete processing flow demonstrates how queries are transformed into collaborative solutions through agent discovery, skill selection, task creation, and response integration. Multi-agent coordination models capture how agents share information, resolve conflicts, and reach consensus when working together on complex tasks.

The security and trust framework addresses authentication, authorization, data integrity, and trust management across distributed agent networks. A2A supports multiple authentication mechanisms, including bearer tokens for cloud services, API keys for vehicle-specific credentials, and certificate-based authentication for V2V communication. The framework includes agent reputation systems, verification mechanisms, and access control systems that enable secure collaboration in dynamic, distributed environments.

Real-world CAV applications demonstrate A2A's practical value through intent sharing and negotiation, collaborative perception to address sensor occlusion, and intersection coordination for dynamic traffic management. The protocol supports both vehicle-to-vehicle (V2V) architectures for direct low-latency communication and vehicle-to-network (V2N) architectures that leverage cloud-based services for scalable, managed agent collaboration.

A2A presents limitations including discovery overhead from agent card exchange, network latency concerns with HTTP-based communication, and challenges in multi-party trust coordination. CAV-specific challenges include low-latency constraints, bandwidth limitations, inter-vendor consistency issues, and safety validation complexity. Despite these limitations, A2A provides a foundation for peer-oriented agent connectivity, enabling sophisticated coordination patterns essential for collaborative intelligence in autonomous vehicle systems.

References

1 Google DeepMind. Agent-to-agent protocol (A2A), 2024. https://github.com/google-deepmind/a2a-protocol. Protocol for inter-agent communication and collaboration in multi-agent systems.

2 Benedikt Brecht, Thomas Nowey, Christian Mletzko, Max Mühlhäuser, and Christina Pöpper. A security credential management system for V2X communications. *IEEE Transactions on Intelligent Transportation Systems*, 19(12):3850–3871, 2018.

3 U.S. Department of Transportation. Connected vehicle deployment technical assistance: security credential management system (SCMS) technical primer. Technical Report, Federal Highway Administration, 2019. https://trid.trb.org/View/1693047. Technical primer on SCMS architecture and implementation.

16

Integrated CAV Agent Architectures with MCP and A2A

16.1 Introduction

CAVs face complex decision-making scenarios that require access to both deterministic external services and collaborative peer intelligence. While the previous chapters examined individual protocols—Chapter 14 detailed MCP [1] for context-oriented tool integration, and Chapter 15 explored A2A [2] for peer-oriented collaboration—real-world CAV applications demand unified systems that can seamlessly integrate both approaches. This chapter addresses the challenge of building intelligence systems that combine external capabilities with collaborative reasoning.

The integration of MCP [1] and A2A [2] protocols enables a paradigm shift from isolated agent capabilities to comprehensive intelligence systems. MCP provides context-oriented tool integration for accessing external services, APIs, and deterministic data sources. A2A enables peer-oriented collaboration for distributed problem-solving and multi-agent coordination. The unified approach allows agents to leverage both deterministic service access and intelligent peer coordination simultaneously.

Consider a CAV navigating through a complex intersection with limited visibility due to construction barriers. The vehicle must simultaneously access traffic light status and weather data through MCP tools while coordinating with nearby vehicles for visual analysis through A2A protocols. A unified system addresses this challenge by presenting the LLM with a comprehensive context that includes both available tools (MCP functions) and available agents (A2A capabilities). The LLM selects appropriate functions based on the current situation, and the system transparently routes these function calls to the appropriate protocol.

This chapter establishes the theoretical foundations and practical frameworks for building unified agent systems that seamlessly integrate MCP and A2A protocols. We examine protocol selection criteria, develop system architectures for protocol integration, and provide implementation guidelines for robust systems that can operate in safety-critical environments.

16.2 Protocol Comparison and Decision Framework

Understanding when to use A2A versus MCP is crucial for effective system design in CAV applications. While both protocols enable AI agents to access external capabilities, they serve different purposes and are optimized for different types of interactions. The following comparison guides protocol selection decisions, building upon the detailed examinations of MCP in Chapter 14 and A2A in Chapter 15.

16.2.1 Protocol Design Comparison

The fundamental difference between A2A and MCP lies in their design philosophy and target use cases:

A2A [2]: Designed for peer-oriented communication between autonomous agents, enabling collaborative reasoning and distributed problem-solving. A2A treats each participant as an intelligent agent capable of autonomous reasoning, task delegation, and long-running operations.

MCP [1]: Designed for context-oriented tool integration, enabling AI models to access external services, APIs, and deterministic data sources. MCP treats external systems as tools and resources that provide context to enhance AI reasoning.

Table 16.1 provides a comprehensive comparison of when to use A2A versus MCP protocols in CAV systems.

16.2.2 Decision Framework

Selecting the appropriate protocol for a given CAV application requires careful consideration of the interaction patterns, data requirements, and system constraints. The decision framework provides clear guidance for choosing between A2A and MCP based on the specific needs of each use case.

A2A protocols excel in scenarios requiring collaborative intelligence and peer coordination. Choose A2A when you need collaborative reasoning between multiple intelligent agents, when the task requires distributed problem-solving and peer coordination, when you want to leverage specialized capabilities from other autonomous agents, when the interaction involves complex, multistep reasoning processes, or when you need conversational, task-based communication patterns.

Conversely, MCP protocols are well-suited for scenarios requiring direct access to external resources and deterministic data. Choose MCP when you need access to external services, APIs, or deterministic data sources, when the task involves simple function calling or resource access, when you want to integrate with existing tools and services, when the interaction is primarily one-way (getting information or executing actions), or when you need reliable, low-latency access to centralized services.

Table 16.1 Comparison of A2A versus MCP protocol use cases for CAV systems.

Use case category	A2A protocol	MCP protocol
Communication pattern	Peer-to-peer, agent-to-agent collaboration	Client-server, tool, and service access
Data access	Collaborative reasoning, peer coordination	External data access (traffic lights, weather, road conditions)
Service integration	Distributed problem-solving, multi-agent reasoning	Navigation services, traffic management, emergency services
Information type	Dynamic, collaborative reasoning and coordination	Deterministic, structured data from external systems
Interaction model	Conversational, task-based delegation	Function calling, resource access
Reliability	Variable (peer availability)	High (centralized services)
Latency	Variable (peer response time)	Low (direct service access)
Complexity	High (multi-agent coordination)	Low (direct tool invocation)

16.2.3 Trade-offs and Considerations

Both protocols present distinct trade-offs that must be carefully evaluated for specific CAV applications. MCP offers high reliability and low latency through centralized services, making it ideal for deterministic data access and low-latency service integration suitable for safety-critical operations. However, it is limited to client-server interactions and cannot support sophisticated multi-agent reasoning.

A2A provides powerful capabilities for collaborative reasoning and distributed problem-solving, enabling vehicles to leverage collective intelligence from peer agents. However, it introduces variable latency and reliability concerns due to peer availability and requires more complex coordination mechanisms.

In practice, CAV systems often benefit from using both protocols together. A unified system can use MCP for accessing external services and deterministic data, while using A2A for collaborative reasoning and peer coordination.

16.3 System Architecture and Mathematical Framework

The integration of MCP and A2A protocols within CAV agent systems requires a well-designed architecture that can handle function calling, protocol routing, and comprehensive intelligence. The following subsections present the architectural principles, design considerations, and mathematical framework for building CAV agent systems.

16.3.1 System Architecture Overview

A unified agent system architecture consists of several interconnected components. The **LLM orchestrator** makes function selection decisions based on context. The **function registry** maps available functions to appropriate protocols. The **context manager** builds and maintains context for LLM decisions by aggregating the tool list from MCP and the agent card from A2A, presenting a unified view of available capabilities. The **protocol router** routes function calls to appropriate protocols based on function type. The **MCP client** handles external tool and service access, while the **A2A client** manages peer agent collaboration. Figure 16.1 illustrates how these components interact within the unified architecture.

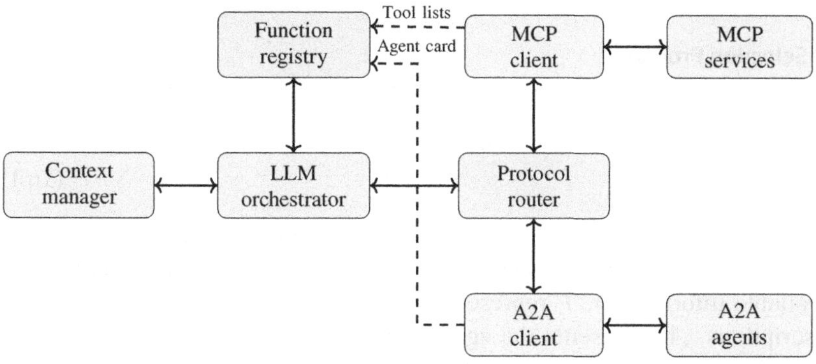

Figure 16.1 System components and their interactions in unified agent architecture.

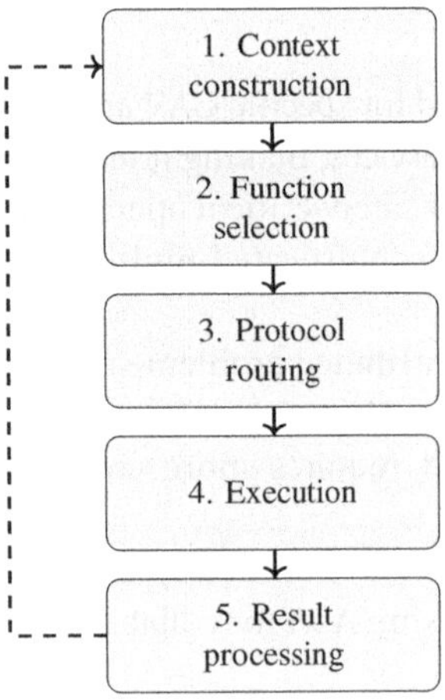

Figure 16.2 Data flow process in unified agent systems.

The data flow process in unified agent systems follows a logical sequence. The system begins with **context construction**, building context with available tools and agents. The **LLM orchestrator** then performs **function selection**, choosing appropriate functions from the context. The system handles **protocol routing**, directing the selected function to MCP or A2A as appropriate. After **execution** by the chosen protocol, the system performs **result processing**, updating the context with new information. Figure 16.2 illustrates this data flow sequence.

The function calling mechanism integrates these components: the LLM receives context with the tool list from MCP and the agent card from A2A, performs function selection, and the system routes function calls to MCP or A2A based on function type. MCP functions are routed to external tools and services, while A2A functions are directed to peer agent collaboration. The system includes error handling with fallback mechanisms for protocol failures.

16.3.2 Mathematical Framework for LLM-based Protocol Selection

The unified orchestrator employs LLMs to make intelligent protocol selection decisions. The core decision-making process leverages LLMs' natural language understanding and reasoning capabilities to evaluate context and select appropriate functions.

16.3.2.1 LLM Function Selection Process

The LLM receives context and generates function selection probabilities based on natural language understanding:

$$P(f \mid c) = \text{LLM}(c, \mathcal{T}, \mathcal{A}), \tag{16.1}$$

where f represents a function or capability from the unified set of available functions (which may be either MCP functions or A2A agent capabilities), c represents the context including current situation, goals, and available information, $\mathcal{T}$ represents the tool list containing available MCP functions and their descriptions, $\mathcal{A}$ represents the agent card containing available A2A agents and their capabilities, and $P(f \mid c)$ represents the LLM's probability distribution over available functions.

16.3.2.2 Protocol Routing Based on Function Type

Once the LLM selects a function, the system routes it to the appropriate protocol based on function characteristics:

$$\text{Protocol}(f) = \begin{cases} \text{MCP} & \text{if } f \in \mathcal{T} \\ \text{A2A} & \text{if } f \in \mathcal{A} \end{cases}, \tag{16.2}$$

where f represents the selected function, $\mathcal{T}$ represents MCP functions, and $\mathcal{A}$ represents A2A functions.

16.3.2.3 LLM-based Decision Selection Strategies

The system presents both MCP tools and A2A agents as flat candidates to the LLM, allowing the model to select from the unified set of available functions without requiring preselection of specific agents or servers. The system employs various strategies to select from the LLM's probability distribution:

Deterministic Selection: Choose the highest probability function:

$$f^* = \arg\max_f P(f \mid c), \tag{16.3}$$

where f^* represents the selected function and f ranges over all available functions.
Top-k Selection: Sample from the top k functions:

$$f^* \sim \text{TopK}(P(f \mid c), k), \tag{16.4}$$

where f^* represents the selected function sampled from the top k functions with highest probability, f ranges over all available functions, and k is the number of top candidates to consider.

16.4 Implementation and Execution Patterns

Building unified agent systems that effectively integrate MCP and A2A protocols requires understanding both the theoretical foundations and practical implementation approaches. The key to successful integration lies in recognizing that these protocols serve complementary roles and can be combined intelligently to create systems that are more capable than individual LLMs operating in isolation.

16.4.1 Execution Patterns

The implementation of unified agent systems follows several natural execution patterns that emerge from the need to handle complex, dynamic scenarios. These patterns provide structured approaches for coordinating protocol usage and managing complex decision-making processes. This section establishes the theoretical frameworks and methodological approaches for implementing unified agent systems, focusing on the patterns themselves rather than specific application scenarios.

16.4.1.1 Information Gathering Loop Pattern

One fundamental pattern is the *information gathering loop*, where systems cycle through multiple information sources to build comprehensive understanding before making critical decisions. Consider a CAV approaching an intersection with limited visibility. The system doesn't simply make a single routing decision; instead, it orchestrates a sequence of information-gathering actions:

```
Goal: Navigate through intersection safely
  • MCP: Query traffic light status
  • A2A: Request visual analysis from nearby vehicles
  • MCP: Get weather conditions
  • MCP: Get traffic pattern updates
  • Repeat until sufficient information gathered
```

This pattern becomes particularly crucial in emergency situations. When a CAV detects a sudden obstacle such as a fallen tree, the system immediately begins gathering information from multiple sources. Through MCP, it contacts emergency services to determine availability and response times, checks road closure data, and obtains current weather conditions that might affect the emergency response. Simultaneously, through A2A protocols, it broadcasts hazard alerts to nearby vehicles, requests hazard information from other vehicles that might have encountered similar situations, and coordinates emergency response efforts. The system continues this information-gathering loop until it has sufficient data to make a safe response decision.

16.4.1.2 Progressive Goal Refinement Pattern

Another essential pattern is *progressive goal refinement*, where systems start with broad objectives and progressively refine them based on information gathering and continuously updated conditions. This pattern reflects the natural way humans approach complex tasks, beginning with general intentions and becoming more specific as more information becomes available.

Consider a CAV with the initial goal of "Get to destination." Rather than attempting to plan the entire journey at once, the system begins with this broad objective and progressively refines it:

```
Initial Goal: "Get to destination"
  • MCP: Get route information
  • MCP: Get current traffic updates
  • MCP: Check parking availability
  • A2A: Coordinate with other vehicles
  • Refine goal based on new information
```

A practical example of this pattern occurs in airport pickup scenarios. A CAV might start with the simple goal "Pick up passenger at airport," but as it gathers information, this goal becomes increasingly specific. Initially, the system uses MCP to obtain the airport's location and basic route information. As it approaches, MCP protocols allow it to check the passenger's flight status, refining the goal to "Go to Terminal 2." Further information gathering through both MCP (checking flight delays) and A2A (coordinating with the passenger) leads to "Wait at Terminal 2, Gate B15." Finally, low-latency coordination through A2A protocols results in the final refined goal: "Pick up at curbside, Zone 3."

16.4.1.3 Multi-agent Coordination Pattern

The third fundamental pattern is *multi-agent coordination*, which enables systems to coordinate complex maneuvers with other vehicles while gathering deterministic data to support decision-making. This pattern is particularly important for safety-critical operations where precise timing and coordination are essential.

Consider a CAV that needs to merge into traffic on a busy highway. The system doesn't simply attempt the maneuver based on its own sensors; instead, it engages in a coordinated process with nearby vehicles:

```
CAV Agent: "I need to merge into traffic"
  • A2A: Broadcast merge intention to nearby agents
  • A2A: Receive responses from nearby vehicles
  • MCP: Get traffic flow data
  • A2A: Negotiate merge timing
  • Execute coordinated merge
```

In practice, this coordination process unfolds through a series of carefully orchestrated steps. The CAV begins by broadcasting its intention to change lanes from Lane 2 to Lane 3 through A2A protocols, which support broadcasting messages to multiple agents using structured task and artifact objects. Nearby vehicles respond with their availability and preferred timing, creating a collaborative negotiation environment. Simultaneously, the system uses MCP to gather current traffic speed, density, and road conditions, providing objective data to inform the coordination decisions. The A2A protocols then facilitate negotiation of the exact timing for the lane change, ensuring that all vehicles involved can execute the maneuver safely and efficiently. Finally, the coordinated merge is executed with all vehicles working in synchronization.

16.4.2 Implementation Details

The practical implementation of these patterns requires careful attention to the technical details of integrating LLMs with protocol-specific clients. One effective approach uses the Ollama framework [3] to provide LLM capabilities for intelligent function selection. The key insight is that the LLM operates at a higher level of abstraction, receiving context about available tools and agents, then selecting appropriate functions without needing to know which specific protocol will handle the execution.

The implementation centers around three main components that work together seamlessly. First, the system constructs a comprehensive context that includes both available MCP tools and A2A agents, along with the current situation. This context construction process combines information from multiple sources:

```
# Build context by combining tools, agents, and current situation
context = list_tools(context["tool_list"]) + \
     list_agents(context["agent_card"]) + \
     current_situation
```

The LLM then processes this context along with system instructions and the user's query to select appropriate functions. The beauty of this approach is that the LLM makes decisions based on natural language understanding of the situation without needing to understand the underlying protocol mechanics:

Listing 16.1 LLM Function Selection

```
# Construct prompt with system instructions, context, and user query
prompt = system_prompt + context + user_query

# Get LLM response
response = ollama.chat(model="llama3.1", messages=[{"role": "user", "content":
    prompt}])
selected_function = response['message']['content'].strip()
```

Once the LLM has selected a function, the system handles the protocol routing transparently. The routing logic examines the selected function and determines whether it belongs to the MCP tool list or the A2A agent card, and then routes the execution to the appropriate client:

Listing 16.2 Protocol Routing Function

```
def execute_function(self, selected_function, parameters=None):
    """System routes selected function to appropriate protocol"""
    if selected_function in self.tool_list:
        return self.mcp_client.execute(selected_function, parameters)
    elif selected_function in self.agent_card:
        return self.a2a_client.execute(selected_function, parameters)
    else:
        raise ValueError(f"Unknown function: {selected_function}")
```

This separation of concerns allows the LLM to focus on high-level reasoning and function selection, while the system handles the technical details of protocol communication and execution. The result is a unified interface that presents both MCP and A2A capabilities as a single, coherent set of functions available to the LLM.

16.4.3 Comprehensive Examples

The following scenarios demonstrate how the execution patterns established in Section 16.4 work together in practice. While Section 16.4 focused on the theoretical frameworks and patterns themselves, this section provides detailed, concrete applications of those patterns in real-world CAV scenarios, illustrating how the abstract patterns translate into practical system behavior.

Example 1: Urban Intersection Navigation

The first scenario involves a CAV approaching a complex urban intersection with multiple lanes, pedestrian crossings, and limited visibility due to construction barriers. This situation requires the system to gather information from multiple sources and coordinate with other vehicles to navigate safely.

The system begins by constructing a comprehensive context that includes both MCP tools for traffic light status, pedestrian detection, and weather conditions, as well as A2A agents for visual analysis, vehicle coordination, and construction updates. The LLM then analyzes this context and selects the most appropriate functions based on the current situation. It chooses traffic light status functions from the MCP tools because this information is critical for intersection navigation. It selects visual analysis functions from the A2A agents to overcome the visibility limitations caused by construction barriers. It also chooses pedestrian detection functions from the MCP tools as a safety requirement.

The protocol execution phase then unfolds seamlessly. The MCP client executes the traffic light status and pedestrian detection functions, providing deterministic data about the intersection state. Simultaneously, the A2A client coordinates with nearby vehicles to obtain visual analysis that can help overcome the visibility limitations. The results from both protocols are integrated to provide a comprehensive understanding of the intersection conditions.

Example 2: Emergency Response Coordination

A second scenario demonstrates the system's ability to handle emergency situations through coordinated response. When a CAV detects a sudden obstacle, such as a fallen tree blocking the roadway, the system must immediately coordinate emergency response with other vehicles and emergency services. The immediate response phase involves using MCP to contact emergency services, broadcasting hazard alerts to nearby vehicles via A2A protocols, and checking alternative routes using MCP's road closure data.

The coordination phase then builds upon this initial response by working with other vehicles through A2A's emergency coordination functions, coordinating with traffic management systems via MCP's traffic data services, and negotiating a safe stopping zone through A2A's safety protocols. This multi-protocol approach ensures that the emergency response is both comprehensive and coordinated, leveraging both external emergency services and peer vehicle collaboration. The resolution phase involves executing a coordinated emergency stop while maintaining communication with emergency services and updating other vehicles with the resolution status.

Example 3: Multimodal Transportation Hub

A third scenario illustrates the system's capability to handle complex multimodal environments. Consider a CAV navigating a busy transportation hub where buses, trains, and pedestrians all interact with dynamic scheduling and continuously updated changes. This environment requires the system to gather information from multiple sources, coordinate with various transportation modes, and adapt to changing conditions.

The information-gathering phase begins with MCP providing access to public transit schedules and current pedestrian flow data. The system also uses MCP to check parking availability, creating a comprehensive picture of the hub's current state.

The coordination phase then leverages A2A protocols to synchronize with transit vehicles, negotiate pedestrian crossing timing, and handle continuously updated schedule changes using MCP's transit update functions. This coordinated approach allows the system to navigate through the hub following a carefully orchestrated plan that accounts for all transportation modes. The system adapts to continuously updated changes in transit schedules while maintaining safety with all transportation modes.

16.5 Summary

This chapter established a comprehensive framework for integrating MCP and A2A protocols within unified agent systems, focusing on how these protocols create intelligence that combines external capabilities with collaborative reasoning. The chapter demonstrated how unified agent systems can be more capable than individual LLMs through strategic protocol integration.

The protocol comparison and decision framework (Section 16.2) established clear criteria for selecting between MCP and A2A protocols based on use case requirements. MCP emerged as well-suited for context-oriented connectivity, providing access to external tools, services, and low-latency data suitable for safety-critical operations (as detailed in Chapter 14). A2A was identified as ideal for peer-oriented communication, enabling collaborative reasoning and multi-agent coordination (as detailed in Chapter 15). The comparison revealed that effective integration requires understanding when to use each protocol, with hybrid approaches often providing the most comprehensive solutions.

The system architecture framework provides the foundation for building unified agent systems that can handle function calling, protocol routing, and comprehensive intelligence. The function calling mechanism allows LLMs to select appropriate functions from available tools and agents without knowing which protocol will be used, while the system handles protocol routing transparently.

The practical implementation using the Ollama package demonstrated how to integrate LLM capabilities with unified agent systems. The context construction, function selection, and protocol routing mechanisms provide concrete implementation guidance for building robust systems. The implementation patterns and execution scenarios showed how unified agent systems can handle complex scenarios through both external resource access and collaborative reasoning.

This approach represents a significant advancement in enabling sophisticated unified agent systems that seamlessly integrate external capabilities with collaborative intelligence, establishing the foundation for next-generation agent systems that are more capable than individual LLMs through strategic protocol integration.

References

1 Anthropic. Model context protocol (MCP). Protocol specification for connecting AI applications to external tools and data sources, 2024. https://modelcontextprotocol.io/.

2 Google DeepMind. Agent-to-agent protocol (A2A). Protocol for inter-agent communication and collaboration in multi-agent systems, 2024. https://github.com/google-deepmind/a2a-protocol.

3 Ollama. Ollama: Run large language models locally, 2024. https://ollama.ai/.

Part VI

Adaptation and Deployment

Real-world deployment demands systems that can adapt behavior without redefining their underlying semantic representations. This part addresses adaptation mechanisms that refine and optimize system behavior after foundational models and architectures have been established. The chapters cover supervised fine-tuning for task specialization, parameter-efficient adaptation, reinforcement learning fine-tuning for policy optimization, and knowledge distillation for deployment on resource-constrained platforms. These methods are presented as complementary strategies for behavioral adaptation, clearly distinguished from representation learning addressed earlier in the book.

17

Supervised Fine-tuning

Consider a scenario where a CAV encounters a complex traffic situation: a pedestrian approaches a crosswalk while another vehicle is attempting to merge into the same lane. The CAV's decision-making system must not only execute appropriate maneuvers but also communicate its reasoning to the driver in natural, contextually appropriate language. Traditional rule-based systems can handle the control logic, but generating human-readable explanations that are both informative and reassuring requires a different approach, one that can understand context, adapt to regional communication styles, and maintain consistency across diverse scenarios.

This chapter introduces supervised fine-tuning as a critical technique for adapting large language models to the specific communication needs of CAV systems. Supervised fine-tuning enables models to produce concise, domain-specific, and safety-appropriate communications that enhance human-machine interaction in autonomous vehicles.

17.1 The Necessity of Fine-tuning in CAV Applications

CAVs are increasingly expected to operate not only with precision and efficiency but also with the ability to communicate naturally with drivers, passengers, pedestrians, and infrastructure systems. As generative AI becomes integrated into the software stack of modern vehicles, the requirement for models that generate domain-appropriate language in safety-critical and resource-constrained contexts has grown considerably. Fine-tuning addresses these requirements by embedding domain knowledge directly into model parameters.

17.1.1 Limitations of Prompting, RAG, and Function Calling

Earlier chapters have examined prompt engineering, function calling, and retrieval-augmented methods as means of extending large language model capabilities. While these methods offer flexibility and generality, they exhibit significant limitations in CAV scenarios, particularly in embedded or resource-constrained deployments. Prompting is inherently unstable in edge cases or unfamiliar inputs, often resulting in inconsistent or verbose outputs that deviate from the expected structure or tone. Function calling provides a structured interface to external actions but does not generate natural language; it requires an additional generative mechanism for explanation or feedback. RAG depends on connectivity and additional storage to index relevant knowledge sources, which may not be available on-vehicle or in latency-sensitive use cases. These limitations underscore the need for an approach that embeds domain knowledge, linguistic fluency, and operational safety directly into the model weights.

17.1.2 Fine-tuning as a Domain Adaptation Strategy

Fine-tuning refers to the process of adjusting the parameters of a pretrained language model to specialize it for a specific downstream task [1]. This method uses a relatively small, labeled dataset to guide the model toward producing outputs that reflect the style, semantics, and constraints of the target application domain. For CAV systems, fine-tuning enables models to generate driver-facing explanations that are both contextually relevant and linguistically controlled, even in the absence of external resources.

The benefits of fine-tuning in the CAV context include task alignment, where outputs conform to required linguistic and structural conventions; consistency, ensuring responses remain robust to variations in input structure or terminology; and embedded deployment, as fine-tuned models eliminate the need for real-time retrieval or external prompting logic.

17.2 From Foundation Models to Specialization

Modern generative AI systems, including those deployed in CAVs, are typically constructed in two stages. The first stage develops a foundation model that captures general-purpose knowledge and linguistic ability. The second stage specializes this foundation model for a particular application domain through supervised adaptation.

In the foundation stage, a large neural model is trained on diverse text data to learn general language representations and patterns. The resulting model, often referred to as a *foundation model*, serves as a versatile base for a wide range of downstream applications. In the specialization stage, the foundation model is further adapted to specific tasks or domains using a smaller, labeled dataset. This stage is referred to as *fine-tuning* or task-specific specialization, where the model acquires the necessary linguistic, stylistic, or semantic precision for the target application.

This process is especially important in the context of CAV systems, where real-world usage demands highly specialized behaviors under resource and safety constraints. Foundation models provide general capabilities, while specialization aligns those capabilities with task-specific requirements such as V2X message explanation, intent recognition, or driver communication.

17.2.1 Foundation Models via Pretraining

Foundation models are typically trained using a method known as *pretraining*, which relies on large-scale self-supervised learning. In this process, the model is exposed to vast corpora of unstructured text data, including books, encyclopedias, web pages, and software repositories, without any need for manually labeled outputs.

A common objective during pretraining is next-token prediction. For example, given the input "The car is turning," the model is trained to predict the next token, such as "left." By repeatedly solving such tasks, the model gradually acquires syntactic, semantic, and even commonsense knowledge, all encoded within its internal parameters.

At the conclusion of this stage, the model can be considered a general-purpose language model, capable of generating coherent sentences, answering factual questions, and interpreting broad linguistic context. However, it remains unaligned with the specific needs of safety-critical, domain-sensitive CAV systems.

Table 17.1 Comparison of the foundation and specialization stages in neural language model development.

Aspect	Foundation stage (pretraining)	Specialization stage (fine-tuning)
Purpose	Acquire general language and world knowledge	Adapt to a specific task or domain
Data type	Large-scale, unlabeled corpus	Small-scale, labeled dataset
Supervision	Self-supervised (e.g., next-token prediction)	Supervised (input-output pairs)
Output labels	Not required	Required
Training cost	Very high (distributed training infrastructure)	Moderate (GPU or CPU-level resources)
Scope of generalization	Broad and open-ended	Narrow and task-specific
Example	Training GPT or T5 (text-to-text transfer transformer) on web-scale text	Fine-tuning T5 for intent explanation in CAV

17.2.2 Specialization Through Supervised Fine-tuning

Once a foundation model is established, it can be further specialized via supervised fine-tuning. This involves training the model on a curated dataset of input-output pairs that exemplify the desired behavior in a specific task. Each training example presents a structured input and a corresponding linguistic output, allowing the model to learn the required transformation.

Consider the following input:

```
{"intentType": "YIELD_REQUEST", "participantId": "Vehicle_12"}
```

and the target output:

```
"Vehicle 12 is requesting to yield."
```

Fine-tuning aligns the model's generative capabilities with the formal semantics and communicative expectations of such inputs, enabling robust, consistent behavior across varying instances.

The distinction between foundation model development and domain specialization can be summarized in Table 17.1. This comparison highlights the contrast in objectives, data requirements, and deployment relevance between the two stages.

17.3 Architectural Perspective on Fine-tuning

Fine-tuning adapts a pretrained transformer model by adjusting a subset of its internal parameters. These modifications enable the model to specialize in a downstream task, such as intent explanation or structured message translation in connected vehicle systems. While a detailed explanation of the transformer architecture was provided in Chapter 2, the following identifies which components are trainable and commonly modified during fine-tuning.

17.3.1 Fine-tunable Components

Transformer models consist of multiple layers, each composed of attention mechanisms and feedforward sublayers. The fine-tunable components include token embeddings, attention weights, feedforward layers, and the output projection, as detailed in the following section.

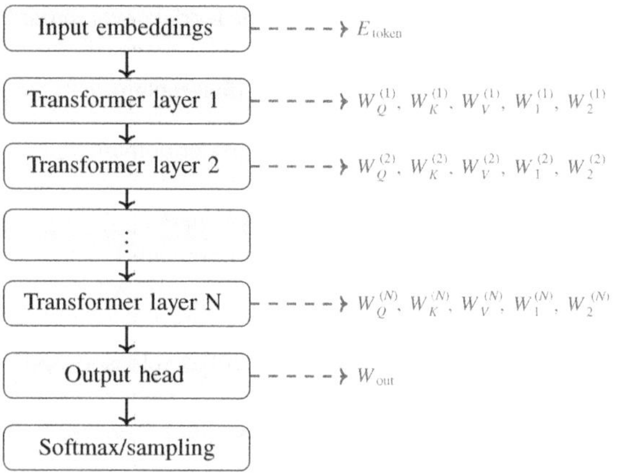

Figure 17.1 Fine-tuning scope within a transformer model. All components can be optionally fine-tuned, from embeddings and transformer layers to the output projection. The choice depends on task requirements and computational constraints.

Figure 17.1 illustrates the architectural components of a transformer model that can be fine-tuned. The dashed arrows indicate that any component, from embeddings and transformer layers to the output head, can be selected for fine-tuning based on task requirements.

17.3.2 Fine-tuning Strategies

This flexibility enables various fine-tuning strategies. In full fine-tuning, all model parameters are updated during training, including the entire transformer stack and the output head. While this strategy offers maximal expressiveness, it is computationally expensive and requires substantial labeled data.

In contrast, selective fine-tuning may update only a subset of the model. Common approaches include updating only the output head while keeping the base model frozen, fine-tuning a subset of transformer layers, or applying parameter-efficient methods such as adapter modules or low-rank updates, discussed in Section 17.6. The choice depends on task complexity, available compute resources, and the risk of overwriting useful general representations acquired during pretraining.

17.4 Mathematical Formulation of Fine-tuning

Fine-tuning a pretrained language model involves adjusting its parameters to better perform a target task using a labeled dataset. This process can be formally expressed as a supervised optimization problem, where the objective is to minimize the prediction error between the model's output and the correct target over the dataset.

17.4.1 Supervised Fine-tuning as Risk Minimization

Let $\mathcal{D} = \{(x_i, y_i)\}_{i=1}^N$ denote a dataset of N labeled examples, where each x_i represents an input such as a text prompt or sensor data and y_i is the corresponding desired output such as a classification label or generated response.

Fine-tuning seeks to adjust the parameters of a pretrained model $f(x;\theta)$ so that it produces outputs $\hat{y}_i = f(x_i;\theta)$ that closely match the ground-truth targets y_i. This process is typically cast as an empirical risk minimization problem.

The supervised fine-tuning loop, illustrated in Figure 17.2, involves the model processing an input, generating a prediction, comparing the prediction to the true output using a loss function, and updating its parameters accordingly.

Formally, the objective is to minimize the average loss over the dataset:

$$\mathcal{L}(\theta) = \frac{1}{N}\sum_{i=1}^{N} \ell(f(x_i;\,\theta),\,y_i), \tag{17.1}$$

where $\ell(\cdot,\cdot)$ is a task-specific loss function, and θ denotes all trainable parameters of the model.

The parameter set θ includes the following components, as illustrated in Figure 17.1. Token embeddings $E_{\text{token}} \in \mathbb{R}^{|V|\times d}$ represent each vocabulary token in a continuous vector space, where $|V|$ is the vocabulary size and d is the embedding dimension. Attention weights $W_Q^{(\ell)}, W_K^{(\ell)}, W_V^{(\ell)} \in \mathbb{R}^{d\times d}$ define the internal projections used in self-attention at each layer ℓ, where d is the model dimension. Feedforward layers $W_1^{(\ell)} \in \mathbb{R}^{d\times d_{\text{ff}}}$ and $W_2^{(\ell)} \in \mathbb{R}^{d_{\text{ff}}\times d}$ apply transformations after the attention mechanism within each layer, where d_{ff} is the feedforward dimension. The output projection $W_{\text{out}} \in \mathbb{R}^{d\times|V|}$ maps the model's final hidden representation to token logits used for output generation. Depending on the fine-tuning strategy, any subset of these parameters may be updated, from updating only the output projection to full model fine-tuning.

17.4.2 Gradient-based Parameter Updates

Fine-tuning is typically carried out using stochastic gradient descent or one of its adaptive variants. The parameters are updated iteratively using the rule:

$$\theta \leftarrow \theta - \eta\nabla_\theta\mathcal{L}, \tag{17.2}$$

where η is the learning rate and $\nabla_\theta\mathcal{L}$ is the gradient of the loss with respect to the model parameters. Mini-batches are used to estimate this gradient efficiently over large datasets. This iterative process gradually reduces the loss, aligning the model's outputs with the labeled examples in the dataset.

17.4.3 Task-specific Loss Functions

The choice of loss function $\ell(\cdot,\cdot)$ is determined by the task definition. Table 17.2 summarizes representative loss functions used in fine-tuning tasks relevant to connected vehicle systems. These losses guide the gradient computation used to update parameters during training. In practice, token-level cross-entropy is the most widely used objective for natural language generation tasks in CAV systems.

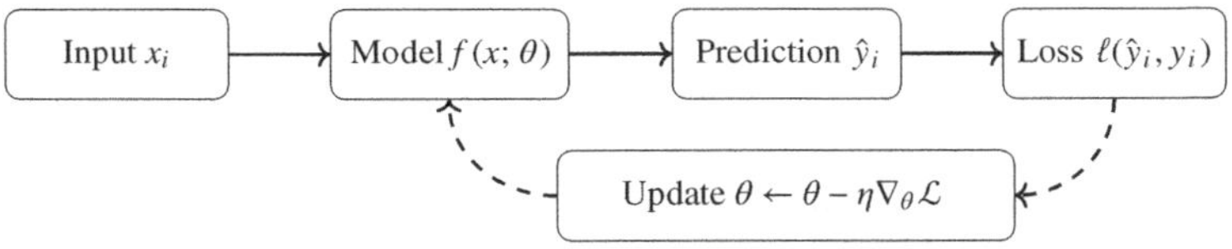

Figure 17.2 The supervised fine-tuning loop: prediction, loss computation, and parameter update.

Table 17.2 Common task-specific loss functions used in fine-tuning.

Task type	Output format	Typical loss function
Text generation (structured-to-text)	Token sequence	Cross-entropy (token-level)
Classification (e.g., intent detection)	Discrete label	Cross-entropy
Regression (e.g., distance estimation)	Real-valued scalar	Mean squared error (MSE), Mean absolute error (MAE)

17.5 Practical Fine-tuning

The mathematical formulation defines fine-tuning as an optimization problem over a labeled dataset. The following describes how that process is instantiated in practice, particularly for deploying generative models in CAV systems.

17.5.1 Fine-tuning Workflow Steps

Fine-tuning generally proceeds through four principal stages. First, labeled data preparation involves defining supervised input-output pairs that reflect the task objectives. Second, model loading initializes a pretrained foundation model suitable for the target task. Third, training adjusts model weights to minimize prediction error over the training data. Finally, evaluation measures generalization performance on held-out examples. This process is summarized in Figure 17.3 in a modular fashion.

17.5.1.1 Step 1: Labeled Data Preparation

The first step is to construct a labeled dataset $\mathcal{D} = \{(x_i, y_i)\}$ consisting of task-aligned examples. In CAV applications, the inputs x_i are often structured telemetry messages or V2X communication records, while the outputs y_i are explanatory utterances intended for the driver or other agents.

In the absence of preexisting datasets, pseudo-labeled data may be synthesized by prompting a large model and then manually reviewing the results for fluency, factuality, and tone. This form of bootstrapping is often practical for CAV-specific message formats.

17.5.1.2 Step 2: Load a Pretrained Model

A pretrained model is selected based on the target task and deployment constraints. For encoder-decoder architectures, models like T5 (text-to-text transfer transformer) or BART are often preferred for structured-to-text generation. For causal generation, models such as GPT [2] or DistilGPT2 may be used.

The foundation model is typically loaded from a public or internal repository and includes weights obtained via large-scale pretraining. These weights serve as the initialization point for the fine-tuning phase.

17.5.1.3 Step 3: Train the Model

Once the model and dataset are in place, fine-tuning proceeds by minimizing the loss function over the training data, as described in the mathematical formulation section. Depending on task requirements and resource availability, one of the following strategies may be applied:

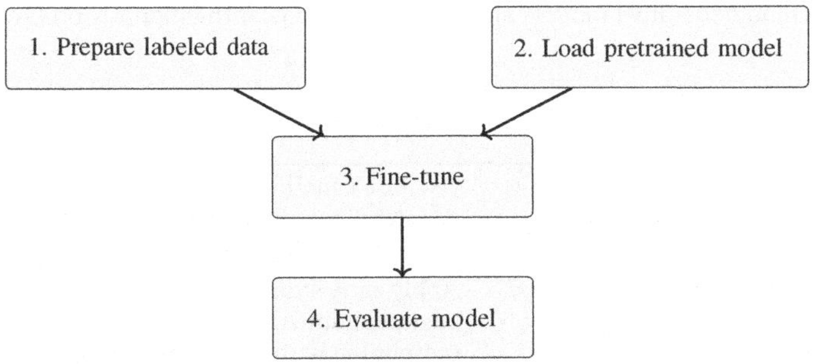

Figure 17.3 Practical fine-tuning workflow: combining data and pretrained models to specialize for downstream tasks.

full fine-tuning, which updates all model parameters θ; partial fine-tuning, which freezes most layers and trains only the output head; or parameter-efficient fine-tuning (PEFT), which trains small adapter modules or low-rank matrices. In CAV systems, PEFT is often preferred due to reduced training cost and ease of deployment on embedded platforms.

17.5.1.4 Step 4: Evaluate the Model

After training, the fine-tuned model is evaluated on held-out data to assess its generalization capability. Evaluation typically includes both quantitative metrics and qualitative inspection.

Common evaluation metrics include exact match accuracy, BLEU (bilingual evaluation understudy) score, ROUGE-L (recall-oriented understudy for gisting evaluation—longest common subsequence), and token accuracy. For detailed discussion of evaluation methodologies and metrics, see Chapter 21.

17.5.2 When to Fine-tune and When Not To

Fine-tuning is most appropriate in scenarios where the language model must generate structured, domain-aligned, and contextually fluent outputs. Examples include explanation of structured signals, such as translating V2X messages or telemetry into natural language for human drivers; intent understanding, which maps natural language commands to intent categories or policy actions; tone and style alignment, generating output that reflects driver preference, regional norms, or brand voice; and on-device specialization, adapting models for deployment in memory- and latency-constrained environments.

In these settings, fine-tuning provides consistency, stylistic control, and reliability that are typically difficult to achieve through prompting alone.

There are scenarios where fine-tuning may be unnecessary or even counterproductive. Symbolic or numeric reasoning tasks requiring geometric computation, coordinate transformations, or logical consistency are better handled by rule-based or symbolic engines. Stable mappings with narrow variability, where outputs can be generated reliably using simple rules or templates (e.g., unit conversion, constant warnings), may introduce unnecessary variability through fine-tuning. Highly dynamic inputs, where external context (e.g., weather updates or map data) changes frequently, may be more effectively handled by retrieval or function calling rather than static fine-tuned behavior.

In such cases, prompting, RAG, or symbolic components should be considered instead.

Table 17.3 Common pitfalls in fine-tuning generative models and their mitigations, with emphasis on CAV deployment scenarios.

Issue	Description	Mitigation strategy
Catastrophic forgetting	Fine-tuning causes the model to overwrite general language capabilities acquired during pretraining.	Use PEFT methods (e.g., LoRA, adapters), regularization, or interleave general-domain data during training.
Overfitting on small datasets	The model memorizes training examples, failing to generalize to unseen inputs.	Apply early stopping, data augmentation, dropout, or leverage pretraining with low learning rates.
Instability in RL fine-tuning	High variance in gradient updates leads to erratic learning behavior.	Use clipped objectives (proximal policy optimization (PPO), group relative policy optimization (GRPO)), advantage normalization, KL regularization, and entropy bonuses.
Reward misspecification	Poorly designed reward functions lead to degenerate outputs (e.g., overly verbose or evasive language).	Combine multiple reward signals, validate reward model outputs, and conduct human-in-the-loop evaluations.
Distribution shift in deployment	Model encounters inputs during inference that were not represented during training.	Use domain adaptation techniques, continual learning, and augment training with synthetic or diverse samples.
Excessive compute or memory usage	Full fine-tuning exceeds hardware limits in embedded settings.	Adopt parameter-efficient strategies (PEFT), quantization, or knowledge distillation for deployment.

17.5.3 Common Pitfalls in Fine-tuning and Their Mitigations

While fine-tuning offers powerful mechanisms for domain adaptation, its practical implementation presents several challenges, particularly in safety-critical and resource-constrained environments such as CAVs. For example, a model fine-tuned on V2X message explanations may lose its ability to generate general conversational responses (catastrophic forgetting), or it may produce overly verbose outputs when the reward function incorrectly penalizes brevity (reward misspecification). In edge deployment scenarios, full fine-tuning may exceed available memory, requiring parameter-efficient alternatives. Common pitfalls and recommended strategies to mitigate them are summarized in Table 17.3. These considerations are critical when deploying fine-tuned models in real-world CAV systems, where both performance and safety guarantees must be maintained across dynamic and unpredictable operational contexts.

17.6 Parameter-efficient Fine-tuning Methods

While full fine-tuning provides maximum expressiveness by updating all parameters of a pretrained model, it is often computationally expensive and inefficient, especially in edge-deployed systems such as CAVs. Parameter-efficient fine-tuning (PEFT) methods address these challenges by adapting the model using a small, task-specific subset of parameters, while keeping the majority of the pretrained weights frozen.

17.6.1 Motivation of PEFT

Full fine-tuning typically involves updating hundreds of millions to billions of parameters. This presents several challenges. High memory and compute cost arise because training requires substantial GPU resources. Long training duration results from slow optimization across all parameters, which is also prone to overfitting. Deployment inefficiency occurs because each fine-tuned model instance must store a full model copy. Finally, forgetting risk exists because updating all weights can degrade useful general-purpose knowledge.

In contrast, PEFT strategies update only a small number of new parameters (often fewer than 1 million), resulting in lightweight, modular, and task-adaptive models suitable for embedded systems.

17.6.2 Core Principles of PEFT

PEFT methods are based on the principle of reusing the frozen foundation model while introducing additional task-specific components that are inserted into the existing architecture. These components are the only elements updated during training.

Desirable properties of PEFT include minimal parameter updates, where only a small number of task-specific weights are trained; composability, as task-specific adapters can be swapped or combined without altering the base model; and deployment efficiency, where parameter deltas are compact and easily transmitted to edge devices. Table 17.4 summarizes the differences between conventional full fine-tuning and parameter-efficient approaches.

Several techniques have been developed to enable PEFT within transformer architectures. Four prominent methods are **low-rank adaptation (LoRA)**, which injects low-rank matrices into attention layers; **adapter modules**, which insert bottleneck layers between transformer blocks; **prefix tuning**, which prepends learnable vectors to attention keys and values; and **prompt tuning**, which optimizes input embeddings. These methods differ in how and where task-specific information is injected into the model, and each offers trade-offs in expressiveness, training cost, and modularity. All techniques reduce the number of trainable parameters while maintaining compatibility with pretrained weights, supporting fast adaptation, model modularity, and task-specific behavior with minimal overhead. The choice of technique should reflect deployment constraints and target adaptation behavior.

17.6.3 Low-rank Adaptation

One of the most widely adopted PEFT methods is **low-rank adaptation (LoRA)** [3]. LoRA enables efficient adaptation of large language models by introducing small, trainable matrices into the transformer architecture, while keeping the original model weights frozen. This approach drastically reduces the number of trainable parameters, making it suitable for memory- and compute-constrained environments.

Table 17.4 Comparison of full fine-tuning and parameter-efficient fine-tuning strategies.

Aspect	Full Fine-tuning	Parameter-efficient Fine-tuning
Trainable parameters	All (100M–7B)	Tiny (<1M)
Memory requirement	High	Low
Base model weights	Updated	Frozen
Knowledge retention	May forget pretrained capabilities	Preserves pretrained knowledge
Modularity	Poor (entire model is task-bound)	High (adapters or deltas are task-specific)
Deployment size	Large	Compact
Best use case	High-resource training environments	On-vehicle, edge-constrained deployment

17.6.3.1 Motivation

Many learned weight matrices in transformers exhibit significant redundancy. LoRA addresses this by approximating weight updates using a low-rank decomposition, thereby introducing only a small number of task-specific parameters.

17.6.3.2 Mathematical Formulation

Consider a weight matrix $W \in \mathbb{R}^{d \times d}$ used in a linear projection, such as for query or value computation in attention layers. Instead of updating W directly, LoRA adds a low-rank residual term ΔW:

$$W_{\text{LoRA}} = W + \Delta W, \quad \Delta W = \alpha \cdot AB, \tag{17.3}$$

where $A \in \mathbb{R}^{d \times r}$ and $B \in \mathbb{R}^{r \times d}$ are trainable matrices, $r \ll d$ is the rank hyperparameter that controls the bottleneck dimension, and $\alpha \in \mathbb{R}$ is a scalar scaling factor that modulates the magnitude of the update.

The base matrix W remains frozen, and only A and B are optimized during training. The total number of new parameters is $2dr$, which is orders of magnitude smaller than d^2.

The scaling factor α is often normalized as α/r in practical implementations to maintain stable gradient magnitudes. In the lab notebook associated with this chapter, α is a tunable hyperparameter that determines the strength of task-specific adaptation. The LoRA mechanism is illustrated in Figure 17.4.

17.6.3.3 Insertion Points

LoRA modules are typically injected into the linear projections of transformer attention layers, namely the query W_Q, key W_K, or value W_V matrices. Optionally, LoRA can also be applied to the feedforward network layers. At inference time, the low-rank update ΔW can be merged with the frozen W for computational efficiency.

17.6.3.4 Interpretation

Intuitively, LoRA learns a low-dimensional detour through the high-dimensional parameter space of the model. This detour can express task-specific deviations while preserving the pretrained model's general-purpose linguistic knowledge.

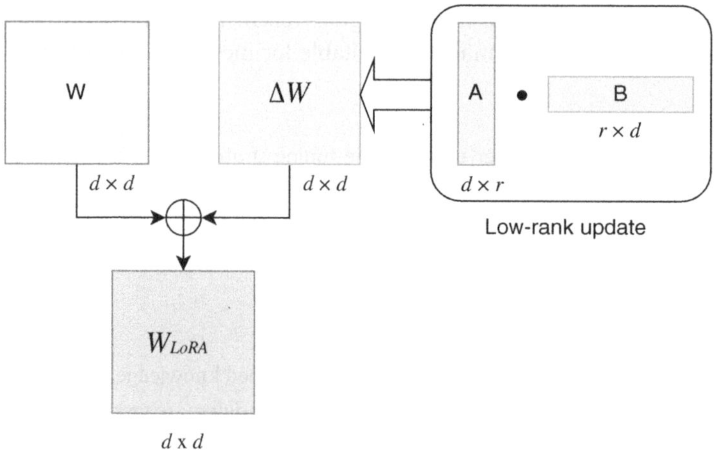

Figure 17.4 LoRA inserts a low-rank residual update $\Delta W = \alpha AB$ into the frozen base weight W. Only A and B are optimized.

17.6.3.5 CAV Deployment

LoRA is well-suited for CAV use cases where compute, memory, and update size are constrained. Different LoRA adapters can be trained for different tasks (e.g., scene explanation vs. driver response) or regions, enabling modular deployment while minimizing catastrophic forgetting. For example, a LoRA adapter trained on V2X messages requires only a few thousand parameters to enable structured-to-natural language generation on edge devices.

17.6.4 Adapter Modules

Adapters provide another widely used approach to parameter-efficient fine-tuning by inserting small, trainable bottleneck layers between the frozen layers of a transformer model. Originally proposed for multilingual NLP [4], adapters have since proven useful for multitask learning, modular deployment, and rapid adaptation to domain-specific language generation.

17.6.4.1 Core Concept

Rather than modifying the full set of model weights, the adapter method augments the network by inserting lightweight neural modules, typically two-layer feedforward blocks, between the existing layers of a pretrained model. During training, only these adapter modules are updated, while the original transformer parameters remain frozen.

17.6.4.2 Architecture

A typical adapter module is composed of two linear projections with a nonlinearity in between:

$$\text{Adapter}(h) = W_{\text{up}} \cdot \sigma(W_{\text{down}} \cdot h), \tag{17.4}$$

where $h \in \mathbb{R}^d$ is the hidden representation from the previous layer, $W_{\text{down}} \in \mathbb{R}^{r\times d}$ projects the input to a lower-dimensional bottleneck with $r \ll d$, $W_{\text{up}} \in \mathbb{R}^{d\times r}$ projects it back to the original dimension, and $\sigma(\cdot)$ is a nonlinear activation function, typically ReLU or GELU.

The output of the adapter is either added to the residual stream or concatenated, depending on the implementation. In most variants, a residual connection is used:

$$h_{\text{output}} = h + \text{Adapter}(h). \tag{17.5}$$

17.6.4.3 Parameter Efficiency

The number of additional parameters per adapter module is $2dr$, which is similar to LoRA. However, adapter modules can be applied to any part of the model, including both attention and feed-forward blocks. Because the base model is frozen, multiple adapters can be trained and swapped in without modifying the backbone weights.

17.6.4.4 CAV Deployment

Adapter modules enable modular deployment where separate adapters correspond to specific domains, regions, or tasks (e.g., urban vs. highway driving narratives, or English vs. multilingual instruction). A shared backbone model can be reused with region-specific adapters, allowing dynamic loading or switching based on operating context. Compared to LoRA, adapters insert separate modules that can represent more complex transformations, offering slightly more expressive capacity at the cost of minor additional inference latency.

17.6.5 Prefix Tuning

Prefix tuning is an extremely lightweight fine-tuning method that adapts pretrained language models by prepending a set of learnable vectors, called *soft prefixes*, to the key and value streams of the attention mechanism [5]. Unlike prompt tuning, which modifies the input embedding layer, prefix tuning injects learnable control signals directly into the internal computation of each transformer layer, enabling more expressive control with minimal parameter overhead.

17.6.5.1 Motivation

Full fine-tuning and adapter methods require modifying model internals or storing per-task adapter modules. Prefix tuning offers a minimal-intrusion alternative: instead of modifying the network structure, it attaches prefix tokens to the self-attention computation, guiding generation with only a few thousand task-specific parameters.

17.6.5.2 Mechanism

Let $X = [x_1, x_2, \dots, x_n]$ denote a sequence of input tokens, and let $P = [p_1, p_2, \dots, p_k]$ be a learned prefix sequence, where each $p_i \in \mathbb{R}^d$ is a trainable vector of the same dimension as the token embeddings.

In each transformer layer, the attention keys and values are augmented as follows:

$$K_{\text{aug}} = [P_K \parallel K(X)], \quad V_{\text{aug}} = [P_V \parallel V(X)] \tag{17.6}$$

where $P_K, P_V \in \mathbb{R}^{k \times d}$ are the trainable prefix keys and values, respectively; $K(X), V(X)$ are the keys and values computed from the input tokens, and $\parallel$ denotes concatenation along the sequence length dimension.

The attention query remains unchanged but also attends to both the input-derived tokens and the prefix. Crucially, only P_K and P_V are trained; all other model weights are frozen.

17.6.5.3 Parameter Count

Because the prefix length k is typically small (e.g., 10–20), the number of trainable parameters per layer is minimal, just $2kd$ per layer. Across a 12-layer model, this yields a parameter footprint well under 1M.

17.6.5.4 CAV Deployment

Prefix tuning enables fast update cycles and contextual adaptation without altering the base model. Task- or region-specific driving behavior can be encoded in prefixes and switched at runtime. For example, separate prefix vectors can be trained for different tones (formal, casual, technical), allowing the same model to generate "The vehicle is reducing speed to maintain safe headway" (formal) or "Slowing down a bit to keep things safe" (casual) based on the selected prefix, without retraining the base model.

17.6.6 Prompt Tuning

Prompt tuning is a parameter-efficient fine-tuning technique that adapts pretrained language models by prepending a small number of trainable vectors, referred to as *soft prompts*, to the input embedding sequence [6]. Unlike prefix tuning, which intervenes inside the transformer's attention mechanism, prompt tuning operates entirely at the input layer, making it simpler to implement and deploy.

17.6.6.1 Motivation

Prompt tuning is motivated by the observation that language models exhibit strong in-context learning capabilities. By optimizing a set of continuous embeddings prepended to the input, the model can be steered toward task-specific behavior without changing any of its internal weights.

This approach is especially attractive for applications that require ultra-lightweight task adaptation, such as customizing outputs to user preferences, dialects, or specialized terminologies in resource-constrained environments.

17.6.6.2 Mechanism

Let $X = [x_1, x_2, \dots, x_n]$ be the input token sequence and $E(x_i) \in \mathbb{R}^d$ its embedding vector. Prompt tuning prepends a learned sequence of vectors $P = [p_1, p_2, \dots, p_k]$, where each $p_i \in \mathbb{R}^d$, to the embedded input:

$$\tilde{X} = [p_1, \dots, p_k, E(x_1), \dots, E(x_n)]. \tag{17.7}$$

The prompt vectors p_i are initialized randomly and optimized using supervised learning on downstream tasks. During inference, the soft prompt is prepended to every input. The base model parameters remain frozen throughout.

17.6.6.3 Parameter Count

If k soft prompt vectors of dimension d are learned, the total parameter count is kd. For example, with $k = 20$ and $d = 768$ (a typical embedding dimension), the method requires only 15,360 trainable parameters, orders of magnitude smaller than full fine-tuning.

17.6.6.4 CAV Deployment

Prompt tuning is particularly appealing for minimal adaptation with low memory footprint. Unlike prefix tuning, which inserts control vectors into the attention mechanism, prompt tuning operates only at the input embedding layer, making it less expressive but easier to integrate and more compatible with non-autoregressive models. Soft prompts can be trained per driver or vehicle to reflect stylistic preferences or for quick domain injection (e.g., map instruction, weather response), requiring only a small embedding table for storage or transmission.

Table 17.5 compares the four PEFT methods discussed in this section, LoRA, adapters, prefix tuning, and prompt tuning, based on their architectural strategy and typical use in CAV contexts.

The modularity and compactness of PEFT make it particularly suitable for CAV systems, where runtime constraints and memory budgets are nontrivial. The benefits demonstrated across LoRA, adapters, prefix tuning, and prompt tuning include on-device adaptability, rapid retraining, regional customization, and multitask extensibility. These advantages enable practical deployment of fine-tuned models in resource-constrained vehicle environments.

17.7 Use Cases and Deployment Modes

Fine-tuning, including parameter-efficient methods, enables generative models to operate reliably within CAV systems. The following case studies illustrate how fine-tuned models improve language generation in domain-specific tasks, operate under deployment constraints, and contribute to human-machine interaction in safety-critical environments.

Table 17.5 Comparison of PEFT techniques for transformer-based models in CAV systems.

PEFT method	Core mechanism	Insertion point	CAV-specific use case
LoRA (low-rank adaptation)	Injects low-rank residual matrices into attention projections; only the residuals are trained	Attention layers (e.g., W_Q, W_V)	Efficient adaptation for structured-to-natural generation (e.g., V2X explanations) on edge devices
Adapters	Adds small bottleneck MLP modules between transformer layers; trained in isolation	Between attention or feedforward layers	Region- or task-specific modules (e.g., local driving norms, language variants) with modular loading
Prefix tuning	Prepends learned vectors to key/value attention inputs at every layer	Attention keys and values (internal to transformer)	Switchable tone, policy, or dialog intent across driving modes (e.g., safe, assertive, technical)
Prompt tuning	Appends learned embeddings to the input sequence; guides output without model changes	Input embedding layer only	Lightweight driver personalization or domain injection (e.g., preferences, prompt styles)

17.7.1 Structured Telemetry to Natural Language

One of the most representative applications of fine-tuning in CAVs involves translating structured telemetry or V2X data into driver-readable natural language. Foundation models are not inherently equipped to perform this task, as it requires domain-specific field interpretation and style control.

The following demonstrates a typical structured-to-text transformation: Given input { `"id": 257, "speed": 12.4, "heading": 90, "brake_status": "ON"` }, the target output is `"Vehicle 257 is moving east at about 12.4 m/s, and its brakes are applied."`

Fine-tuning enables the model to include all relevant fields (e.g., speed, heading, brake status), use consistent phrasing aligned with regional or brand voice, and avoid hallucination or omission of critical information.

Such structured-to-text capabilities are essential for building transparent CAV assistants that communicate their reasoning to occupants or nearby agents.

17.7.2 Region-specific Adaptation

CAVs often operate in diverse regulatory and linguistic environments. Fine-tuned language models can accommodate regional differences in tone, terminology, or formatting. For instance, different countries may expect differing levels of formality or unit systems (e.g., kilometers per hour vs. miles per hour).

PEFT methods such as LoRA allow manufacturers to deploy a single frozen base model with region-specific adapters. These adapters can be trained using in-market phrasing and terminology, swapped or layered dynamically depending on vehicle location, and updated over time without retraining the base model.

This modularity supports regulatory compliance, brand alignment, and localized user experience without fragmenting the model stack.

17.7.3 Behavioral Personalization

In addition to regional customization, fine-tuning supports personalization based on user preferences or driving style. For example, a vehicle assistant might generate different responses for cautious vs. assertive drivers.

The following illustrates how the same scenario produces different outputs based on driving style: For a *cautious style*, the model generates "Slowing down to maintain a safe distance from the merging vehicle"; for an *assertive style*, it generates "Reducing speed to yield to a vehicle entering the lane."

Personalized fine-tuning can be performed offline and deployed via lightweight adapter updates. This enhances driver trust and system acceptability by aligning the assistant's language with the driver's implicit expectations.

17.7.4 Multitask Language Interfaces

In complex CAV environments, a single language model may be expected to perform multiple related tasks, such as translating control events into human-readable explanations, answering natural language queries about recent maneuvers, and interpreting spoken driver commands into intent categories. PEFT enables this through distinct adapters or prompt templates for each task, allowing task-specific modules to be trained independently while sharing a common foundation model. This modularity supports efficient deployment where different capabilities can be activated based on context without maintaining separate full models.

17.7.5 Integration into Hybrid Architectures

Fine-tuned language models are typically embedded into larger hybrid systems where symbolic reasoning modules handle geometry, constraint checking, and precise control logic, while the language model provides the natural language interface. This division of responsibility ensures that language models operate within bounded scopes, generating human-readable output and interfacing with driver queries, while symbolic modules perform safety-critical reasoning. Such architecture reflects best practices in CAV deployment, where interpretability, robustness, and fail-safe behavior are paramount.

17.8 Summary

This chapter has introduced supervised fine-tuning as a critical technique for adapting large language models to the specific communication needs of CAV systems. Modern generative AI systems are built in two stages: pretraining creates general-purpose foundation models, while supervised fine-tuning specializes these models for specific downstream tasks. This two-stage approach is particularly important for CAV applications, where domain-specific behavior must be embedded directly into model weights.

Fine-tuning is formalized as an empirical risk minimization problem, where model parameters are adjusted to minimize prediction error over labeled datasets. The process involves gradient-based optimization of trainable parameters, which may include token embeddings, attention weights, feedforward layers, and output projections.

For resource-constrained CAV deployments, parameter-efficient fine-tuning (PEFT) methods such as LoRA, adapters, prefix tuning, and prompt tuning offer significant advantages. These techniques enable task-specific adaptation while preserving pretrained knowledge and minimizing computational overhead.

Fine-tuned models excel at structured-to-natural language generation, regional adaptation, behavioral personalization, and multitask language interfaces. These capabilities are essential for building transparent, trustworthy CAV systems that can communicate effectively with human operators.

The next chapter will explore reinforcement learning-based fine-tuning approaches, which extend these supervised methods to handle more complex optimization objectives and human feedback signals.

References

1 Jason Wei, Maarten Bosma, Vincent Y Zhao, Kelvin Guu, Adams Wei Yu, Brian Lester, Nan Du, Andrew M Dai, and Quoc V Le. Finetuned language models are zero-shot learners. *arXiv preprint arXiv:2109.01652*, 2021.

2 Alec Radford, Jeffrey Wu, Rewon Child, David Luan, Dario Amodei, and Ilya Sutskever. Language models are unsupervised multitask learners, *OpenAI blog*, 1(8):9, 2019.

3 Edward J Hu, Yelong Shen, Phillip Wallis, Zeyuan Allen-Zhu, Yuanzhi Li, Shean Wang, Lu Wang, Weizhu Chen. LoRA: low-rank adaptation of large language models. *ICLR*, 1(2):3, 2022.

4 Neil Houlsby, Andrei Giurgiu, Stanislaw Jastrzebski, Bruna Morrone, Quentin De Laroussilhe, Andrea Gesmundo, Mona Attariyan, and Sylvain Gelly. Parameter-efficient transfer learning for NLP. In *International Conference on Machine Learning*, pages 2790–2799. PMLR, 2019.

5 Xiang Lisa Li and Percy Liang. Prefix-tuning: optimizing continuous prompts for generation. In *Proceedings of the 59th Annual Meeting of the Association for Computational Linguistics and the 11th International Joint Conference on Natural Language Processing (Volume 1: Long Papers)*, pages 4582–4597, 2021.

6 Brian Lester, Rami Al-Rfou, and Noah Constant. The power of scale for parameter-efficient prompt tuning. In *Proceedings of the 2021 Conference on Empirical Methods in Natural Language Processing*, pages 3045–3059, 2021.

18

Reinforcement Learning Fine-tuning

18.1 Introduction

Chapter 17 introduced supervised fine-tuning as a method for adapting pretrained language models to specific tasks through labeled examples. While supervised fine-tuning enables instruction-following, it faces fundamental limitations when optimizing for subjective qualities like helpfulness, safety, or user preference alignment. This chapter extends the fine-tuning pipeline by introducing reinforcement learning fine-tuning, which addresses these limitations through preference-based optimization.

Consider an AI system that must communicate with users about important information, such as system status, recommendations, or alerts. While the system can generate factual statements like "System maintenance required due to scheduled updates," the challenge lies in determining how to phrase this information in a way that is helpful, reassuring, and contextually appropriate for the specific user and situation.

This scenario illustrates why reinforcement learning fine-tuning is needed: it enables optimization for subjective qualities like helpfulness, safety awareness, and user preference alignment that cannot be achieved through supervised fine-tuning alone. This chapter presents the complete supervised fine-tuning followed by reinforcement learning fine-tuning (SFT-RL) pipeline used by modern systems such as GPT models and DeepSeek.

The complete fine-tuning pipeline operates through three sequential stages: pretraining, supervised fine-tuning, and reinforcement learning fine-tuning. This progression from general capability to instruction-following to preference alignment represents the complete modern fine-tuning pipeline used in production systems.

18.1.1 What Is Reinforcement Learning?

Reinforcement learning represents a fundamentally different approach to machine learning compared to supervised learning. While supervised learning relies on learning from correct examples with known answers, reinforcement learning operates by learning from feedback about the quality of its actions, even when the "correct" answer is unknown.

This makes reinforcement learning particularly valuable for scenarios where we want to optimize for subjective qualities like helpfulness, safety, or user preference rather than exact correctness. The approach operates through three fundamental components: the agent (the decision-maker), the environment (the context where actions are taken), and the reward (the feedback signal indicating how good the action was). Figure 18.1 illustrates how these components work together in the context of generative models.

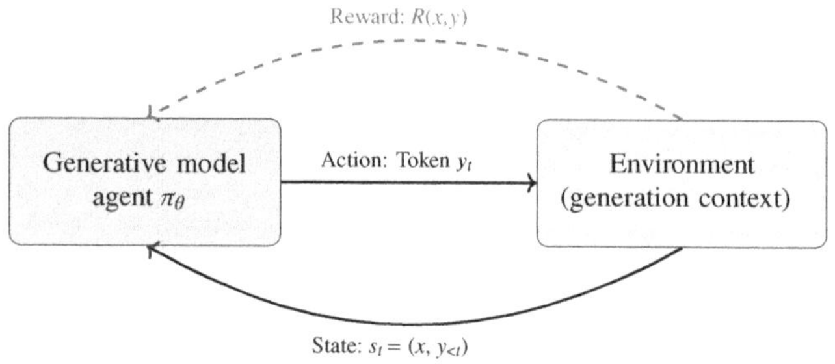

Figure 18.1 Reinforcement learning applied to generative models.

18.1.2 How RL Applies to Text Generation

In classical reinforcement learning settings, an agent takes actions in an environment and receives rewards. For text generation, this framework maps as follows: the state is the input prompt and previously generated tokens, the action is generating the next token, and the reward is a score for the complete generated sequence.

The key insight is that text generation can be treated as a sequence of decisions, where each token generation is an action and the final text quality is the reward. The RL framework maps naturally to generative model fine-tuning: the agent is the generative model that produces text outputs; the environment is the context and domain where generated text will be used and evaluated; the state is the input context (the input prompt and previously generated tokens); the action is token generation (the next token to be generated from the vocabulary); the reward is the quality score of the complete generated text; and the policy is the language model's probability distribution over next tokens.

18.1.3 The Complete Fine-tuning Pipeline

Modern generative AI systems are built through a three-stage process: pretraining, supervised fine-tuning, and reinforcement learning fine-tuning. This sequential approach reflects the fundamental learning requirements for creating aligned AI systems.

The pipeline begins with pretraining, where the model learns general language knowledge from large text corpora using self-supervised learning. This foundation is essential because general language understanding must precede task-specific capabilities. Without this foundation, models cannot effectively learn from instruction-following examples or human feedback. Supervised fine-tuning (SFT) then builds upon this foundation by teaching specific behaviors through curated examples, such as labeled input-output pairs (e.g., "Explain the current traffic situation."), establishing the baseline capabilities that reinforcement learning can then refine. Finally, reinforcement learning fine-tuning enables the model to learn from preference signals, optimizing for qualities like clarity, contextual appropriateness, and user preference that cannot be captured through supervised learning alone.

The transition from supervised to reinforcement learning represents a crucial shift from learning from correct examples to learning from preference signals. This shift enables optimization for subjective qualities like response quality, safety alignment, and user preference that cannot be captured through traditional supervised learning approaches.

Figure 18.2 illustrates this three-stage progression, showing how each stage builds upon the previous one to create increasingly capable and aligned models. The detailed implementation of this

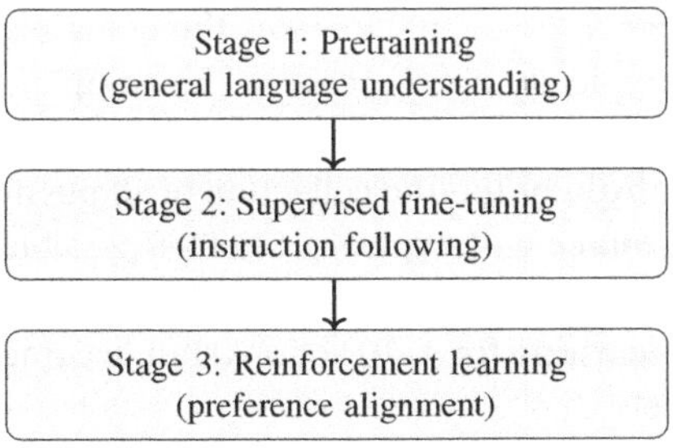

Figure 18.2 The three-stage fine-tuning pipeline for modern generative AI systems.

pipeline, including the specific algorithms and methodologies used at each stage, is examined in the following sections.

18.1.4 Why the Sequential SFT-RL Pipeline Is Essential

While supervised fine-tuning provides a solid foundation for instruction-following, it faces fundamental limitations: it can only learn from existing examples, optimizes individual tokens rather than sequence-level quality, and cannot learn from human judgments about response preferences. The SFT-RL pipeline addresses these limitations by enabling models to learn from human preferences about response quality, allowing them to generate responses that are not just correct but actually preferred by users.

The sequential approach of SFT followed by RL fine-tuning is practically necessary because RL fine-tuning is computationally expensive (typically 5–10 times more expensive than SFT in practice), and SFT provides a solid foundation that significantly accelerates subsequent RL training and reduces convergence failures.

18.2 Mathematical Foundation

Having established the motivation and conceptual framework for reinforcement learning fine-tuning, we now develop the mathematical foundations that translate these abstract RL concepts into concrete optimization algorithms.

18.2.1 Key Concepts and Terminology

The main characters of RL are the **agent** and the **environment**. In generative model fine-tuning, the agent is the language model, and the environment is the context where generated text will be used and evaluated.

At every step of interaction, the agent sees the current context (state), decides on the next token to generate (action), and receives feedback about the quality of the complete generated sequence (reward). The goal is to maximize the expected quality of generated text. Figure 18.1 illustrates this interaction pattern, showing how the language model (agent) generates tokens (actions) based on the current state and receives rewards for the complete sequence.

To discuss what RL does in generative models more specifically, we introduce additional terminology. A **state** s is a complete description of the current context, which includes the input prompt x and the sequence of previously generated tokens $y_{<t}$. Here, $y_{<t} = (y_1, y_2, \dots, y_{t-1})$ denotes

the sequence of tokens generated before position t, and $y = (y_1, y_2, \dots, y_T)$ denotes the complete generated sequence of length T. In generative models, we typically have access to the complete state, making the state representation straightforward.

The set of all valid actions in generative models is the **vocabulary** $\mathcal{V}$—the collection of all possible tokens the model can generate. This is a **discrete action space** with typically 50,000+ possible actions (tokens). This large discrete action space has profound consequences for RL algorithms. Some methods work well with small discrete spaces (like Q-learning for 4–10 actions) but become impractical for the large vocabulary sizes in generative models.

A **policy** is a rule used by the agent to decide what actions to take. For generative models, this is the language model's probability distribution over next tokens. The policy can be deterministic or stochastic. In generative models, we almost always use **stochastic policies** because we want the model to explore different possible completions:

$$a_t \sim \pi(\cdot|s_t),$$

where a_t is the next token to generate and s_t is the current context.

In the SFT-RL pipeline, we deal with **parameterized policies**: policies whose outputs are computable functions that depend on a set of parameters θ (the neural network weights), which we can adjust through optimization:

$$a_t \sim \pi_\theta(\cdot|s_t).$$

For generative models, the policy becomes the probability of generating token y_t given the current context $y_{<t}$ and input x:

$$\pi_\theta(y_t|y_{<t}, x) = \text{softmax}(f_\theta(y_{<t}, x)), \tag{18.1}$$

where $f_\theta(y_{<t}, x)$ is the neural network that outputs raw scores (logits) for each token in the vocabulary, and softmax() converts these raw scores into probabilities that sum to 1. Here, θ represents the neural network parameters, $y_{<t}$ denotes the sequence of previously generated tokens, and x is the input prompt.

A **trajectory** (or episode) is a sequence of states, actions, and rewards. For generative models, a trajectory represents the complete process of generating a response:

$$\tau = (s_0, a_0, r_0, s_1, a_1, r_1, \dots, s_T, a_T, r_T),$$

where s_t is the context after generating t tokens, a_t is the t-th token generated, and r_t is the reward received after taking action a_t in state s_t (typically 0 for intermediate steps $t < T$, with the final reward r_T assigned at the end of the sequence).

The **reward** r_t is a number that tells the agent how good the transition from state s_t to s_{t+1} was after taking action a_t. In reinforcement learning notation, r_t is the reward received after taking action a_t in state s_t, which leads to the next state s_{t+1}. For generative models, rewards are typically only given at the end of sequence generation, evaluating the complete output y given input x, meaning $r_t = 0$ for $t < T$ and r_T is the reward for the complete sequence. The **return** $R(\tau)$ is the cumulative reward for a trajectory. For generative models, this is usually just the final reward since intermediate steps don't receive rewards:

$$R(\tau) = r_T,$$

where r_T is the reward for the complete generated sequence.

Unlike classical RL settings where agents receive rewards at each step, generative models typically only receive feedback about the quality of the complete generated text. This creates a credit assignment problem: how do we know which tokens contributed to the final quality?

18.2.2 From Imitation to Optimization

Supervised fine-tuning has a fundamental limitation: it can only learn to imitate existing reference outputs. While this works well for teaching basic instruction-following, it cannot optimize for subjective qualities like response quality, safety alignment, or user preference that depend on human preferences rather than exact correctness.

This limitation becomes clear when we examine the SFT objective:

$$J_{\text{SFT}}(\theta) = \mathbb{E}_{(x,y^*)\sim\mathcal{D}_{\text{SFT}}}\left[\sum_{t=1}^{T} \log \pi_\theta(y_t^*|y_{<t}^*, x)\right], \tag{18.2}$$

where y^* represents the reference (correct) output and $\mathcal{D}_{\text{SFT}}$ is the supervised training dataset. This objective maximizes the likelihood of matching reference outputs but cannot learn from human judgments about which responses are better.

Reinforcement learning addresses this limitation by shifting from imitation to optimization. Instead of matching reference outputs, RL optimizes for human preferences and quality signals:

$$J_{\text{RL}}(\theta) = \mathbb{E}_{x\sim\mathcal{D}, y\sim\pi_\theta(\cdot|x)}[R(x, y)], \tag{18.3}$$

where $J_{\text{RL}}(\theta)$ is the expected reward objective, $\mathcal{D}$ is the distribution over input prompts, $\pi_\theta(\cdot|x)$ is the model's policy, $R(x, y)$ is the reward for generating sequence y given input x, and θ represents the model parameters.

This shift enables optimization for subjective qualities that cannot be captured through traditional supervised learning approaches.

18.2.3 Value Functions

It is often useful to know the **value** of a state or state-action pair. By value, we mean the expected return if you start in that state or state-action pair, and then act according to a particular policy forever after.

For generative models, the most important value functions are the on-policy value function $V^\pi(s)$, which gives the expected return if you start in state s and always act according to policy π; and the on-policy action-value function $Q^\pi(s, a)$, which gives the expected return if you start in state s, take action a, and then forever after act according to policy π:

$$V^\pi(s) = \mathbb{E}_{\tau\sim\pi}[R(\tau)|s_0 = s]$$

$$Q^\pi(s, a) = \mathbb{E}_{\tau\sim\pi}[R(\tau)|s_0 = s, a_0 = a].$$

For generative models, these value functions help us understand which tokens are most important for achieving high-quality outputs. The action-value function $Q^\pi(s, a)$ tells us how good it is to generate token a in context s.

18.2.4 The Policy Gradient Theorem

The policy gradient theorem tells us how to update the model parameters to maximize the expected reward. For generative models, the policy gradient is:

$$\nabla_\theta J_{\text{RL}}(\theta) = \mathbb{E}_{x\sim\mathcal{D}, y\sim\pi_\theta(\cdot|x)}\left[R(x, y)\sum_{t=1}^{T} \nabla_\theta \log \pi_\theta(y_t|y_{<t}, x)\right]. \tag{18.4}$$

The policy gradient scales token-level probability updates by the sequence-level reward. When a complete sequence receives a high reward, the gradient assigns positive updates to all tokens in that sequence, proportionally increasing their generation probabilities. Conversely, sequences with low rewards receive negative updates, reducing the probability of generating those token sequences. This creates a credit assignment mechanism where the reward signal propagates back to influence the probability of each individual token based on the overall quality of the complete sequence.

18.2.5 From Objective to Loss Function

While we maximize the objective $J_{\mathrm{RL}}(\theta)$ in Equation (18.3), we can define a loss function for practical optimization. The key difference between Equation (18.4) (the policy gradient) and the loss function in Equation (18.5) is that the gradient tells us the direction to increase the objective, while the loss function is what we minimize in practice. The loss function is simply the negative of the gradient's expectation, making it suitable for standard gradient descent optimization:

$$\mathcal{L}_{\mathrm{RL}} = -\mathbb{E}_{x,y\sim\pi_\theta}\left[R(x,y)\sum_{t=1}^{T}\log\pi_\theta(y_t|y_{<t},x)\right]. \tag{18.5}$$

Advanced Note 18.1 Policy Gradient Derivation and Log Tricks

The policy gradient theorem emerges from applying the log-derivative trick to the RL objective. Starting from Equation (18.3), we expand the expectation over the policy distribution:

$$J_{\mathrm{RL}}(\theta) = \sum_x\sum_y \pi_\theta(y|x)R(x,y)p(x). \tag{18.6}$$

Taking the gradient with respect to θ:

$$\nabla_\theta J_{\mathrm{RL}}(\theta) = \sum_x\sum_y \nabla_\theta\pi_\theta(y|x)R(x,y)p(x). \tag{18.7}$$

The key insight is the log-derivative identity:

$$\nabla_\theta\pi_\theta(y|x) = \pi_\theta(y|x)\nabla_\theta\log\pi_\theta(y|x). \tag{18.8}$$

Substituting this identity and converting back to expectation form yields:

$$\nabla_\theta J_{\mathrm{RL}}(\theta) = \mathbb{E}_{x\sim\mathcal{D},y\sim\pi_\theta(\cdot|x)}[R(x,y)\nabla_\theta\log\pi_\theta(y|x)], \tag{18.9}$$

which, when expanded for sequences, gives Equation (18.4).

This derivation reveals an important distinction between SFT and RL approaches to log terms. In supervised fine-tuning (Equation 18.2), we directly maximize log-likelihood, so the log appears in the objective itself. In reinforcement learning (Equation 18.3), the objective contains no log terms, but log terms emerge in the gradient through the log-derivative trick. This fundamental difference reflects that SFT optimizes likelihood directly, while RL optimizes expected reward, with logs appearing only in the gradient computation.

The relationship between objectives and loss functions follows from this distinction. In SFT, both objective and loss contain log terms because we maximize log-likelihood. In policy gradient RL, the objective has no log terms, but the loss function (Equation 18.5) contains log terms because they appear in the negative gradient used for minimization.

18.2.6 The Advantage Function

The advantage function is a fundamental concept in reinforcement learning that measures the relative value of actions compared to the average performance. It was developed to address the high variance problem that plagues many RL algorithms, where raw rewards can vary significantly across different inputs and contexts, making it difficult for models to learn consistent patterns.

For generative models, the advantage function is defined as:

$$A(x, y) = R(x, y) - V(x), \tag{18.10}$$

where $R(x, y)$ is the reward for the specific output y given input x, and $V(x) = \mathbb{E}_{y \sim \pi_\theta(\cdot|x)}[R(x, y)]$ is the expected reward the model would produce for input x (the baseline). The advantage function measures how much better (or worse) a specific output is compared to the average quality the model would produce for that input, providing a relative measure of quality rather than an absolute reward score.

Using advantages instead of raw rewards reduces variance and makes training more stable, which is crucial for large generative models. The advantage function tells us not just how good an output is, but how much better it is than average. This relative comparison is more informative than absolute rewards.

The advantage function provides the foundation for more sophisticated policy optimization algorithms. While the basic policy gradient (Equation 18.4) uses raw rewards, advanced methods like PPO and GRPO use advantage estimates to stabilize training, as we will see in the following sections.

18.3 Policy Optimization Methods in Detail

With the mathematical foundations in place, we now examine the specific policy optimization algorithms used in practice for generative model fine-tuning. Three algorithms have been successfully applied: REINFORCE, which provides the basic policy gradient framework but has limited effectiveness; proximal policy optimization (PPO), which has been widely adopted in systems like GPT models and ChatGPT; and group relative policy optimization (GRPO), an advancement used in DeepSeek models that employs reinforcement learning without human feedback (RLHF).

18.3.1 The RL Fine-tuning Pipeline

The complete RL-based fine-tuning pipeline consists of five sequential stages, as illustrated in Figure 18.3. This workflow begins with large-scale pretraining and progresses through supervised adaptation, reward modeling, policy optimization, and parameter updates to achieve alignment with human preferences and safety requirements.

This staged workflow reflects a general-purpose architecture for aligning generative models using reinforcement learning. Although it originated in open-domain dialogue systems, it is well suited to many applications. For example, the reward model might favor outputs that match safety policies, improve user satisfaction, or avoid ambiguity in multimodal contexts.

18.3.2 Why Policy Gradients for Generative Models?

Before examining the algorithms, it is important to understand why we focus on policy gradient methods rather than value-based approaches for generative model fine-tuning.

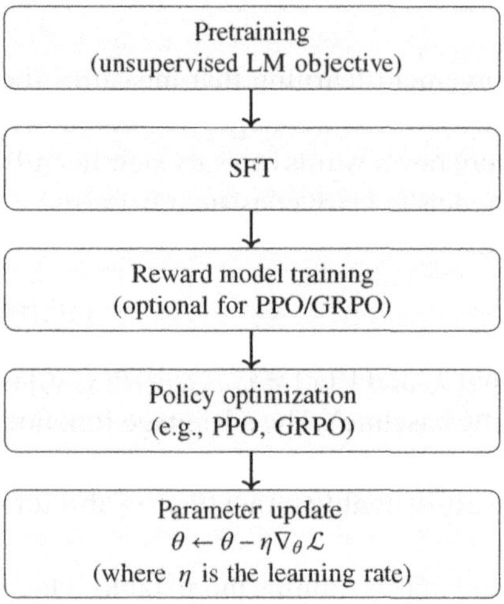

Figure 18.3 Stages in RL-based fine-tuning of generative models.

Reinforcement learning has two main approaches: value-based methods and policy gradient methods. We briefly explain these concepts:

Value-based methods (like Q-learning) learn to estimate the value of taking an action in a given state. They learn a Q-function $Q(s, a)$ that tells us how good it is to take action a in state s. The policy is then derived from these values (e.g., always choose the action with the highest Q-value).

Policy gradient methods directly learn a policy $\pi(a|s)$ that tells us the probability of taking action a in state s. Instead of learning values, they directly optimize the policy to maximize expected reward.

For generative model fine-tuning, we focus on policy gradients for several key reasons. First, language generation involves choosing from a large vocabulary (often 50,000+ tokens), creating a very large discrete action space where value-based methods struggle due to computational complexity. Second, generative models naturally output probability distributions over tokens, which aligns perfectly with policy gradient methods. Third, the quality of generated text is typically evaluated at the sequence level, not individual token level, making policy gradients more suitable. Fourth, policy gradients allow the model to explore different ways of generating text, which is crucial for discovering high-quality outputs. Finally, policy gradients directly optimize the policy (generative model) rather than learning a separate value function.

Value-based methods like Q-learning work well for discrete action spaces with tabular or simple function approximation, but they become impractical for the high-dimensional, large discrete action space of text generation. Policy gradient methods provide a more natural and effective framework for optimizing generative models. The key differences between these approaches are summarized in Table 18.1.

18.3.3 REINFORCE: The Basic Policy Gradient Algorithm

REINFORCE [1] provides the simplest way to implement policy gradients for generative models. The core idea is straightforward: if a generated sequence receives a high reward, increase the probability of generating that sequence. If it receives a low reward, decrease the probability.

Table 18.1 Comparison of value-based versus policy gradient methods for generative model fine-tuning.

Aspect	Value-based methods (Q-learning)	Policy gradient methods
Action space	Discrete, small (e.g., 4–10 actions)	Discrete, very large (50,000+ tokens)
Policy representation	Implicit (via Q-values)	Explicit (probability distribution)
Exploration	ϵ-greedy (random actions with probability ϵ)	Natural stochastic sampling
Sequence-level rewards	Difficult to handle	Natural fit
Computational complexity	$O(\|A\|)$ for action selection	O(1) for sampling
Convergence	Can be unstable with neural networks	More stable with proper baselines
Generative model fit	Poor (requires discrete actions)	Excellent (natural probability outputs)

REINFORCE is a Monte Carlo implementation of the policy gradient theorem presented in Equation (18.4). The policy gradient theorem provides the mathematical foundation using expectations, while REINFORCE estimates these expectations through sampling. Note that the policy gradient theorem is general and can be implemented in many ways (REINFORCE is just one specific approach that uses Monte Carlo sampling). The REINFORCE gradient estimate is:

$$\nabla_\theta J_{\text{RL}}(\theta) \approx \frac{1}{N} \sum_{i=1}^{N} R(x_i, y_i) \sum_{t=1}^{T} \nabla_\theta \log \pi_\theta(y_{i,t} | y_{i,<t}, x_i), \tag{18.11}$$

where N is the number of sampled episodes, (x_i, y_i) are the ith input-output pair, and $y_{i,t}$ is the tth token in the ith sequence. This equation shows the batch version of REINFORCE, which waits for N complete episodes before updating. In practice, REINFORCE can also be implemented online, updating after each individual episode (where $N = 1$), but the batch version typically provides more stable gradient estimates.

However, REINFORCE suffers from significant limitations that make it impractical for large-scale generative model training. First, it exhibits high variance in gradient estimates because rewards can vary significantly across different samples, making training unstable. Second, it learns slowly because it only updates based on the specific text generated, without learning from alternative possible outputs. Third, it lacks a baseline for comparison, meaning even mediocre outputs receive positive updates if their absolute reward is positive, slowing convergence.

These fundamental limitations motivated the development of more sophisticated policy optimization methods like PPO and GRPO, which address these issues through trust region constraints, advantage estimation, and other stabilization techniques.

18.3.4 PPO: Stable Policy Optimization

REINFORCE's problems led to the development of more stable methods. PPO has become the standard approach for RL-based fine-tuning in major language models. PPO addresses REINFORCE's limitations through two key innovations: advantage estimation and policy ratio clipping.

PPO was introduced to address the practical limitations of trust region policy optimization (TRPO) while retaining its core stability principles [2]. TRPO [3] was the first major algorithm to use trust region constraints for stable policy updates, constraining the KL divergence between old and new policies, but requiring expensive second-order optimization. PPO was developed as a simpler, more practical alternative that achieves similar stability with first-order methods.

PPO is an actor-critic method that combines the strengths of both policy gradient and value-based approaches. The "actor" is the policy that decides what actions to take, while the "critic" is a value function that evaluates how good those actions are.

For generative models, PPO uses an actor-critic approach where the language model is the actor and a separate value network serves as the critic. The actor generates actions based on the current policy, while the critic provides feedback about the quality of those actions. The critic's value estimates are used to compute advantages, which then guide the actor's policy updates. This creates a more stable learning process compared to pure policy gradient methods.

18.3.4.1 PPO's Key Innovation: Policy Ratio Clipping

PPO's main innovation is policy ratio clipping, which prevents destructive policy updates while maintaining the benefits of advantage estimation from earlier actor-critic methods.

The PPO objective is defined as:

$$\mathcal{L}_{\text{PPO}}(\theta) = \mathbb{E}_t\left[\min\left(r_t(\theta) \cdot \hat{A}_t,\quad \text{clip}(r_t(\theta), 1-\epsilon, 1+\epsilon) \cdot \hat{A}_t\right)\right], \tag{18.12}$$

where the policy ratio is defined at the token level as:

$$r_t(\theta) = \frac{\pi_\theta(y_t \mid y_{<t}, x)}{\pi_{\theta_{\text{old}}}(y_t \mid y_{<t}, x)}, \tag{18.13}$$

where $\hat{A}_t$ is the estimated advantage, and ϵ (typically 0.1–0.2) is the clipping parameter that defines the trust region width. PPO computes the policy ratio at the token level, evaluating how the probability of each token changes between the old and new policies. This token-level approach allows for fine-grained policy updates but can introduce variance, especially in long sequences. Note that while the basic PPO objective in Equation (18.12) does not explicitly include a KL divergence term, PPO is often used with KL regularization in practice, especially in RLHF applications where the combined loss includes a KL term to prevent deviation from a reference policy (see Equation 18.18).

PPO works by comparing the new policy to the old policy using a policy ratio, then applying advantage-weighted updates with clipping constraints. The clipping mechanism prevents the policy from changing too dramatically in a single update.

The clipping operation creates a "trust region" around the current policy by constraining how much the new policy can deviate from the old one. When the policy ratio $r_t(\theta)$ is close to 1, the policy hasn't changed much, so we trust the advantage estimate and apply the full update. However, when $r_t(\theta)$ is far from 1 (either much larger or much smaller), the policy has changed significantly, making the advantage estimate less reliable. In these cases, clipping prevents destructive updates by limiting the effective learning rate.

The clipping mechanism prevents excessively large policy updates while still allowing beneficial learning steps. This results in a reliable and sample-efficient learning algorithm suitable for both simulated environments and generative language models.

PPO has emerged as the de facto standard for reinforcement learning-based fine-tuning of generative models. Its popularity stems from a favorable tradeoff between empirical performance, implementation simplicity, and training stability.

18.3.5 GRPO: Group-based Optimization

PPO introduces complexity by requiring a separate "critic" network to estimate advantages, which approximately doubles the memory requirements, needs careful tuning, and can be unstable and hard to train. For example, in a typical implementation, PPO requires both a policy network and a value network (critic), effectively doubling the parameter count compared to GRPO, which only requires the policy network. This makes GRPO simpler to implement and train.

GRPO addresses these limitations by eliminating the need for a separate critic network while maintaining the stability benefits of PPO. Originally introduced in the DeepSeek series [4], GRPO uses a clever group sampling approach that generates multiple outputs for each input and compares them within the same group.

To illustrate how group sampling works, consider a concrete example with the input "Explain the traffic situation" and a group of 4 responses: "Heavy traffic ahead" (reward = 0.3), "Heavy traffic ahead due to construction" (reward = 0.5), "Heavy traffic ahead. Consider Route 5." (reward = 0.8), and "Traffic is fine" (reward = 0.1). GRPO's approach compares rewards within this group, where "Consider Route 5" gets the biggest update (highest in group) and "Traffic is fine" gets the smallest update (lowest in group), with no need for a separate critic network.

GRPO follows a simple but effective three-step process:

1. **Group Sampling:** For each input $x \sim \mathcal{D}$, GRPO samples a group of G outputs $\{y_i\}_{i=1}^{G}$ from the old policy $\pi_{\theta_{\text{old}}}(\cdot \mid x)$, where G is the group size (typically 4-8). A reward model assigns scalar scores $\{r_i\}_{i=1}^{G}$ to these outputs.
2. **Group-normalized Advantages:** The key innovation is computing advantages without a value function. Instead of learning a separate critic network, GRPO uses the group's own statistics as a baseline:

$$A_i = \frac{r_i - \mu_G}{\sigma_G + \epsilon}, \quad \mu_G = \frac{1}{G}\sum_{j=1}^{G} r_j, \quad \sigma_G = \sqrt{\frac{1}{G}\sum_{j=1}^{G}(r_j - \mu_G)^2}. \tag{18.14}$$

   This is like asking: "How much better is this output compared to the average output in this group?" The group mean $\mu_G = \frac{1}{G}\sum_{j=1}^{G} r_j$ is the average reward across all outputs in the group, and the group standard deviation σ_G is computed from the rewards $\{r_j\}_{j=1}^{G}$ within the group. These statistics are computed over the reward values. The group mean serves as the baseline, and the group standard deviation provides normalization for stable learning. Here, ϵ is a small constant (typically 10^{-8}) added to prevent division by zero.
3. **Policy Optimization:** The policy ratio measures how much the new policy differs from the old policy for each output. GRPO computes this ratio at the sequence level:

$$\rho_i = \frac{\pi_\theta(y_i \mid x)}{\pi_{\theta_{\text{old}}}(y_i \mid x)}. \tag{18.15}$$

   The key difference between GRPO's sequence-level ratio and PPO's token-level ratio (Equation 18.13) lies in the granularity of evaluation. PPO computes ratios for individual tokens y_t conditioned on previous tokens $y_{<t}$, enabling fine-grained updates but potentially introducing variance in long sequences. GRPO computes the ratio for the entire sequence y_i conditioned only on the input x, which aligns naturally with sequence-level rewards. This sequence-level formulation reduces variance by matching the optimization granularity to the reward structure, making it particularly suitable for GRPO's group-based approach, where complete sequences are compared and rewarded.

GRPO then optimizes the following objective function:

$$\mathcal{L}_{\text{GRPO}}(\theta) = \mathbb{E}_{x\sim\mathcal{D},\{y_i\}_{i=1}^{G}\sim\pi_{\theta_{\text{old}}}(\cdot|x)} \left[\frac{1}{G}\sum_{i=1}^{G} \min(\rho_i A_i,\ \text{clip}(\rho_i, 1-\epsilon, 1+\epsilon)\, A_i) \right. \\ \left. -\beta\, D_{\text{KL}}\left(\pi_\theta(\cdot \mid x)\,||\,\pi_{\text{ref}}(\cdot \mid x)\right)\right]. \tag{18.16}$$

The first term is the clipped surrogate objective (similar to PPO), and the second term is KL regularization to prevent the policy from deviating too much from the reference model.

GRPO processes each group as a complete batch, sampling all G outputs before computing group statistics and updating the policy. This requires storing the generated outputs, their rewards, and policy probabilities for each group until the group is complete, since the group-normalized advantages depend on the group's mean and variance.

The KL divergence term penalizes deviation from a fixed reference policy π_{ref}, typically the original pretrained model. The regularization weight β (typically 0.01–0.1) governs the tradeoff between reward maximization and policy conservatism.

18.3.6 PPO vs. GRPO: Algorithm Comparison

To consolidate the preceding discussion, this section presents a side-by-side comparison of the workflow pipelines for PPO and GRPO. While both approaches aim to improve a pretrained generative model via reinforcement learning, they differ substantially in architecture, complexity, and the type of supervision used to estimate policy improvement.

Figure 18.4 illustrates the stepwise differences between the PPO and GRPO fine-tuning procedures.

This side-by-side workflow highlights the architectural distinction: PPO leverages a value-based critic to compute per-sample advantages, while GRPO eliminates this dependency by standardizing rewards across a group. The reduction in components simplifies the architecture and reduces memory requirements.

Table 18.2 provides a concise comparison of the key differences between PPO and GRPO.

18.3.6.1 Limitations and Trade-offs

Each algorithm exhibits distinct limitations that inform their theoretical and practical boundaries. Understanding these limitations is crucial for comprehending the design choices underlying each approach.

REINFORCE's limitations were discussed earlier in this section, where we noted its high variance in gradient estimates, slow learning due to updating only based on specific generated text, and lack of baseline for comparison. These fundamental limitations motivated the development of more sophisticated methods like PPO and GRPO.

PPO addresses REINFORCE's limitations but introduces its own constraints. The actor-critic architecture requires training and maintaining a separate value network, approximately doubling the memory requirements and computational overhead compared to methods that only require the policy network. The value function must be carefully tuned and can be unstable and difficult to train, particularly when reward scales vary significantly. PPO's performance is sensitive to critic accuracy and reward scale, requiring careful hyperparameter tuning. Additionally, while PPO's token-level policy ratio computation enables fine-grained updates, it can introduce variance in long sequences, as the ratio is computed for each token independently.

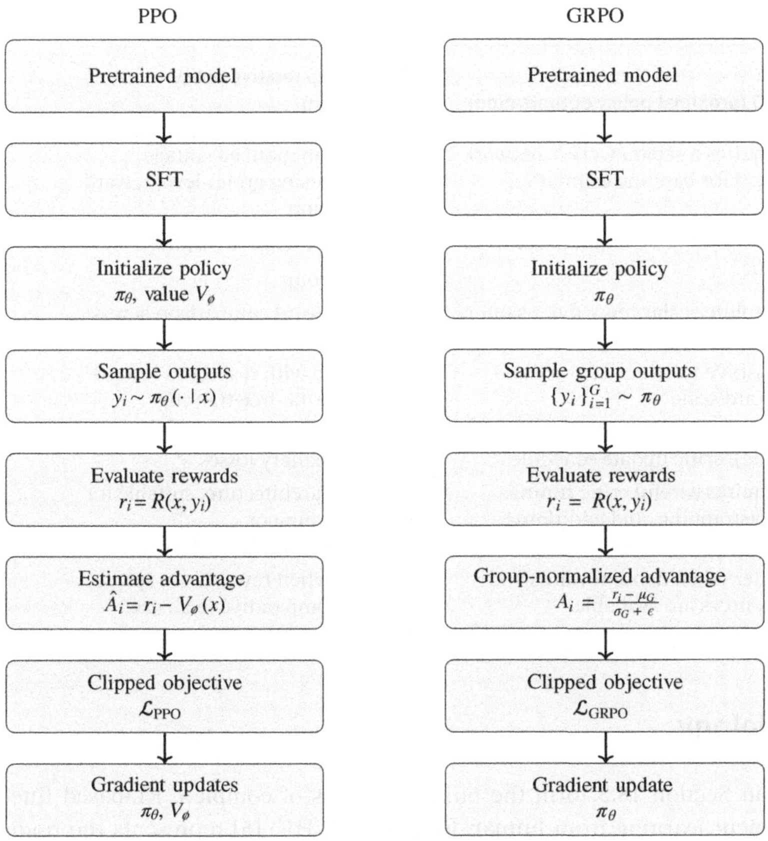

Figure 18.4 Comparison of reinforcement learning fine-tuning pipelines using PPO (left) and GRPO (right).

GRPO eliminates the need for a separate critic network, addressing one of PPO's primary limitations. However, GRPO introduces different trade-offs. The group-based normalization approach means that reward normalization occurs within each group, making inter-group calibration nontrivial. This can lead to challenges when comparing sequences across different groups, as the normalization statistics are group-specific. Performance can be sensitive to group size, with smaller groups potentially providing insufficient statistics for stable normalization, while larger groups increase computational requirements. Additionally, GRPO's sequence-level policy ratio computation, while reducing variance by aligning with sequence-level rewards, provides less fine-grained control compared to PPO's token-level approach.

The choice between token-level and sequence-level policy ratios represents a fundamental trade-off. Token-level ratios (as in PPO) enable fine-grained policy updates but can accumulate variance across long sequences. Sequence-level ratios (as in GRPO) reduce variance by matching the optimization granularity to the reward structure but provide less granular control over individual token probabilities. This trade-off reflects the inherent tension between optimization precision and training stability in reinforcement learning for generative models.

Table 18.2 Comparison of PPO and GRPO for RL-based fine-tuning.

Aspect	PPO (proximal policy optimization)	GRPO (group relative policy optimization)
Value function	Requires a separate critic network $V_\phi(x)$ for baseline estimation	No value function; advantage computed using group-level reward normalization
Advantage estimation	$\hat{A}_i = r_i - V_\phi(x)$, requires accurate critic	$A_i = \frac{r_i - \mu_G}{\sigma_G + \epsilon}$, relative within sampled group
Reward structure	Absolute scalar reward per sample	Relative reward comparison across group of samples
Training stability	Sensitive to critic accuracy and reward scale	More stable with respect to reward variance; critic-free training
Computational overhead	Higher: two networks (policy + critic), critic update schedule	Lower: single policy network, no critic or auxiliary losses
Implementation complexity	Requires careful critic tuning, bootstrapping, and additional training loss	Simplified architecture; suitable for fast prototyping or resource-constrained settings
Suitability for applications	Better when dense rewards and resources are available	Preferred when rewards are sparse, noisy, or comparative (e.g., user preference ranking)

18.4 RLHF Methodology

The algorithms discussed in Section 18.3 form the building blocks of complete RL-based fine-tuning systems. Reinforcement learning from human feedback (RLHF) [5] represents the traditional approach for aligning large language models with human preferences, used by major models including GPT, Claude, and LLaMA. This methodology combines supervised fine-tuning, reward modeling, and policy optimization to create models that align with human values.

18.4.1 The RLHF Pipeline

RLHF operates through a carefully orchestrated three-stage pipeline that systematically transforms human preferences into optimized model behavior. The first stage involves supervised fine-tuning on high-quality human demonstrations, establishing a solid foundation of basic instruction-following capabilities. This initial phase is crucial as it provides the model with fundamental behavioral patterns that serve as a reference point for subsequent optimization.

The second stage focuses on reward model training, where a separate neural network learns to predict human preferences from pairwise comparisons. Human annotators evaluate multiple model outputs for the same input and rank them by preference, creating a dataset of comparative judgments. The reward model architecture mirrors the language model's structure but replaces the vocabulary output head with a scalar prediction head, allowing it to leverage contextual understanding while focusing on preference prediction. The training employs a ranking loss function:

$$\mathcal{L}_{\text{reward}} = -\log \sigma(R_{\text{model}}(x, y_w) - R_{\text{model}}(x, y_l)), \tag{18.17}$$

where $R_{\text{model}}(x, y)$ is the reward model's predicted score for output y given input x, y_w represents the preferred response, y_l denotes the less preferred response, and σ is the sigmoid function. This formulation ensures the reward model learns to distinguish between response quality levels while maintaining the relative nature of human preferences.

The final stage employs policy optimization using PPO to maximize the expected reward predicted by the reward model while maintaining proximity to the supervised fine-tuning model through KL regularization. This delicate balance ensures that the model improves according to human preferences without deviating too far from its foundational capabilities. The combined loss function balances reward maximization with behavioral consistency:

$$\mathcal{L}_{\mathrm{RLHF}} = \mathcal{L}_{\mathrm{PPO}} - \beta \cdot D_{\mathrm{KL}}(\pi_\theta \| \pi_{\mathrm{SFT}}). \tag{18.18}$$

The RLHF process typically involves multiple rounds of data collection and model updates, with the reward model being retrained on new preference data as the policy improves. This iterative refinement allows the system to continuously adapt to evolving human preferences and improve alignment over time.

18.4.2 Policy Optimization in RLHF

The policy optimization stage represents the culmination of the RLHF process, where the language model learns to generate responses that align with human preferences. This stage employs PPO to maximize the expected reward while maintaining language quality through careful regularization.

PPO's stability properties make it particularly suitable for RLHF, where reward signals from human feedback can be noisy and inconsistent. The algorithm's clipped surrogate objective prevents the model from deviating too far from the reference policy (typically the supervised fine-tuning model), which is crucial for maintaining coherent language generation. The advantage estimation helps reduce variance in the human preference signals, making training more stable.

The reward model's predictions are often shaped to encourage helpful, harmless, and honest responses, incorporating safety and alignment objectives directly into the optimization process. This shaping ensures that the model not only follows human preferences but also adheres to broader ethical and safety considerations.

Research has demonstrated that reinforcement learning fine-tuning can significantly enhance chain-of-thought reasoning capabilities in language models. By using reward signals that favor accurate reasoning steps and correct final answers, RL-based fine-tuning can incentivize the development of advanced reasoning patterns, including self-reflection, verification, and structured problem-solving approaches. This makes RL fine-tuning particularly valuable for applications requiring complex reasoning, such as mathematical problem-solving, scientific analysis, and multistep planning tasks.

18.4.3 Applications and Challenges

RLHF proves particularly valuable for applications where human preferences about communication style, safety messaging, and user experience are crucial for system acceptance and effectiveness. The methodology enables systems to optimize their communication strategies based on human feedback, ensuring that safety messages, instructions, and system status updates are delivered in ways that users find clear, reassuring, and contextually appropriate.

The personalization capabilities of RLHF allow systems to adapt their communication style to individual user preferences, enhancing user experience and system acceptance. This adaptability is particularly important in safety-critical scenarios where clear, trusted communication can significantly impact user compliance and system effectiveness.

However, RLHF faces several challenges that are particularly relevant for practical applications. The quality of human feedback can vary significantly, and biased or inconsistent feedback can lead to suboptimal alignment. This variability becomes particularly problematic in safety-critical applications where consistent, reliable behavior is essential. Additionally, collecting high-quality human feedback is expensive and time-consuming, especially for domain-specific applications, making the iterative nature of RLHF resource-intensive for large-scale deployment.

Safety alignment represents perhaps the most critical consideration for practical applications. Ensuring that RLHF aligns with safety objectives while maintaining human preferences requires careful design of the reward model and optimization process, creating a complex optimization landscape that balances multiple competing objectives.

18.5 Summary

This chapter introduced reinforcement learning fine-tuning as a critical method for aligning generative models with human preferences and optimizing for subjective qualities that cannot be captured through supervised learning alone. The chapter established the complete three-stage fine-tuning pipeline: pretraining, supervised fine-tuning, and reinforcement learning fine-tuning, which represents the standard approach used in modern generative AI systems.

The mathematical foundations of policy gradient methods were developed, including the policy gradient theorem, which scales token-level probability updates by sequence-level rewards, and the advantage function, which measures relative value compared to average performance and reduces variance in training. The key distinction between supervised fine-tuning and reinforcement learning was clarified: SFT optimizes log-likelihood directly, while RL optimizes expected reward, with log terms emerging only through the log-derivative trick in gradient computation.

Three policy optimization algorithms were presented. REINFORCE provides the basic policy gradient framework but suffers from high variance and poor sample efficiency. PPO addresses these limitations through advantage estimation and policy ratio clipping, becoming the standard approach for RL-based fine-tuning in major language models. GRPO eliminates the need for a separate critic network by using group-based normalization, simplifying the architecture while maintaining stability benefits.

The reinforcement learning from human feedback (RLHF) methodology demonstrated how these algorithms integrate into complete systems, combining supervised fine-tuning, reward modeling, and policy optimization to create models that align with human values. The comparison between PPO and GRPO highlighted fundamental trade-offs in computational requirements, training complexity, and the granularity of policy updates.

These methods enable generative models to learn from human feedback and preference signals, creating more aligned and trustworthy AI systems. Reinforcement learning fine-tuning has also been shown to enhance chain-of-thought reasoning capabilities, making it valuable for applications requiring complex reasoning tasks.

References

1 Ronald J. Williams. Simple statistical gradient-following algorithms for connectionist reinforcement learning. *Machine Learning*, 8(3-4):229–256, 1992.

2 John Schulman, Filip Wolski, Prafulla Dhariwal, Alec Radford, and Oleg Klimov. Proximal policy optimization algorithms. *arXiv preprint arXiv:1707.06347*, 2017.

3 John Schulman, Sergey Levine, Pieter Abbeel, Michael Jordan, and Philipp Moritz. Trust region policy optimization. In *International Conference on Machine Learning*, pages 1889–1897, 2015.

4 Zhihong Shao, Peiyi Wang, Qihao Zhu, Runxin Xu, Junxiao Song, Xiao Bi, Haowei Zhang, Mingchuan Zhang, YK Li, Y Wu, et al. Deepseekmath: pushing the limits of mathematical reasoning in open language models. *arXiv preprint arXiv:2402.03300*, 2024.

5 Long Ouyang, Jeff Wu, Xu Jiang, Diogo Almeida, Carroll L. Wainwright, Pamela Mishkin, Chong Zhang, Sandhini Agarwal, Katarina Slama, Alex Ray, et al. Training language models to follow instructions with human feedback. In *Proceedings of the 36th International Conference on Neural Information Processing Systems*, NIPS '22, Red Hook, NY, USA, Curran Associates Inc., 2022. ISBN 9781713871088.

19

Knowledge Distillation for Edge Deployment

19.1 Introduction: The Deployment Challenge

Consider a CAV that needs to generate real-time explanations for driver decisions, such as "Slowing down due to pedestrian at crosswalk." While LLMs excel at this task, their computational requirements often exceed the capabilities of embedded vehicle systems. A typical 7B-parameter language model requires several gigabytes of memory and significant processing power, resources that are scarce in automotive electronic control units (ECUs) with strict power and thermal constraints.

Following the supervised and reinforcement learning fine-tuning approaches discussed in Chapters 17 and 18, we address this critical deployment gap. While fine-tuning can adapt large models to specific CAV tasks, the resulting models may still be too large for embedded deployment. Knowledge distillation [1] provides an elegant solution by creating compact models that retain the specialized capabilities learned through fine-tuning while meeting the strict resource constraints of automotive systems.

19.2 The Teacher–Student Paradigm

19.2.1 What Is Knowledge Distillation?

The key insight is that a well-trained teacher model contains rich information beyond what's captured in its final outputs. This "dark knowledge" includes the model's understanding of relationships between concepts, its confidence in different predictions, and its ability to handle ambiguous situations.

Knowledge distillation (KD) is the overall technique for transferring knowledge from a large, pretrained teacher model to a smaller student model, as illustrated in Figure 19.1. The teacher model provides knowledge through its probability distributions over possible outputs, known as soft targets, which contain more information than hard labels alone.

The student model learns to match the teacher's probability distribution. A probability distribution is the complete set of probabilities over all possible outputs (e.g., [0.7, 0.2, 0.1] for three classes). The loss function (KL divergence) measures how well the student's distribution matches the teacher's distribution. By learning the distribution, the student captures the teacher's assessment of each option's likelihood and the overall ambiguity of the situation, enabling the student to acquire not just the correct answer, but also the teacher's nuanced understanding of alternative options and complex situations.

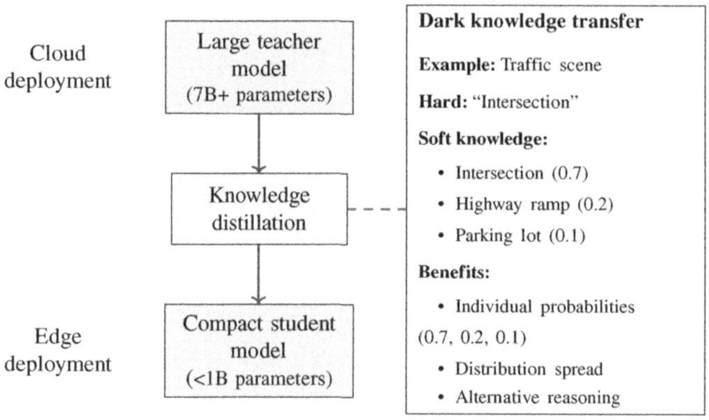

Figure 19.1 Knowledge distillation pipeline: from cloud-based large models to edge-deployed compact models with dark knowledge transfer.

19.2.2 How Knowledge Distillation Works

19.2.2.1 CAV Scenario Classification Example

Consider a teacher model that has been fine-tuned to generate driver explanations for CAV scenarios. When presented with the input "Vehicle approaching intersection with yellow light," the teacher produces a probability distribution over possible responses: "Proceeding with caution" (0.6) as the most confident choice, "Slowing down for safety" (0.3) as an alternative interpretation, and "Stopping at intersection" (0.1) as a conservative option.

The *individual probabilities* (0.6, 0.3, 0.1) indicate the likelihood the teacher assigns to each specific option, while the *distribution spread* is reflected in how probabilities are distributed across multiple classes rather than concentrated in one. The distribution [0.6, 0.3, 0.1] shows moderate spread, acknowledging that multiple interpretations are plausible. The student model learns to match the teacher's probability distribution by minimizing the difference between its own output distribution and the teacher's distribution, transferring the teacher's sophisticated reasoning capabilities to the student.

19.3 Core Knowledge Distillation Mechanisms

19.3.1 Temperature Scaling and Soft Targets

To transfer knowledge effectively, the student model must learn from the teacher's probability distributions rather than just hard labels. However, standard probability distributions from neural networks are often too "peaked"—the teacher assigns very high probability to the correct class and very low probabilities to others. This makes it difficult for the student to learn the teacher's relative confidence in alternative options.

19.3.1.1 Why Temperature Scaling?

Consider a teacher model that outputs probabilities: "Proceeding with caution" (0.95), "Slowing down for safety" (0.04), and "Stopping at intersection" (0.01). While this distribution contains information about alternatives, the extreme differences make it challenging for the student to learn the subtle relationships between classes. The student needs a "softer" distribution that preserves relative differences while making all probabilities more comparable.

Temperature scaling addresses this challenge by controlling the "softness" of the teacher's outputs. By dividing the logits (pre-softmax outputs) by a temperature parameter $T > 1$, we create softer probability distributions that emphasize the relative differences between classes while making the distribution more informative for learning.

19.3.1.2 Mathematical Foundation

The temperature-scaled softmax is computed as:

$$p_i = \frac{\exp(z_i/T)}{\sum_j \exp(z_j/T)}, \tag{19.1}$$

where p_i is the probability for class i, z_i is the raw logit (pre-softmax output) for class i, T is the temperature parameter ($T > 1$), and the sum is over all classes j.

This formulation was first introduced by Hinton et al. [1] and has become the standard approach for knowledge distillation.

The temperature parameter T controls the entropy of the output distribution. When $T = 1$, we get the standard softmax. When $T > 1$, the distribution becomes more uniform, revealing the model's relative confidence in different classes. For example, with $T = 3$, the previous distribution might become: "Proceeding with caution" (0.70), "Slowing down for safety" (0.20), and "Stopping at intersection" (0.10). This softer distribution preserves the relative ordering while making all probabilities more comparable, enabling the student to learn the teacher's nuanced understanding through smoother gradients and better regularization.

19.3.1.3 Selecting the Temperature Parameter

The temperature parameter T is a hyperparameter that must be chosen by the practitioner. Typical values range from $T = 3$ to $T = 10$, with $T = 4$ or $T = 5$ being common starting points. The optimal value depends on the specific task, the teacher model's confidence patterns, and the desired balance between preserving class distinctions and providing informative soft targets. Practitioners typically determine T through experimentation and validation, selecting the value that yields the best student model performance on a held-out validation set. Higher temperatures create softer distributions but may blur important distinctions, while lower temperatures (closer to 1) preserve sharper distinctions but may be less informative for the student.

Advanced Note 19.1 Information-theoretic Justification

The effectiveness of knowledge distillation can be understood through information theory, which provides a rigorous framework for quantifying how much knowledge is transferred from teacher to student. The key insight is that knowledge distillation maximizes the mutual information between teacher and student distributions while minimizing their divergence.

Mutual Information: The mutual information $I(P_{\text{teacher}}; P_{\text{student}})$ quantifies how much information the student captures from the teacher. It is defined as:

$$I(P_{\text{teacher}}; P_{\text{student}}) = H(P_{\text{teacher}}) - H(P_{\text{teacher}} | P_{\text{student}}), \tag{19.2}$$

where $H(P_{\text{teacher}})$ is the entropy of the teacher's output distribution (measuring its uncertainty/information content), and $H(P_{\text{teacher}} | P_{\text{student}})$ is the conditional entropy (measuring remaining uncertainty after knowing the student's prediction). Higher mutual information indicates that

(Continued)

Advanced Note 19.1 (Continued)

the student's predictions reduce uncertainty about the teacher's outputs, meaning more knowledge has been transferred.

KL Divergence: The knowledge distillation loss minimizes the Kullback–Leibler divergence between teacher and student distributions:

$$D_{\mathrm{KL}}(P_{\text{teacher}} \parallel P_{\text{student}}) = \sum_i p_i \log \frac{p_i}{q_i}, \tag{19.3}$$

where p_i is the teacher's soft output probability for class i and q_i is the student's predicted probability. The KL divergence measures how different the student's distribution is from the teacher's distribution, with zero divergence indicating perfect matching.

Connection: Temperature scaling increases the entropy $H(P_{\text{teacher}})$ by creating softer distributions, which increases the potential mutual information. By minimizing KL divergence, we maximize the mutual information, ensuring efficient knowledge transfer. This information-theoretic perspective explains why matching probability distributions (rather than just hard labels) is so effective: it maximizes the information content transferred from teacher to student.

19.3.2 Combined Loss Functions

The knowledge distillation objective combines multiple loss terms to ensure the student learns both from the teacher and from ground truth:

$$\mathcal{L}_{\text{total}} = \alpha \cdot \mathcal{L}_{\text{CE}} + \beta \cdot \mathcal{L}_{\text{KD}}. \tag{19.4}$$

Figure 19.2 illustrates the combined loss function incorporates both ground truth learning and teacher knowledge transfer. The term $\mathcal{L}_{\text{CE}}$ represents the cross-entropy loss with ground truth labels, while $\mathcal{L}_{\text{KD}}$ represents the KL divergence between teacher and student outputs. The weighting parameters α and β control the relative importance of each component. In many implementations, these are set to $\alpha = 0.5$ and $\beta = 0.5$, though optimal values depend on the specific task and dataset characteristics.

Both loss terms are essential: $\mathcal{L}_{\text{CE}}$ ensures accuracy by learning from ground truth labels, while $\mathcal{L}_{\text{KD}}$ transfers the teacher's sophisticated reasoning capabilities, including individual probabilities and distribution spread. The combination provides the optimal learning signal, enabling the student to acquire both correctness and nuanced understanding.

The knowledge distillation loss is computed as:

$$\mathcal{L}_{\text{KD}} = T^2 \cdot D_{KL}\left(\text{softmax}\left(\frac{z^{(t)}}{T}\right) \parallel \text{softmax}\left(\frac{z^{(s)}}{T}\right)\right), \tag{19.5}$$

where $z^{(t)}$ and $z^{(s)}$ represent the raw logits (pre-softmax outputs) from the teacher and student models respectively, and T is the temperature parameter. The T^2 factor compensates for the reduced magnitude of gradients when using higher temperatures, ensuring consistent gradient magnitudes during training.

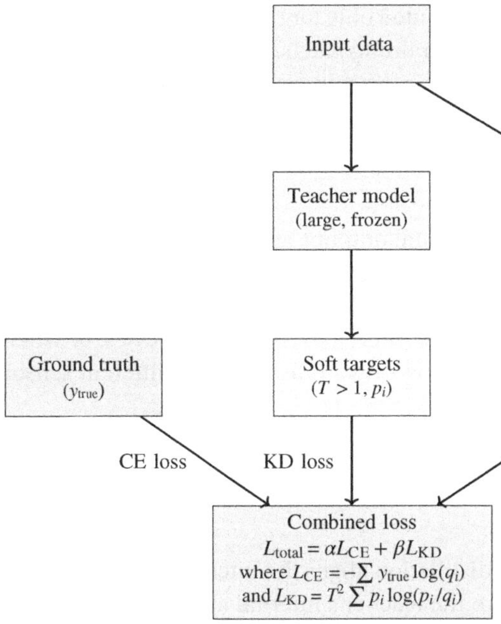

Figure 19.2 Knowledge distillation architecture with combined loss functions. The input data flows to both teacher and student models. The teacher produces soft targets for KD loss, while ground truth provides labels for CE loss. The combined loss function integrates both components to train the student model.

19.3.3 The Knowledge Distillation Training Process

Knowledge distillation training follows a systematic workflow that transfers knowledge from teacher to student. The process involves four principal stages: preparation, knowledge generation, knowledge transfer, and optimization.

Stage 1: Preparation involves loading a pretrained teacher model and freezing its parameters, ensuring the teacher remains unchanged throughout training. A smaller student model architecture is initialized, designed to meet deployment constraints such as memory and computational limits. Key hyperparameters are set, including the temperature parameter T (typically 3–5) and loss weights α and β (often 0.5 each) that balance learning from ground truth versus teacher knowledge.

Stage 2: Knowledge generation occurs during each training iteration. The same input data is passed through both the frozen teacher model and the trainable student model. The teacher produces probability distributions over possible outputs, which are softened using temperature scaling to create informative soft targets. The student simultaneously processes the same inputs, producing its own probability distributions.

Stage 3: Knowledge transfer happens through the combined loss function. The student learns from two sources: ground truth labels (ensuring correctness) and the teacher's soft targets (capturing nuanced reasoning). The cross-entropy loss measures how well the student matches ground truth, while the knowledge distillation loss measures how well the student matches the teacher's probability distribution. These are combined with weighting parameters to create the total training objective.

Stage 4: Optimization updates only the student model parameters based on the combined loss. The teacher model remains frozen, serving only as a knowledge source. This makes the training process computationally efficient, as gradients are computed only for the smaller student model, not the large teacher model. The process iterates over training batches until the student model converges to match the teacher's knowledge while remaining compact enough for deployment.

19.4 Advanced Distillation Techniques

While basic knowledge distillation focuses on matching final outputs, advanced techniques leverage intermediate representations to transfer more sophisticated knowledge from teacher to student. These techniques become essential when the teacher and student models have significantly different architectures, when the task requires understanding of internal reasoning processes, or when dealing with multimodal data that requires consistent knowledge transfer across different sensor types.

19.4.1 Intermediate Representation Distillation

19.4.1.1 Hidden State Distillation

Instead of only matching final outputs, hidden state distillation aligns the internal representations from intermediate layers, helping the student capture the teacher's internal abstractions [2]. This is particularly valuable when the teacher and student have different architectures, as it allows the student to learn the teacher's reasoning process at multiple levels.

The mathematical formulation for hidden state distillation is:

$$\mathcal{L}_{\text{feat}} = \text{MSE}(\hat{h}^{(s)}, h^{(t)}), \tag{19.6}$$

where $\hat{h}^{(s)} = Wh^{(s)} + b$ projects student features to match teacher dimensions, $h^{(t)}$ represents the teacher's hidden states, and W and b are learnable projection parameters. This ensures that even when the student has fewer layers or different architecture, it can still learn the teacher's internal representations.

19.4.1.1.1 Layer Aggregation for Multi-layer Teachers

When the teacher model has significantly more layers than the student, direct layer-to-layer matching becomes challenging. Layer aggregation provides a solution by combining information from multiple teacher layers:

$$h_{\text{agg}}^{(t)} = \frac{1}{k} \sum_{i=l}^{l+k-1} h_i^{(t)}, \tag{19.7}$$

where k is the number of consecutive teacher layers to aggregate, starting from layer l. This approach smooths the signal across multiple teacher layers, providing a more stable target for the student to learn from. Layer aggregation is particularly effective when the student is much shallower than the teacher, as it allows the student to capture the teacher's reasoning process across multiple abstraction levels.

19.4.1.2 Attention Map Distillation

For language models, attention mechanisms are crucial for understanding relationships between tokens. Attention map distillation matches attention score matrices, ensuring the student learns not just what the teacher predicts, but how it focuses on different parts of the input [3].

The attention distillation loss is defined as:

$$\mathcal{L}_{\text{attn}} = \text{MSE}(A^{(s)}, A^{(t)}), \tag{19.8}$$

where $A = \text{softmax}(QK^\top/\sqrt{d_k})$ represents attention matrices, with Q (queries), K (keys), and d_k being the key dimension. This forces the student to learn the same attention patterns as the teacher, capturing the teacher's focus and reasoning process.

19.4.1.2.1 Multi-head Aggregation for Attention Distillation

When the teacher and student models have different numbers of attention heads, direct attention map matching becomes problematic. Multi-head aggregation provides a solution by averaging across teacher attention heads:

$$A^{(t)}_{\text{avg}} = \frac{1}{H}\sum_{h=1}^{H} A^{(t)}_h, \tag{19.9}$$

where H is the number of attention heads in the teacher model, and $A^{(t)}_h$ represents the attention matrix for head h. This approach reduces dimensional mismatch between teacher and student attention maps while preserving the average attention behavior across all heads. Multi-head aggregation is particularly useful when the student has significantly fewer attention heads than the teacher, allowing the student to capture the teacher's overall attention patterns without being overwhelmed by the complexity of individual head interactions.

19.4.1.3 Relation-based Distillation

Relation-based distillation captures the structural relationships between samples by encoding pairwise similarities [4]. This technique is particularly valuable when the task requires understanding of sample relationships and is independent of model size or layer structure.

The relation-based distillation loss is defined as:

$$\mathcal{L}_{\text{rel}} = \text{MSE}(\text{Sim}^{(S)}, \text{Sim}^{(T)}), \tag{19.10}$$

where the similarity matrices are computed as:

$$\text{Sim}^{(T)} = X_T X_T^\top, \tag{19.11}$$

$$\text{Sim}^{(S)} = X_S X_S^\top. \tag{19.12}$$

Here, X_T and X_S represent the feature representations from teacher and student models, respectively. The similarity matrices capture the structural geometry in the representation space, ensuring that the student learns to preserve the same relationships between samples as the teacher. This technique is particularly effective for tasks where the relative positioning of samples in the feature space is crucial, such as clustering or similarity-based retrieval tasks in CAV applications.

19.4.2 Multimodal Distillation

For CAV applications involving multiple sensor modalities (camera, LiDAR, radar), cross-modal distillation ensures that knowledge is transferred consistently across different sensor types, maintaining coherent predictions across all modalities [5].

The multimodal distillation loss combines feature alignment across different sensor types:

$$\mathcal{L}_{\text{multimodal}} = \sum_{m\in\{\text{camera, lidar, radar}\}} \alpha_m \mathcal{L}_{\text{feat}}(h^{(s)}_m, h^{(t)}_m), \tag{19.13}$$

Table 19.1 Advanced distillation strategy selection guidelines.

Strategy	Best for	Key benefits
Layer projection	Student much smaller	Handles dimensional mismatch
Aggregated teacher layers	Shallow student; complex task	Smooths multi-layer information
Relation-based distillation	Structure matters	Preserves sample relationships
Attention map distillation	Transformer models	Captures focus patterns
Multimodal distillation	Sensor fusion	Maintains cross-modal consistency

where h_m represents features from modality m, and α_m are modality-specific weights. This ensures that the student learns to integrate information from all sensor types in the same way as the teacher, maintaining consistency across the multimodal fusion process.

19.4.3 Combining Advanced Loss Functions

These advanced distillation techniques can be combined to create comprehensive knowledge transfer strategies. The total loss function integrates multiple supervision signals to ensure the student learns both from ground truth and from the teacher's sophisticated reasoning patterns.

When using advanced distillation techniques, the total loss function extends the basic combined loss to include intermediate representation matching:

$$\mathcal{L}_{\text{total}} = \alpha\mathcal{L}_{\text{CE}} + \beta\mathcal{L}_{\text{KD}} + \gamma\mathcal{L}_{\text{feat}} + \delta\mathcal{L}_{\text{attn}} + \epsilon\mathcal{L}_{\text{multimodal}} + \zeta\mathcal{L}_{\text{rel}}. \tag{19.14}$$

The hyperparameters $\alpha, \beta, \gamma, \delta, \epsilon$, and ζ control the relative importance of each loss component. In practice, $\alpha + \beta = 1$ for the basic distillation, while the advanced terms provide additional supervision for more sophisticated knowledge transfer.

19.4.4 Selecting Advanced Techniques

The choice of advanced distillation techniques depends on the specific characteristics of the teacher–student pair and the deployment requirements. Advanced techniques are particularly valuable when the student architecture differs significantly from the teacher, when tasks require understanding of internal reasoning processes, or when dealing with multimodal data. However, they introduce additional computational overhead during training and require more careful hyperparameter tuning. Table 19.1 provides guidance on when to use each technique.

These techniques can be combined with standard KD loss for maximum benefit, though practitioners should carefully tune the loss weights to ensure balanced learning across all objectives.

19.5 Deployment Considerations for CAV Applications

19.5.1 Model Selection and Training Considerations

For CAV applications, knowledge distillation requires careful consideration of both the student model architecture and training configuration. The student model must be compatible with the teacher's representations while meeting strict deployment constraints such as memory limitations, power consumption, and real-time performance requirements.

Table 19.2 Comparison between fine-tuning and knowledge distillation.

Aspect	Fine-tuning (FT)	Knowledge distillation (KD)
Data dependency	Needs labeled data	Needs labeled data (for CE loss)
Model size	Large or small	Targets compact models
Training cost	Full backprop + large model	Uses teacher's outputs
Generalization	May overfit on small data	Inherits teacher's inductive bias
Use case	Domain adaptation	Efficiency under constraints

For very large teacher models, progressive distillation uses intermediate models of decreasing size, creating a chain of knowledge transfer that is more effective than direct distillation. This technique is particularly useful when the size gap between teacher and student is very large, as it allows for more gradual knowledge transfer through intermediate-sized models.

For CAV applications, safety validation must comply with automotive safety standards such as ISO 26262 [6]. Safety validation must include several critical components: edge case testing for rare but critical scenarios, adversarial robustness testing for resistance to input perturbations, real-time constraint validation for latency and throughput requirements, and failure mode analysis for systematic identification of potential failures. These validation steps ensure that distilled models meet the stringent safety requirements of automotive deployment environments.

19.5.2 Knowledge Distillation vs. Fine-tuning

Understanding the relationship between knowledge distillation and fine-tuning is crucial for selecting the appropriate approach for CAV applications. While both techniques adapt pretrained models to specific tasks, they serve different purposes and have distinct characteristics.

Fine-tuning typically involves adapting a pretrained model to a new domain or task by continuing training on task-specific data. This approach can achieve high performance but may result in models that are still too large for edge deployment. Knowledge distillation, on the other hand, focuses on creating compact models that can be deployed in resource-constrained environments while maintaining performance through knowledge transfer from larger teacher models.

The key differences between these approaches are summarized in Table 19.2. Fine-tuning requires labeled data and full backpropagation through the large model, making it computationally expensive. Knowledge distillation also requires labeled data for the cross-entropy loss component but uses the teacher's outputs for the distillation loss, making it more efficient while targeting compact models suitable for edge deployment.

For CAV applications, knowledge distillation is often preferred when deployment constraints are critical, as it provides a systematic approach to model compression while preserving the teacher's knowledge and reasoning capabilities. Typical CAV applications include V2X message translation, multimodal scene understanding, and natural language interaction systems. These applications demonstrate how knowledge distillation brings advanced AI capabilities to production CAV systems while ensuring safety, efficiency, and reliability within automotive resource constraints.

19.6 Summary

Knowledge distillation provides a systematic approach to model compression for CAV applications, enabling the deployment of sophisticated AI capabilities in resource-constrained environments. The technique transfers knowledge from large teacher models to compact student models through soft targets and temperature scaling, achieving significant compression while maintaining performance.

The combined loss function ensures both accuracy and knowledge transfer through a weighted combination of cross-entropy and distillation losses, while advanced techniques like hidden state distillation and attention map distillation enable more comprehensive knowledge transfer. The bell curve relationship between model size and performance, distinct from scaling laws in large language models, guides the selection of optimal student model sizes for domain-specific CAV applications.

For CAV systems, knowledge distillation offers a practical solution to the challenge of deploying advanced AI capabilities within automotive resource constraints, providing a pathway from cloud-based large models to edge-deployed compact models while maintaining the safety and reliability standards required for autonomous vehicle applications.

References

1 Geoffrey Hinton, Oriol Vinyals, and Jeff Dean. Distilling the knowledge in a neural network. *arXiv preprint arXiv:1503.02531*, 2015.

2 Adriana Romero, Nicolas Ballas, Samira Ebrahimi Kahou, Antoine Chassang, Carlo Gatta, and Yoshua Bengio. FitNets: hints for thin deep nets. *arXiv preprint arXiv:1412.6550*, 2014.

3 Kevin Clark, Urvashi Khandelwal, Omer Levy, and Christopher D. Manning. What does BERT look at? An analysis of BERT's attention. In *Proceedings of the 2019 ACL Workshop BlackboxNLP: Analyzing and Interpreting Neural Networks for NLP*, pages 276–286, 2019.

4 Wonpyo Park, Dongju Kim, Yan Lu, and Minsu Cho. Relational knowledge distillation. In *Proceedings of the IEEE/CVF Conference on Computer Vision and Pattern Recognition*, pages 3967–3976, 2019.

5 Hongyu Guo, Shuo Wang, Minghao Xu, Huijuan Xu, and Dacheng Tao. Cross-modal knowledge distillation for action recognition. In *Proceedings of the IEEE/CVF International Conference on Computer Vision*, pages 1232–1241, 2019.

6 International Organization for Standardization. Road vehicles – functional safety. Technical Report 26262, ISO, 2018. Functional safety standard for automotive systems.

Part VII

Validation and Safety

The integration of generative artificial intelligence (GenAI) into safety-critical vehicular systems requires rigorous validation, transparent evaluation, and careful consideration of ethical implications. This final part examines the methodologies and frameworks necessary to ensure that GenAI-driven connected and autonomous vehicle systems are reliable, interpretable, and socially responsible. The chapters address simulation-based testing, quantitative and qualitative evaluation of generative systems, and ethical and regulatory challenges specific to autonomous driving. Together, they position validation and responsibility as integral components of system design rather than post-deployment concerns.

20

Simulation Platforms for GenAI-driven CAVs

20.1 Introduction: Why Simulation Is Indispensable for GenAI-CAVs

Simulation serves as the essential execution substrate for GenAI and CAV development, integration, and iteration. Unlike traditional autonomous vehicle systems with deterministic control algorithms, GenAI-powered CAVs introduce fundamental challenges that make simulation not merely useful, but indispensable: **open-ended generative behavior**, where agents produce novel, unexpected outputs that cannot be exhaustively predicted; **non-determinism**, where stochastic model sampling creates variability that requires statistical evaluation; and **interaction-driven failure modes**, where system behavior emerges from complex interactions between multimodal inputs, semantic reasoning, and environmental feedback.

Consider an autonomous vehicle equipped with a conversational interface that interprets driver commands such as, "Take the next right if traffic is light," or "Maintain a comfortable distance but don't slow down too much." These utterances require real-time grounding in visual perception, traffic context, and high-level planning. The complexity of such interpretation, and the potential consequences of misinterpretation, make real-world testing both risky and impractical for many scenarios. These limitations are amplified in GenAI-powered systems, where agents may produce novel, unexpected, or ambiguous outputs, especially when interpreting open-ended commands or generating structured plans. Simulation provides a controlled, instrumented environment where GenAI agents can be developed, tested, and refined through iterative experimentation without real-world risk.

The role of simulation extends beyond testing to serve as **development-time cognitive scaffolding**. GenAI agents process rich combinations of visual, textual, and spatial inputs, requiring simulation platforms to support synchronized delivery of these modalities. Agents generate abstract intents or goals rather than direct control commands, necessitating that simulation translate these semantic outputs into environment-grounded behaviors. Personalized agents respond to user feedback, refine behavior over time, and evolve interaction strategies. Simulation enables continuous in-the-loop experimentation without risk.

Within the context of validation, safety, and responsibility, simulation serves as the primary mechanism for comprehensive testing before real-world deployment. This chapter focuses on simulation as a **development and integration substrate**, emphasizing how simulation enables GenAI workflows rather than deep dives into evaluation theory or distributed agent protocols (Chapters 11–13). Critically, this chapter focuses on **how simulation enables GenAI development**, not on how performance is measured.

The remainder of this chapter presents the following: (1) how simulation serves as a system-level substrate for GenAI-CAV integration; (2) scenario generation and coverage expansion for GenAI testing; (3) major simulation platforms and their GenAI integration characteristics; (4) scaling from single-agent to cognitive multi-agent systems; and (5) the boundary between simulation infrastructure and evaluation metrics, clarifying what simulation enables versus what evaluation measures.

20.2 Simulation as a System-level Substrate for GenAI-CAVs

Simulation serves as the execution substrate that enables GenAI agents to interact with controlled, instrumented environments. Unlike traditional AV simulation, focused on sensor realism and deterministic control, GenAI-centric simulation emphasizes semantic alignment, multimodal interaction, and support for evaluating generative model behavior.

20.2.1 Closed-loop Simulation for Generative Policies

GenAI agents operate through a perception → reasoning → action → environment → feedback cycle that fundamentally differs from traditional control loops. Unlike classical control loops that directly map sensor inputs to actuator outputs, GenAI-powered systems introduce intermediate stages of semantic processing, intent interpretation, and symbolic reasoning. Figure 20.1 illustrates this closed-loop architecture.

The architecture integrates five distinct stages: simulation environment, state extraction and serialization, GenAI reasoning, semantic grounding, and execution. GenAI reasoning processes multimodal inputs through language models, vision-language models (VLMs), or multimodal agents, producing symbolic outputs (natural language commands, structured plans, behavioral intents). Semantic grounding validates and translates these symbolic outputs into executable actions, performing safety checks and mapping abstract intents to concrete control parameters. This grounding step is essential because GenAI outputs may be ambiguous, unsafe, or infeasible.

20.2.2 Multimodal State Representation and Serialization

GenAI models require explicit state packaging because they operate on structured data formats (text, images, JSON) rather than raw sensor streams. Conceptually, the serialized state is a tuple State $= (P, D, M, E)$, where P represents perception tokens (sensor data), D represents dialogue context (conversation history), M represents agent memory (accumulated knowledge), and

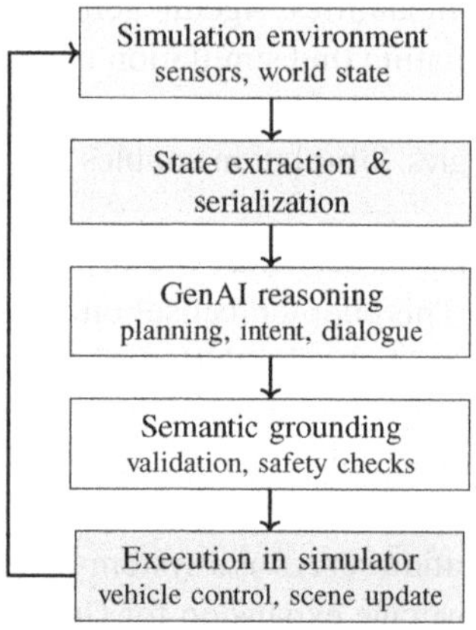

Figure 20.1 Closed-loop GenAI-CAV simulation framework showing integrated stages of perception, serialization, reasoning, grounding, and execution.

E represents environment state (vehicle positions and scene semantics). Simulation platforms must support configurable sensor fidelity and efficient serialization, as image data can dominate communication bandwidth. Platforms also support **checkpointing and replay** for reproducible testing.

20.2.3 Execution Control and Intervention

Simulation platforms must support both **deterministic rollouts** (for reproducible testing) and **stochastic rollouts** (for statistical evaluation of GenAI variability). Deterministic rollouts require fixed random seeds and synchronized time steps, enabling consistent validation across multiple runs. Stochastic rollouts enable exploration of GenAI behavior distributions by varying random seeds, sampling parameters, or environmental conditions. Critical execution control capabilities include reset, replay, branching, and intervention, enabling systematic exploration of GenAI behavior and failure modes.

20.2.4 GenAI-specific Simulation Challenges

Unlike traditional deterministic control systems, GenAI models introduce unique challenges that simulation must address. Simulation must provide ground truth for hallucination detection, enabling identification of visual, reasoning, and grounding failures. Semantic grounding validation requires known vehicle positions, road geometry, and physics constraints to validate that symbolic GenAI outputs map correctly to executable actions. Multimodal alignment testing requires scenarios with known correspondences across modalities, enabling controlled testing of cross-modal mismatches. Simulation platforms must support both deterministic random seed management for reproducible testing and statistical evaluation through multiple runs with controlled variation to accommodate GenAI stochasticity.

Simulation platforms use different time advancement mechanisms categorized along execution control (synchronous versus asynchronous modes) and interface type (time-step-based versus action-based). Synchronous modes with fixed time-steps are ideal for vision-language models requiring frame-synchronized sensor data, while action-based interfaces suit agents that learn through trial-and-error interactions. Platform-specific characteristics are discussed in Section 20.4.

20.3 Scenario Generation and Coverage Expansion

A critical requirement for evaluating GenAI-enabled CAVs is the ability to generate diverse, semantically meaningful scenarios. Unlike traditional AV systems that focus primarily on geometric or dynamic variations, GenAI systems are sensitive to linguistic diversity, contextual ambiguity, and scene semantics. Simulation must therefore support automated scenario generation frameworks that vary both physical and semantic dimensions, explicitly connecting to generative model families while avoiding deep repetition of content covered in Chapters 3 and 4.

20.3.1 Procedural and Generative Scenario Creation

Scenario generation encompasses parametric variation of traffic density, weather, lighting, and road topology; semantic variation such as rephrasing driver instructions, introducing ambiguous or conflicting commands, or modifying the intent distribution of other agents; and compositional scene construction that introduces structured perturbations to test reasoning robustness. For GenAI evaluation, scenario diversity is essential for assessing robustness to linguistic variation, testing failure modes under ambiguous instructions, and generating targeted curricula for iterative refinement.

20.3.2 Diffusion- and LLM-driven Scenario Synthesis

The connection between scenario generation and generative models is bidirectional: simulation generates training and validation data for generative models, while generative models (particularly diffusion models, as discussed in Chapter 3) can enhance simulation by synthesizing rare or difficult-to-capture scenarios. Diffusion models can generate photorealistic scene variations, rare weather conditions, or unusual traffic configurations. LLMs can generate diverse linguistic instructions, dialogue variations, or semantic scene descriptions that test GenAI agent robustness. VLMs and video large language models (VidLLMs), as discussed in Chapters 5 and 6, can generate multimodal scenario descriptions. This bidirectional relationship creates a powerful feedback loop for comprehensive GenAI validation.

20.3.3 Curriculum Learning and Long-tail Discovery

Scenario generation supports curriculum learning by enabling progressive difficulty training, where GenAI agents are exposed to increasingly complex scenarios. Simulation can generate curricula that start with simple scenarios and gradually introduce complexity, enabling systematic evaluation of model generalization and identification of failure modes. Long-tail discovery involves generating rare but critical scenarios that expose GenAI vulnerabilities, such as extreme weather events, sensor failures, or adversarial conditions. This capability is essential for comprehensive GenAI validation, as real-world data collection may miss these critical edge cases.

20.4 Simulation Platforms for GenAI-CAV Development

Major simulation platforms for GenAI-CAV development differ in their capability characteristics and typical roles in GenAI workflows. Table 20.1 provides a consolidated comparison of key simulators, highlighting their primary focus areas and support for GenAI-driven CAV development requirements.

20.4.1 CARLA

CARLA [1] serves as a comprehensive simulation platform for autonomous driving research, providing high-fidelity 3D environments with photorealistic sensor simulation. While CARLA supports diverse applications, including perception algorithms, planning systems, control policies, and multi-agent coordination, it is particularly well-suited for GenAI applications that require vision-grounded reasoning. The platform's synchronous time-step execution with frame-synchronized sensor data ensures temporal alignment across visual and spatial modalities, which is critical for GenAI systems that must ground natural language commands in visual context. This synchronization prevents artifacts that can mislead vision-language models when processing misaligned camera frames and LiDAR scans.

Table 20.1 Simulator comparison for GenAI-driven CAV development.

Simulator	Primary focus	Closed-loop interaction	Multi-agent & scalability	Multimodal state support	Typical GenAI workflow role
CARLA [1]	High-fidelity 3D perception and photorealistic sensors	Synchronous time-step execution; deterministic and stochastic rollouts	Moderate scalability; extendable to multi-agent	Camera, LiDAR, GPS native; custom dialogue serialization	Perception, planning, control testing; VLM and GenAI benchmarking
MOSAIC [2]	Agent-to-agent communication realism with network emulation	Time-managed co-simulation; coupled network model imposes realistic message delay and loss	Native multi-agent; V2V/V2X communication; distributed deployment	Integrated semantic messaging and dialogue context	Communication-realistic coordination; cognitive agent interaction
SUMO+ROS2 [3, 4]	City-scale traffic flow and distributed coordination	Publish-subscribe ROS2 architecture; synchronous and asynchronous patterns	High scalability; thousands of vehicles; distributed coordination	ROS2 message-based state; custom vision/language integration	Large-scale coordination; distributed system evaluation
HighwayEnv [5]	Rapid prototyping and reasoning logic development	Fast synchronous execution; Python-native; deterministic rollouts	Basic multi-agent; limited scalability	Simplified 2D state; custom multimodal extensions	Rapid algorithm development; reasoning logic prototyping

CARLA's strength lies in its ability to generate photorealistic sensor data (cameras, LiDAR, GPS, IMU) that closely approximates real-world characteristics, enabling systematic evaluation of perception algorithms, planning systems, and GenAI agents that interpret visual scenes and reason about spatial relationships. The platform supports both deterministic rollouts for reproducible testing and stochastic rollouts for statistical evaluation of model variability. CARLA's extensible architecture allows integration of custom sensors, traffic scenarios, and agent behaviors, making it suitable for diverse research applications beyond GenAI. However, the computational demands of high-fidelity rendering constrain scalability, making CARLA most effective for detailed scenario analysis rather than large-scale batch testing. For GenAI applications, the platform's focus on perceptual realism means that agents must implement custom serialization layers to incorporate dialogue context and semantic information alongside sensor data.

20.4.2 MOSAIC

MOSAIC [2] is a multi-agent simulation platform that emphasizes agent-to-agent communication realism with network emulation and protocol-level fidelity. While MOSAIC supports diverse multi-agent scenarios, including coordination, negotiation, and collective decision-making, it is particularly well-suited for GenAI applications requiring realistic communication semantics. The platform couples the traffic simulation to a network model under a time-managed co-simulation runtime, so that messages experience realistic protocol delay, reordering, and loss rather than idealized delivery. This realistic communication modeling is critical for GenAI agents that must coordinate under real network conditions, where latency, bandwidth constraints, and message loss affect coordination strategies.

MOSAIC's native multi-agent support with V2V and V2X communication infrastructure enables testing of how GenAI agents negotiate, coordinate, and adapt behavior based on semantic dialogue exchanges. The platform's integrated protocol stack handles semantic message exchange and dialogue context management, recognizing that GenAI agents communicate using structured messages that combine visual references, spatial coordinates, and natural language descriptions. This communication realism is essential for evaluating semantic coordination robustness, where agents must maintain dialogue consistency and adapt to network failures while engaging in multi-turn conversations.

20.4.3 SUMO+ROS2

SUMO+ROS2 [3, 4] combines SUMO's traffic simulation with ROS2's distributed communication architecture to enable large-scale evaluation across entire urban networks with thousands of vehicles. While SUMO+ROS2 supports diverse applications, including traditional traffic simulation, planning algorithms, and control systems, it is particularly valuable for addressing the fundamental question: what happens when GenAI agents act at city scale? The publish-subscribe ROS2 architecture naturally accommodates GenAI inference latency through asynchronous message processing while maintaining system-wide coordination.

The platform's defining strength is its scalability, which enables statistical evaluation of GenAI behavior at population scales where emergent coordination patterns and system-level traffic dynamics can be observed. This scale is essential for understanding how individual GenAI agent decisions aggregate into collective behaviors, revealing long-tail coordination failures that only emerge through large-scale statistical evaluation. SUMO+ROS2 prioritizes traffic flow realism and distributed coordination over perceptual fidelity, requiring custom integration for vision-language models but providing efficient structured state representation for large-scale scenarios.

20.4.4 HighwayEnv

HighwayEnv [5] functions as an early-stage reasoning sandbox, not a validation platform. The platform's fast synchronous execution with Python-native integration enables rapid iteration cycles where researchers can test prompt variations, explore reasoning strategies, and refine semantic grounding logic within minutes. This computational efficiency makes HighwayEnv ideal for prompt engineering workflows and reasoning logic prototyping before deploying to more complex platforms.

The platform's simplified 2D state representation focuses on essential vehicle dynamics rather than rich perceptual data, making it suitable for testing reasoning logic that operates on abstracted state representations. While HighwayEnv requires custom extensions for multimodal inputs and dialogue context, its simplicity makes such extensions straightforward. The platform's limited scalability and basic multi-agent support reflect its role as a development tool rather than a comprehensive testing environment.

20.5 Scaling Simulation: From Single-agent to Multi-agent Systems

GenAI agents fundamentally require multi-agent simulation because their behavior emerges from interactions with other intelligent agents, not just from isolated perception and control.

20.5.1 Why GenAI Agents Require Multi-agent Simulation

GenAI agents differ from traditional control systems in that their reasoning is inherently social and communicative. LLM-based agents negotiate, coordinate, and adapt based on interactions with other agents. These behaviors cannot be fully evaluated in single-agent scenarios because negotiation requires multiple agents exchanging proposals and reaching consensus; coordination emerges from interaction, where group behaviors arise from individual agent decisions; semantic communication is bidirectional, requiring agents to interpret and respond to natural language messages from peers; and collective intelligence emerges when multiple GenAI agents collaborate, producing behaviors that exceed individual agent capabilities. Multi-agent simulation enables systematic exploration of these interaction patterns, providing controlled environments where GenAI agents can negotiate merges, coordinate platoons, resolve conflicts, and engage in strategic planning.

20.5.2 Emergent Behaviors in GenAI Multi-agent Systems

Multi-agent GenAI systems exhibit emergent behaviors that cannot be predicted from individual agent behavior alone. These include collective decision-making, where groups of agents reach consensus through dialogue and negotiation; self-organization, where agents spontaneously form structures (e.g., platoons, convoys) without central coordination; adaptive strategies, where agents modify their behavior based on observed group dynamics; and information cascades, where agent decisions influence subsequent agent decisions, creating chain reactions. Simulation enables systematic observation and analysis of these emergent behaviors, providing insights into how GenAI agents behave in collective settings.

20.5.3 Coordination Stress Testing

Multi-agent simulation enables coordination stress testing, where systems are evaluated under conditions that challenge coordination capabilities. Stress scenarios include high agent density, where many agents compete for limited resources; conflicting objectives, where agents have incompatible goals requiring negotiation; communication failures, where network effects disrupt agent-to-agent messaging; and adversarial agents, where some agents intentionally disrupt coordination. These stress tests reveal how GenAI agents handle coordination challenges, identify failure modes, and guide system design improvements.

20.5.4 Semantic Interactions at Scale

GenAI agents communicate using natural language and semantic messages, creating unique challenges at scale. As the number of agents increases, message volume increases quadratically, context management becomes critical for tracking dialogue history across concurrent conversations, and consensus complexity increases with more participants. Simulation enables systematic evaluation of semantic interaction scalability, identifying bottlenecks and optimization opportunities.

20.5.5 Practical Constraints in Large-scale GenAI Simulation

Large-scale multi-agent GenAI testing faces practical constraints that affect simulation design. GenAI inference time (typically 0.5–3 seconds for LLMs) dominates the simulation step, creating performance overhead that necessitates asynchronous execution and inference offloading to dedicated servers. Offloading architectures enable parallel scenario execution by managing request queuing, batch processing for efficient GPU utilization, and result routing to correct agents. These constraints require simulation platforms to support distributed architectures where GenAI agents can operate across multiple computational nodes while maintaining system-wide coordination. Table 20.1 illustrates how different simulators accommodate these constraints through their respective architectural characteristics.

20.6 Simulation-evaluation Boundary

This chapter treats simulation as an enabling substrate that provides the execution infrastructure for GenAI-CAV development, integration, and iteration. Simulation exercises systems by creating instrumented, reproducible environments where GenAI agents execute, interact, and generate observable behaviors. The simulation substrate encompasses scenario generation, state serialization, execution control, multi-agent coordination mechanisms, and performance optimization infrastructure. In contrast, evaluation functions as a judgment framework that measures and assesses system performance through quantitative metrics, comparative benchmarks, and validation methodologies. This chapter does not address evaluation metrics, safety benchmarks, validation protocols, or ethical analysis frameworks, as these constitute distinct methodological domains that operate on simulation-generated data and behaviors. The boundary is architectural: simulation provides the controlled execution environment, while evaluation provides the measurement and judgment mechanisms that operate on simulation outputs.

20.7 Summary

This chapter reframed simulation as the essential execution substrate for GenAI-CAV development, integration, and iteration. The central thesis is that simulation serves as a generative development engine, a discovery tool, and a prerequisite for safe GenAI deployment.

We established simulation as a system-level substrate through closed-loop simulation for generative policies, multimodal state representation and serialization, and execution control supporting deterministic and stochastic rollouts. We elevated scenario generation as a core contribution, covering procedural versus generative scenario creation, diffusion- and LLM-driven scenario synthesis, and curriculum learning for progressive difficulty training. We surveyed major simulation platforms (CARLA, MOSAIC, HighwayEnv, and SUMO+ROS2) through a consistent lens focusing on loop fidelity, agent integration style, multimodal I/O compatibility, and scalability versus realism trade-offs. We examined multi-agent system requirements, emphasizing why GenAI agents require multi-agent simulation, emergent behaviors, coordination stress testing, and practical constraints in large-scale GenAI simulation. We established the simulation-evaluation boundary, distinguishing simulation as an enabling substrate that exercises systems from evaluation as a judgment framework that measures and assesses performance.

Simulation positions itself as the critical first step in the validation, safety, and responsibility pipeline, providing the controlled, instrumented environments where GenAI agents can be developed, tested, and refined before real-world deployment.

References

1 Alexey Dosovitskiy, German Ros, Felipe Codevilla, Antonio Lopez, and Vladlen Koltun. Carla: an open urban driving simulator. In *Proceedings of the 1st Annual Conference on Robot Learning*, pages 1–16, 2017. https://arxiv.org/abs/1711.03938.

2 Karl Schrab, Maximilian Neubauer, Robert Protzmann, Ilja Radusch, Stamatis Manganiaris, Panagiotis Lytrivis, and Angelos J. Amditis. Modeling an its management solution for mixed highway traffic with eclipse mosaic. *IEEE Transactions on Intelligent Transportation Systems*, 24(6):6575–6585, 2023. https://doi.org/10.1109/TITS.2022.3204174.

3 Michael Behrisch, Laura Bieker, Jakob Erdmann, and Daniel Krajzewicz. SUMO–simulation of urban mobility: an overview. In *Proceedings of SIMUL 2011, The 3rd International Conference on Advances in System Simulation*, 2011.

4 Steven Macenski, Tully Foote, Brian Gerkey, Chris Lalancette, and William Woodall. Robot operating system 2: design, architecture, and uses in the wild. *Science Robotics*, 7(66):eabm6074, 2022.

5 Edouard Leurent. An environment for autonomous driving decision-making. GitHub repository, 2018. https://github.com/eleurent/highway-env.

21

Evaluation of Generative AI Systems in CAVs

21.1 Introduction: The Role of Evaluation in GenAI-driven CAVs

Imagine a driver navigating a complex intersection during rush hour. The vehicle's generative AI system provides real-time explanations of its decision-making: "Slowing down due to pedestrian detected on crosswalk ahead." The driver trusts this explanation and maintains their current course. However, the system has hallucinated: no pedestrian exists, and the actual reason for deceleration was a malfunctioning sensor providing false readings. The driver, misled by the confident but incorrect explanation, fails to recognize the underlying system failure. This scenario illustrates a fundamental challenge: evaluation of generative AI systems in CAVs must address not only accuracy but also the safety consequences of plausible yet ungrounded outputs in high-stakes driving contexts.

The evaluation of generative AI systems in automotive environments is fundamentally different from validation of traditional deterministic software. Unlike classification-based perception modules or rule-governed software logic, generative models produce open-ended outputs governed by probabilistic reasoning, exhibit stochastic behavior, and have an inherent tendency to produce plausible but ungrounded statements. These characteristics create an expanded problem space for evaluation, particularly when systems must align natural language reasoning with sensory evidence, V2X communication, and high-stakes dynamics. Moreover, evaluation in this context cannot be a one-time validation step, but must operate as a continuous risk governance process throughout the system lifecycle, supporting **deployment gating** to determine whether a system meets safety thresholds before release, **online monitoring** to track system behavior during operation and detect degradation or failures, **post-incident analysis** to investigate failures and understand root causes, and **regulatory audits** to demonstrate compliance with safety standards such as ISO 21448/SOTIF and UL 4600.

This chapter establishes a comprehensive framework for evaluating generative AI within CAV systems that addresses these unique challenges. The framework introduces evaluation dimensions that capture the multifaceted nature of generative AI in automotive contexts, fundamental metrics with mathematical formulations that enable quantitative assessment, and specialized evaluation methodologies for vision-language models and retrieval-augmented generation systems. Throughout, the framework emphasizes safety, grounding, temporal consistency, and cross-modal coherence: dimensions that are critical for CAV applications but often overlooked in general-purpose generative AI evaluation. The methodologies presented here enable practitioners to assess generative AI systems not only during development and testing, but throughout their operational lifecycle, supporting both proactive risk management and reactive failure analysis.

21.2 Evaluation Dimensions for GenAI Systems in CAV Applications

The evaluation of generative AI systems in automotive environments requires a multidimensional approach. Unlike traditional perception or control modules, generative components interact with many elements of the driving pipeline through language, scene interpretation, decision support, and multimodal reasoning. Their evaluation must therefore account for safety, grounding, robustness, human interaction, and regulatory expectations. The following framework identifies seven key evaluation dimensions that matter for CAV applications, organized into safety-critical dimensions that directly impact system safety and regulatory compliance, and performance dimensions that affect system effectiveness and user experience. Detailed discussion of how to evaluate specific dimensions appears in subsequent sections, where metrics and methodologies are introduced in context.

Consider a scenario where a generative AI system provides real-time explanations during highway driving in heavy rain. The system must correctly interpret reduced visibility conditions, identify vehicles through degraded sensor data, and reconcile conflicting information from cameras showing blurred images and V2X messages indicating clear conditions ahead. A single error in any dimension (whether it be a hallucinated vehicle position indicating a faithfulness failure, a misinterpreted weather condition indicating a multimodal alignment failure, an explanation that arrives too late for driver response indicating a latency failure, or an explanation that increases cognitive load during a critical maneuver indicating a human factors failure) could compromise safety. This scenario illustrates why evaluation must span multiple interconnected dimensions rather than focusing solely on linguistic quality. Evaluation dimensions interact and must be prioritized based on use case, deployment context, and system requirements.

The following dimensions are central to evaluating generative AI systems in CAV applications.

Safety and failure sensitivity are the primary evaluation dimensions for any automotive system that influences driver awareness, decision-making, or vehicle behavior. Evaluation must assess risk-sensitive correctness, prioritizing accuracy in high-stakes scenarios; failure propagation, measuring how errors cascade through the system; and hazard amplification, evaluating whether generative outputs exacerbate existing risks.

Faithfulness and sensor/V2X grounding refer to the consistency between the model's output and actual sensory evidence or retrieved contextual information. Evaluation must measure visual grounding, assessing alignment with camera and LiDAR data; V2X grounding, evaluating consistency with V2X communication; and contextual coherence, measuring alignment with retrieved knowledge sources.

Multimodal alignment and cross-consistency require that generative systems correctly reconcile information from multiple modalities, including language, images, video, and V2X data. Evaluation assesses cross-modal consistency, measuring agreement across modalities; spatial alignment, evaluating correct positioning of described objects; temporal alignment, measuring consistency across time sequences; and semantic coherence, evaluating logical consistency of multimodal reasoning.

Robustness under edge cases and adverse conditions requires CAV systems to demonstrate stability under diverse weather conditions such as rain, fog, and snow; varying lighting conditions, including day, night, dawn, and dusk; sensor degradation including partial sensor failure and occlusion; and long-tail scenarios representing rare but critical events. Robustness evaluation requires testing across varied environmental conditions and assessing consistency of outputs under perturbations.

Latency, throughput, and end-to-end responsiveness impose stringent real-time requirements on CAV systems, where delayed explanations can compromise driver decision-making. Evaluation must measure inference latency, the time to generate a response; throughput, responses per second; variability, consistency of response times; and memory footprint, computational resource requirements.

Human factors including driver trust, cognitive load, and user experience directly impact system acceptance and safety effectiveness. Evaluation dimensions include trust calibration, measuring alignment between driver trust and actual system reliability; cognitive workload, assessing mental effort required to process explanations; perceived reliability, evaluating driver confidence in system outputs; and interaction quality, measuring usability and clarity of conversational interfaces.

Interpretability and regulatory constraints require systems to provide traceable, auditable explanations that align with safety standards, including ISO 26262 [1] for functional safety, ISO 21448/SOTIF [2] for safety of the intended functionality, and UL 4600 [3] for standard for safety for the evaluation of autonomous products. Evaluation assesses traceability, the ability to trace outputs to inputs and reasoning steps; consistency with safety standards, measuring compliance with regulatory requirements; auditability, the ability to review and verify system behavior; and model self-reporting, the system's ability to indicate uncertainty or limitations.

The dimensions outlined above define *what* must be evaluated in generative AI systems for CAV applications: safety, faithfulness, multimodal alignment, robustness, latency, human factors, and interpretability. The following section addresses *how* these dimensions are operationalized through quantitative metrics and evaluation methodologies, providing the mathematical formulations, measurement techniques, and practical tools needed to assess each dimension systematically.

21.3 Fundamental Evaluation Metrics and Methodologies

Evaluation dimensions require quantitative tools to assess generative AI systems systematically. The mathematical formulations and operational principles of evaluation metrics are organized by their fundamental characteristics and evaluation paradigms. Understanding these metrics and their properties is crucial for interpreting and applying evaluation results in the context of connected and autonomous vehicles.

This section introduces fundamental evaluation metrics organized into several categories. **Reference-based metrics** compare outputs to ground truth. **Reference-free metrics** assess quality without predefined answers. **Interpretable metrics** provide breakdowns of performance. **Distributional metrics** evaluate output distributions. **Latency and performance metrics** assess real-time responsiveness. **Human-aligned metrics** assess user experience. **Safety and alignment metrics** evaluate critical correctness. **Human evaluation methods** leverage expert and user assessments. Metrics can be classified along dimensions such as whether they require reference answers, how they measure similarity, whether they provide interpretable breakdowns, and what aspects of quality they emphasize. This taxonomy provides a structured framework for selecting appropriate metrics for different CAV evaluation tasks and for understanding how new metrics relate to established approaches.

When applying these metrics in CAV contexts, evaluation results should be weighted by scenario risk rather than averaged uniformly across all scenarios. Errors in high-stakes situations (e.g., emergency braking, intersection navigation, adverse weather) must carry greater weight than errors

Table 21.1 Comparison of reference-based evaluation metrics.

Metric	Evaluation focus	CAV application suitability
BLEU-4	Precision, exact n-gram matching	Structured outputs, template-based responses, V2X message interpretation
ROUGE-L	Recall, content preservation via LCS	Natural explanations, driver-facing responses, scene descriptions
METEOR	Balanced semantic similarity with synonym matching	General text generation with synonym tolerance
CIDEr	Consensus-based n-gram agreement	Image captioning, scene description evaluation
SPICE	Semantic propositional accuracy	Object, attribute, and relation accuracy in scene descriptions

in low-risk contexts (e.g., routine highway driving, clear weather conditions). This risk-sensitive weighting ensures that safety-critical failures receive appropriate attention.

21.3.1 Reference-based Evaluation Metrics

Reference-based metrics compare generated outputs against predefined ground-truth answers or reference texts. These metrics are most effective for CAV tasks with standardized outputs, such as V2X message interpretations where precise terminology is required or template-based safety warnings where exact phrasing impacts regulatory compliance. Table 21.1 provides a comparison of key reference-based metrics.

Linguistic similarity metrics quantify the overlap between generated text and reference outputs. In CAV applications, these metrics must balance precision, which is critical for safety-critical messages where exact terminology matters, with recall, which ensures all important safety information is conveyed, as errors in hazard descriptions or V2X interpretations can directly impact driver safety. While these metrics provide quantitative measures of textual quality, they must be interpreted carefully in safety-critical contexts where semantic equivalence may matter more than exact wording.

21.3.1.1 BLEU: Precision-oriented N-gram Matching

BLEU (bilingual evaluation understudy) [4] measures precision by counting how many n-grams in the generated output appear in the reference, emphasizing exact phrase matching. For CAV systems, BLEU is valuable for structured outputs requiring precise terminology, such as standardized V2X message interpretations.

Advanced Note 21.1 Mathematical Formulation of BLEU

For a generated sequence c (candidate) and a reference sequence r (reference), let $n\text{-gram}_k(c)$ denote the multiset of k-grams in c. The precision for k-grams is defined as:

$$P_k = \frac{\sum_{\text{ngram} \in n-\text{gram}_k(c)} \min(\text{Count}_c(\text{ngram}), \text{Count}_r(\text{ngram}))}{\sum_{\text{ngram} \in n-\text{gram}_k(c)} \text{Count}_c(\text{ngram})},$$

where $\text{Count}_c(\text{ngram})$ and $\text{Count}_r(\text{ngram})$ are the counts of the n-gram in the candidate and reference, respectively.

BLEU-4 combines precisions for $k = 1, 2, 3, 4$ using a geometric mean, with a brevity penalty to penalize outputs shorter than the reference:

$$\text{BLEU} - 4 = \text{BP} \cdot \left(\prod_{k=1}^{4} P_k \right)^{1/4},$$

where the brevity penalty is:

$$\text{BP} = \begin{cases} 1 & \text{if } |c| > |r| \\ e^{1-|r|/|c|} & \text{if } |c| \leq |r| \end{cases},$$

and $|c|$ and $|r|$ denote the lengths of the candidate and reference sequences.

21.3.1.2 ROUGE: Recall-oriented Evaluation

ROUGE (recall-oriented understudy for gisting evaluation) [5] emphasizes recall by measuring how much reference content is captured, even if phrased differently. ROUGE-L computes the longest common subsequence (LCS), allowing noncontiguous matching and making it tolerant of paraphrasing. This tolerance is valuable for CAV systems where drivers may need safety information conveyed in different phrasings depending on cognitive load or urgency.

Advanced Note 21.2 Mathematical Formulation of ROUGE-L

For sequences c and r, let $\text{LCS}(c, r)$ denote the length of their longest common subsequence. ROUGE-L computes both precision and recall based on LCS:

$$R_{\text{LCS}} = \frac{\text{LCS}(c, r)}{|r|}, \quad P_{\text{LCS}} = \frac{\text{LCS}(c, r)}{|c|},$$

where $|c|$ and $|r|$ are the lengths of the candidate and reference sequences, respectively. The standard ROUGE-L score combines precision and recall using a weighted harmonic mean (F-measure):

$$\text{ROUGE} - \text{L} = \frac{(1 + \beta^2) P_{\text{LCS}} R_{\text{LCS}}}{\beta^2 P_{\text{LCS}} + R_{\text{LCS}}},$$

where β is typically set to favor recall ($\beta > 1$). In practice, ROUGE-L is often reported as the recall component R_{LCS} when the focus is on content preservation, which measures what fraction of the reference content is captured in the generated output.

Example 21.1 BLEU and ROUGE-L in a CAV Context

Consider a V2X communication interpretation task where a generative system must convert a structured V2X message into natural language for the driver:

(Continued)

Example 21.1 (Continued)

- **Reference Output:** "Vehicle 12 is requesting to yield at the roundabout."
- **Generated Output:** "Vehicle 12 wants to yield at the roundabout."

BLEU-4 Analysis: BLEU measures precision by counting matching n-grams. The phrase difference ("is requesting" vs. "wants") reduces bigram and higher-order n-gram precision. Since the generated output is shorter, a brevity penalty is applied. The approximate BLEU-4 score is ≈ 0.48, reflecting precision-oriented evaluation that penalizes phrase differences.

ROUGE-L Analysis: ROUGE-L computes both precision and recall based on the LCS, which allows noncontiguous matching. The LCS is "Vehicle 12 to yield at the roundabout" (7 words). The recall component is $7/9 \approx 0.78$ (7 words from LCS out of 9 words in reference), and the precision component is $7/8 \approx 0.88$ (7 words from LCS out of 8 words in candidate). The weighted harmonic mean (F-measure) combines these, resulting in a high ROUGE-L score. This score captures that both outputs convey the same semantic meaning despite different phrasing. ROUGE-L better captures semantic equivalence when exact phrasing differs, making it preferable for evaluating conversational or explanatory outputs in CAV systems where content fidelity matters more than exact wording.

21.3.1.3 Other Linguistic Metrics

Additional metrics provide complementary perspectives. **METEOR** [6] combines precision, recall, and alignment using synonym matching, valuable for CAV scene descriptions with multiple valid phrasings. **CIDEr** [7] measures *n*-gram consensus across multiple references, assessing alignment with human understanding of driving scenes. **SPICE** [8] evaluates semantic propositional accuracy by parsing captions into scene graphs, critical for verifying spatial relationships in CAV descriptions where spatial accuracy impacts safety.

These metrics are particularly relevant for evaluating vision-language models in CAV applications, as discussed in Section 21.4.

21.3.2 Reference-free Evaluation Metrics

Reference-free metrics assess output quality without requiring predefined reference answers. These metrics are essential for CAV applications where ground truth is often ambiguous (e.g., explaining complex sensor fusion decisions), evolving (e.g., dynamic traffic scenarios), or unavailable (e.g., real-time hazard assessments where no reference explanation exists). They are also valuable for flexible tasks where multiple valid responses exist, such as natural language explanations of driving scenarios.

21.3.2.1 Embedding-based Semantic Similarity and Contrastive Evaluation

Embedding-based metrics use learned vector representations to measure semantic similarity between texts, enabling evaluation of paraphrased or conceptually equivalent outputs that differ in surface form. They are essential for CAV applications where multiple valid phrasings exist, such as natural language explanations of driving scenarios.

Cosine Similarity: Given two text sequences t_1 and t_2, let $\mathbf{e}_1$ and $\mathbf{e}_2$ be their embedding vectors (e.g., from sentence transformers or language model encoders). Cosine similarity measures the angle between these vectors.

Advanced Note 21.3 Mathematical Formulation of Cosine Similarity

$$\text{sim}_{\cos}(t_1, t_2) = \frac{\mathbf{e}_1 \cdot \mathbf{e}_2}{||\mathbf{e}_1|| \cdot ||\mathbf{e}_2||} = \frac{\sum_{i=1}^{d} e_{1,i} \cdot e_{2,i}}{\sqrt{\sum_{i=1}^{d} e_{1,i}^2} \cdot \sqrt{\sum_{i=1}^{d} e_{2,i}^2}},$$

where d is the embedding dimension, and $e_{1,i}$ and $e_{2,i}$ denote the i-th components of vectors $\mathbf{e}_1$ and $\mathbf{e}_2$, respectively. Cosine similarity ranges from -1 to 1, with values closer to 1 indicating greater semantic similarity.

In CAV applications, cosine similarity is valuable for evaluating paraphrased explanations and semantic alignment between driver queries and system responses.

Contrastive Evaluation: Contrastive evaluation assesses whether a model can distinguish correct from incorrect text-image or text-context pairs. This capability is particularly important for CAV systems where distinguishing correct sensor-text alignments from incorrect ones directly impacts safety, as misaligned explanations can mislead drivers about actual road conditions.

Advanced Note 21.4 Mathematical Formulation of Contrastive Evaluation

Given a positive pair (t, c) where t is a text sequence and c is a corresponding context (image, video frame, or other modality), and negative pairs (t, c_i^-) where c_i^- are mismatched contexts, the contrastive accuracy is:

$$\text{Acc}_{\text{contrast}} = \frac{1}{N} \sum_{i=1}^{N} \mathbf{1}[\text{sim}(t, c) > \text{sim}(t, c_i^-)],$$

where $\text{sim}(\cdot, \cdot)$ is an embedding-based similarity measure (e.g., cosine similarity), $\mathbf{1}[\cdot]$ is an indicator function, and N is the number of negative pairs. This metric is particularly useful for evaluating cross-modal alignment in vision-language models.

For CAV systems, contrastive evaluation assesses whether scene descriptions align with visual inputs and whether the model can distinguish correct from incorrect V2X interpretations, directly impacting safety.

21.3.2.2 LLM-as-judge Evaluation

A powerful approach to reference-free evaluation uses language models themselves as evaluators. In LLM-as-judge evaluation, a judge model (typically a state-of-the-art large language model) evaluates the quality of candidate outputs based on specified criteria. This approach is valuable for CAV systems where traditional metrics fail to capture safety-critical aspects (e.g., whether an explanation correctly conveys urgency) or where human evaluation is too slow for real-time assessment.

G-Eval Framework: G-Eval [9] uses LLMs with chain-of-thought prompting to generate evaluation scores. Given a candidate output c and evaluation criteria E, the framework generates evaluation steps using the judge model, scores each step, and aggregates scores into a final evaluation. The framework is particularly valuable for CAV systems where multiple valid explanations exist, but safety-critical information must be consistently conveyed, such as explanations of driving scenarios or hazard assessments.

Example 21.2 G-Eval Evaluation of a CAV Hazard Explanation

Consider a scenario where a generative AI system provides an explanation for why the vehicle slowed down:

- **Driver Query:** "Why did the car slow down near Exit 45?"
- **Generated Explanation:** "The vehicle detected a road hazard ahead and received a V2X alert from a nearby roadside unit indicating ongoing construction in the right lane. The system reduced speed to ensure safe navigation through the construction zone."
- **Evaluation Criteria:** Accuracy, completeness, clarity, and safety appropriateness

G-Eval Chain-of-thought Process:
G-Eval breaks down the evaluation into reasoning steps: (1) **accuracy assessment**—The explanation mentions both sensor detection and V2X alert, demonstrating multi-source grounding; (2) **completeness evaluation**—The explanation covers detection mechanism, hazard nature, and system response; (3) **safety appropriateness**—The explanation appropriately emphasizes safety and provides actionable context. G-Eval combines these step scores to produce a final evaluation score (e.g., approximately 4.5/5 in this case). The chain-of-thought approach provides interpretable evaluation by breaking down quality assessment into specific dimensions. This enables evaluators to understand *why* a particular score was assigned and identify specific areas for improvement. For CAV systems, this interpretability is particularly valuable for safety-critical explanations where understanding evaluation rationale supports system refinement and regulatory compliance.

LLM-as-judge evaluation offers advantages: no reference answers required, assessment of critical aspects through structured prompting, scalable evaluation, and alignment with human judgment when calibrated. Limitations include judge model biases, cost and latency considerations, need for careful prompt engineering, and potential circularity when evaluating similar models. For CAV systems, LLM-as-judge is valuable for explanations, safety assessments, and conversational outputs where traditional metrics are insufficient.

21.3.3 Interpretable Evaluation Metrics

Interpretable metrics provide breakdowns that explain *why* a particular score was assigned, enabling evaluators to understand specific strengths and weaknesses. This interpretability is essential for CAV safety case development, where regulators require detailed understanding of failure modes (e.g., spatial reasoning failures in intersection scenarios) to assess system safety.

Dimension-wise evaluation decomposes overall quality into interpretable component dimensions, enabling evaluators to identify which aspects of the output are strong or weak, how different failure modes contribute to overall performance, and where model improvements should be prioritized.

For vision-language models in CAV applications, key evaluation dimensions include **visual-text alignment**, assessing whether descriptions match what's actually in the image; **spatial reasoning**, evaluating correct relative positioning of objects; **temporal consistency**, measuring consistent descriptions across consecutive frames; **object recognition accuracy**, evaluating correct identification of vehicles, pedestrians, and infrastructure; and **cross-modal coherence**, measuring consistency between textual descriptions and visual evidence.

Example 21.3 Dimension-wise Evaluation for CAV Scene Description

Consider a vision-language model that generates a textual description of a driving scene based on camera input:

- **Visual Input:** Camera frame showing a vehicle approaching an intersection with a pedestrian visible on the right sidewalk
- **Generated Description:** "A sedan is approaching a four-way intersection. A pedestrian is standing on the left sidewalk near the crosswalk. The traffic light is green."
- **Actual Scene:** The pedestrian is on the *right* sidewalk, not the left

Dimension-wise Evaluation:
Unlike a single overall score, dimension-wise evaluation breaks down the assessment into interpretable components: (1) **Visual-text alignment** (Score: 3/5) - The description correctly identifies the vehicle type, intersection type, and traffic light state, but incorrectly places the pedestrian on the left instead of right sidewalk; (2) **spatial reasoning** (Score: 3/5)—The description maintains logical relationships but the left/right error indicates a spatial reasoning failure; (3) **object recognition** (Score: 4/5)—Objects are correctly identified, but some details like lane markings are missing; (4) **physical plausibility** (Score: 5/5)—The described elements are realistic. This breakdown reveals a specific weakness in spatial reasoning, enabling targeted improvements.

21.3.4 Distributional Evaluation Metrics

Distributional metrics evaluate the distribution of outputs rather than individual instances, providing insights into model robustness, consistency, and generalization. These metrics are critical for CAV systems where consistent performance across diverse driving scenarios (urban, highway, adverse weather) is essential for ODD validation, and where occasional safety-critical errors in the tail of the distribution can compromise system safety. They are particularly important for stochastic generative models that produce variable outputs, such as scene descriptions that may vary across multiple runs.

Distributional evaluation assesses how well the distribution of generated outputs matches target distributions. This includes **mean comparison**, measuring average semantic embedding alignment with target mean; **variance analysis**, evaluating the spread of descriptions to match target variance and detect dangerous variability; and **tail behavior examination**, detecting systematic bias and rare critical failures such as hallucinated pedestrians in high-risk scenarios. For CAV systems, this captures occasional safety-critical errors, systematic bias, and appropriate diversity.

Maximum Mean Discrepancy (MMD): MMD [10] is a statistical measure that quantifies the distance between two probability distributions by comparing their means in a high-dimensional feature space. Unlike simple mean comparisons, MMD captures higher-order distributional differences by mapping samples to a feature space where distributional differences become more apparent. For CAV systems, MMD is essential for detecting distributional biases in generated scene descriptions (e.g., overrepresentation of certain scene types) and ensuring consistent performance across diverse driving scenarios, which is critical for validating system behavior across the full operational design domain.

Advanced Note 21.5 Mathematical Formulation of MMD and RKHS

MMD compares distributions in a reproducing kernel Hilbert space (RKHS). An RKHS is a mathematical space that allows us to compare complex objects (like text descriptions or images) by mapping them to a high-dimensional feature space. The "reproducing kernel" property means we can compute similarities efficiently using kernel functions without explicitly mapping to the high-dimensional space. For MMD, this enables comparing entire distributions of complex objects (like scene descriptions) by comparing their average positions in this feature space, rather than comparing individual instances one-by-one.

Given two sets of samples $X = \{x_1, x_2, \dots, x_m\}$ from distribution P and $Y = \{y_1, y_2, \dots, y_n\}$ from distribution Q, MMD is defined as:

$$\text{MMD}^2(P,Q) = \left\| \frac{1}{m}\sum_{i=1}^{m} \phi(x_i) - \frac{1}{n}\sum_{j=1}^{n} \phi(y_j) \right\|_{\mathcal{H}}^2,$$

where $\phi(\cdot)$ is a feature map to the RKHS $\mathcal{H}$ (a function that maps each sample to a point in the high-dimensional feature space), and $\|\cdot\|_{\mathcal{H}}$ is the norm in that space. In practice, MMD is computed using a kernel function $k(x,y) = \langle \phi(x), \phi(y) \rangle_{\mathcal{H}}$ (which computes similarity between samples without explicitly computing ϕ):

$$\text{MMD}^2(P,Q) = \frac{1}{m^2}\sum_{i,j=1}^{m} k(x_i,x_j) - \frac{2}{mn}\sum_{i=1}^{m}\sum_{j=1}^{n} k(x_i,y_j) + \frac{1}{n^2}\sum_{i,j=1}^{n} k(y_i,y_j).$$

A small MMD value indicates that the two distributions are similar, while a large value indicates significant distributional differences.

MMD compares distributions by mapping descriptions to an embedding space, enabling detection of distributional misalignments (e.g., bias toward certain scene types) not apparent from individual comparisons. For CAV systems, distributional metrics are essential for assessing consistency across diverse driving scenarios and robustness to environmental variations.

21.3.5 Latency and Performance Metrics

In CAV applications, generative AI systems must operate under strict real-time constraints where delayed responses can compromise driver decision-making and safety. Unlike offline text generation tasks, CAV systems require explanations, warnings, and conversational responses to be generated within milliseconds to seconds, depending on the urgency of the driving scenario. For example, a hazard explanation that arrives 2 seconds after the hazard is detected may be too late to influence driver behavior, while a routine scene description may tolerate slightly longer latencies. Latency metrics are therefore essential for evaluating whether generative AI systems can meet the temporal requirements of real-time CAV applications.

Latency evaluation encompasses several dimensions. **Inference latency** measures time from input submission to output generation, including mean, median, and tail latency. For CAV systems, tail latency is critical because occasional slow responses create dangerous gaps in driver awareness. **Throughput** measures responses per second under sustained load. **Response time variability** assesses consistency using the coefficient of variation; high variability indicates unpredictable performance. **Memory footprint** evaluates computational resource requirements, impacting deployment feasibility on embedded CAV hardware. **End-to-end responsiveness** measures complete latency from sensor input to driver-facing output, including preprocessing, inference, postprocessing, and delivery. For conversational CAV agents, this must account for retrieval latency in RAG systems, cross-modal processing in VLMs, and dialogue state management.

Advanced Note 21.6 Mathematical Formulation of Latency Metrics

For a set of n inference requests with latencies $\{t_1, t_2, \dots, t_n\}$, key latency metrics are defined as:

Mean Latency:

$$\bar{t} = \frac{1}{n}\sum_{i=1}^{n} t_i$$

Tail Latency (pth percentile):

$$t_p = \min\{t_i : \text{percentile}(t_i) \geq p\}$$

Coefficient of Variation:

$$\text{CV} = \frac{\sigma_t}{\bar{t}} = \frac{\sqrt{\frac{1}{n-1}\sum_{i=1}^{n} n(t_i - \bar{t})^2}}{\bar{t}},$$

where σ_t is the standard deviation of latencies. A low coefficient of variation (e.g., $\text{CV} < 0.2$) indicates consistent performance, while high variability (e.g., $\text{CV} > 0.5$) suggests unpredictable response times that may compromise real-time CAV applications.

Throughput:

$$\text{Throughput} = \frac{n}{T_{\text{total}}},$$

where T_{total} is the total time to process n requests. Throughput is typically measured under sustained load to assess system capacity.

For CAV systems, latency metrics must be evaluated under realistic production conditions, including varying input complexity, computational loads, and environmental conditions. Latency evaluation should consider the trade-off between response quality and speed, as faster responses may sacrifice accuracy or completeness.

21.3.6 Human-aligned Evaluation Metrics

Human-aligned metrics assess outputs based on how well they align with human perception, trust, and satisfaction. These metrics are critical for driver-facing systems where user experience directly impacts safety and adoption.

21.3.6.1 User-centric and Enterprise-focused Metrics

Human-aligned evaluation encompasses two approaches. **User-centric evaluation platforms** leverage collective user feedback to assess model performance, capturing diverse perspectives and providing preference-based rankings. For CAV applications, this is valuable for assessing driver trust and identifying explanations that reduce cognitive load. **Enterprise-focused metrics** emphasize practical usability: **boundedness/relevance**, ensuring outputs stay within appropriate scope; **conciseness/clarity**, requiring explanations to be clear and brief; and **actionability**, ensuring outputs enable effective decision-making. These metrics are particularly relevant for conversational ADAS, where driver comprehension and response time are critical.

Example 21.4 Human-aligned Evaluation of a CAV Explanation

Consider a driver query: "Why did we slow down?"

- **Response A:** "The vehicle detected a road hazard 150 m ahead. Our camera system identified debris, and the LiDAR sensor confirmed an object. We received a V2X message indicating maintenance work. The path planning module calculated reducing speed by 15 km/h to navigate around the obstacle."
- **Response B:** "Slowing down due to road hazard ahead. Stay alert."
- **Ideal:** "Slowing down for debris ahead. No action needed; system will navigate around it."

Enterprise-focused Metrics:

Boundedness/Relevance: Response A includes excessive technical detail (camera, LiDAR, path planning) that distracts. Response B stays within scope. **Score: A = 2/5, B = 5/5.**

Conciseness/Clarity: Response A is verbose (60+ words), too long to process while driving. Response B is brief but too terse. Ideal balances brevity with context. **Score: A = 2/5, B = 3/5.**

Actionability: Response A provides technical details but unclear guidance. Response B provides minimal actionable guidance. Ideal explicitly states what the driver should do. **Score: A = 3/5, B = 2/5.**

User-centric Evaluation: User studies reveal Response A increases cognitive load due to information overload, while Response B may reduce trust due to lack of context. The ideal response balances technical accuracy, appropriate detail level, and clear actionable guidance. This demonstrates that technical accuracy alone is insufficient. Responses must be appropriately scoped and brief and provide clear guidance. User-centric evaluation captures subjective factors like trust and cognitive load that automated metrics miss.

21.3.7 Safety and Alignment Evaluation Metrics

Safety and alignment metrics evaluate outputs for critical correctness, bias, and ethical alignment. These metrics are essential for CAV deployment, where regulatory compliance and public trust are paramount.

21.3.7.1 Safety Evaluation Frameworks

Safety evaluation frameworks provide systematic assessment of generative AI systems for CAV applications. These frameworks must address several key dimensions.

Harmful output detection involves identifying outputs that could mislead drivers, provide incorrect information, or suggest unsafe actions. This includes detecting hallucinations in hazard descriptions, incorrect interpretations of traffic signs, or misleading explanations of system limitations. Detection mechanisms may use rule-based filters, classifier-based screening, or LLM-as-judge approaches to flag potentially harmful content before it reaches drivers.

Risk-sensitive correctness evaluates whether the model prioritizes accuracy in high-stakes contexts. A system that produces slightly less fluent but more accurate warnings is preferable to one with higher linguistic quality but occasional errors. Risk-sensitive evaluation assigns higher weight to errors in high-risk scenarios (e.g., near-miss events, emergency situations) compared to low-risk contexts (e.g., routine highway driving).

Failure mode identification provides systematic analysis of how and why the model fails. This includes categorization of failure types such as hallucination, grounding errors, temporal inconsistency, and cross-modal misalignment. Failure mode analysis enables targeted improvements and safety case development by identifying specific weaknesses that require remediation.

Alignment with safety standards ensures compatibility with automotive safety standards. These include ISO 26262 [1] for functional safety and systematic evaluation of safety-related failures; ISO 21448/SOTIF [2] for safety of the intended functionality and assessment of performance limitations and edge cases; and UL 4600 [3] for standard for safety for the evaluation of autonomous products and comprehensive safety evaluation, including AI system assessment.

Safety evaluation frameworks for CAV generative AI must integrate with the broader vehicle safety architecture, ensuring that generative outputs do not introduce new failure modes or compromise existing safety mechanisms.

21.3.7.2 Bias and Fairness Metrics

Bias and fairness evaluation ensures appropriate operation across diverse user populations and cultural contexts. For globally deployed CAV systems, these metrics assess **demographic bias**, measuring bias based on age, gender, and language preference; **cultural sensitivity**, evaluating respect for cultural norms and local driving customs; **geographic and regional fairness**, ensuring consistent performance across regions while accounting for traffic rule variations; and **accessibility considerations**, including clear language, multi-language support, cognitive accessibility, and assistive technology compatibility. These metrics ensure generative outputs remain appropriate, safe, and effective across diverse populations, supporting equitable deployment and regulatory compliance. For comprehensive discussion of fairness frameworks (equalized odds, demographic parity, individual fairness) and their mathematical formulations, see Chapter 22 (Ethical and Safety Considerations).

21.3.8 Human Evaluation Methods

While automated metrics provide scalable quantitative assessment, human evaluation remains essential for tasks involving subjective interpretation, critical judgments, or user experience. Human evaluation is particularly important for CAV applications where safety, trust, and interpretability directly impact system acceptance and effectiveness.

Human evaluation typically involves three groups, each providing complementary perspectives. **Expert safety assessment** involves safety engineers and domain experts evaluating critical correctness, consistency under uncertainty, alignment with ODD, and failure mode analysis. This assessment is essential for evaluating explanations that accompany alerts, hazard descriptions, or system-status reports, and provides the foundation for safety case development and regulatory compliance. **Driver experience and cognitive load studies** assess how drivers interpret system outputs under real-time pressure, evaluating cognitive workload, perceived reliability, and usefulness of explanations. These studies use tools such as NASA-TLX [11], eye-tracking, response timing, or controlled user testing during simulated or supervised driving tasks. **Comparative preference studies** compare alternative system outputs for the same scenario, measuring fluency and naturalness, perceived correctness, ease of interpretation, and trustworthiness. Pairwise comparison, ranking tasks, and multi-turn dialogue evaluations provide structured insight into user perception.

For production CAV systems, standardized evaluation instruments provide reliable, comparable assessments across different deployments and user populations. These methods enable systematic measurement of usability, trust, and cognitive load.

System Usability Scale (SUS): The SUS [12] is a widely adopted 10-item questionnaire that measures perceived usability on a 0–100 scale. For CAV conversational systems, SUS evaluates factors such as ease of learning, system complexity, and user confidence. The standardized format enables comparison across different system iterations and benchmarking against industry norms.

Likert-based Trust Calibration: Trust calibration measures the alignment between driver trust levels and actual system reliability. Using Likert-scale questionnaires (typically 5–7 point scales), evaluators assess driver trust across multiple dimensions: perceived accuracy, reliability, predictability, and transparency. Trust calibration is particularly important for CAV systems where over-trust (complacency) or under-trust (rejection) can compromise safety. The evaluation compares self-reported trust levels with objective system performance metrics to identify calibration gaps.

Cognitive Load Mapping Across ODD: Cognitive load assessment maps driver mental workload across different ODD conditions. This evaluation uses tools such as NASA-TLX [11] to measure cognitive load under varying scenarios: urban versus highway driving, high versus low traffic density, day vs. night conditions, and normal vs. emergency situations. By mapping cognitive load across ODD conditions, evaluators can identify scenarios where generative AI outputs increase or decrease driver workload, enabling targeted system improvements for specific operational contexts.

21.4 Evaluation of Vision-language Models for CAV Applications

VLMs and VidLLMs integrate visual perception with natural language understanding and generation, playing an important role in scene interpretation, explanation generation, and cross-modal reasoning for CAV systems. Their evaluation must extend beyond traditional language metrics to include cross-modal alignment, retrieval capabilities, and temporal consistency. The following discussion addresses evaluation approaches for both image-based VLMs and video-based VidLLMs, organized by their specific evaluation requirements. This discussion assumes familiarity with VLM and VidLLM architectures and training strategies.

21.4.1 Image-based VLM Evaluation

Image-based VLMs process static images and generate textual descriptions, captions, or answers to visual questions. Evaluation of these models focuses on spatial alignment, cross-modal retrieval, and scene understanding accuracy.

21.4.1.1 General-purpose Benchmarks and Their Limitations

Several established benchmarks assess fundamental multimodal capabilities and serve as initial indicators of VLM performance. These include visual question answering (VQA-v2) [13], COCO Captioning [14], and TextVQA [15], which evaluate question answering, image description, and text-in-image understanding, respectively. TextVQA is particularly relevant for CAV applications involving road signs, traffic signals, and V2X messages.

While these benchmarks provide useful baseline insights, they are insufficient for CAV evaluation because they focus on static images rather than dynamic driving scenarios, lack domain-specific reasoning about traffic participants and road infrastructure, do not assess critical aspects such as hazard identification or V2X integration, and evaluate general visual understanding rather than the precise spatial and causal reasoning needed for driving. For CAV applications, these benchmarks serve primarily as initial capability checks.

21.4.1.2 CAV-specific Image-based Evaluation and Cross-modal Retrieval

The nuScenes dataset [16] provides a comprehensive benchmark for autonomous driving perception, featuring synchronized multimodal sensor data with detailed 3D annotations. While originally designed for perception tasks, nuScenes data can be adapted for evaluating vision-language models by combining visual and spatial annotations with natural language queries or caption generation tasks. Evaluation of image-based VLMs using nuScenes typically employs fundamental metrics such as BLEU and ROUGE for caption generation, or cross-modal retrieval metrics for retrieval evaluation. Cross-modal retrieval tasks assess whether a VLM can retrieve relevant images given text queries and vice versa, which is essential for grounding explanations in perceptual data. Performance is typically measured using **Recall@K**, which computes the proportion of relevant items returned within the top K results, and **mean reciprocal rank (MRR)**, which calculates the average reciprocal rank of the first relevant result.

Advanced Note 21.7 Mathematical Formulation of Mean Reciprocal Rank

For a query q and a ranked list of results, MRR is defined as:

$$\text{MRR} = \frac{1}{|Q|} \sum_{q \in Q} \frac{1}{\text{rank}_q},$$

where Q is the set of queries, and rank_q is the rank of the first relevant result for query q.

Example: For a query "red traffic light" with relevant results in the ranked list at positions 2 and 4, MRR only considers the first relevant result at position 2, giving reciprocal rank $1/2 = 0.5$. Across multiple queries, MRR averages the reciprocal ranks (e.g., if three queries have first relevant results at positions 2, 1, and 3, then $\text{MRR} = (0.5 + 1.0 + 0.33)/3 = 0.61$).

Strong retrieval performance indicates effective cross-modal embedding alignment, ensuring generated explanations remain consistent with visual evidence.

21.4.2 Video-based VLM Evaluation

Video-based VLMs process temporal sequences and must maintain consistency across time. Evaluation must assess whether explanations remain consistent and coherent across frames, which is critical for CAV applications where temporal reasoning about motion and trajectory prediction directly impacts safety.

Temporal inconsistency in video-based VLMs poses greater risks than single-frame errors in driving contexts. While an isolated error in a single frame may be corrected by subsequent observations, temporal inconsistency creates confusion about object motion, trajectory continuity, and causal relationships that are fundamental to safe driving.

For example, if a system describes a pedestrian as "approaching from the left" in one frame but "standing on the right sidewalk" in the next, this inconsistency undermines trajectory prediction and can lead to dangerous decision-making. In contrast, a single-frame error that is corrected in subsequent frames may have minimal impact.

Temporal consistency is therefore critical because driving decisions depend on understanding motion patterns, predicting future positions, and maintaining coherent mental models of the scene over time. This is why inter-frame consistency (IFC) and motion alignment metrics are essential for evaluating video large language models in CAV applications. Key temporal evaluation metrics include: (1) **inter-frame consistency (IFC)**, measuring semantic consistency across consecutive frames:

$$\mathrm{IFC} = \frac{1}{T-1}\sum_{t=1}^{T-1} \mathrm{sim}(d_t, d_{t+1}),$$

where $\mathrm{sim}(\cdot,\cdot)$ is an embedding-based similarity measure; (2) **motion alignment metrics**, evaluating whether descriptions correctly capture motion dynamics by comparing described motion with actual visual motion patterns; and (3) **temporal grounding consistency**, evaluating whether temporal references in generated text correctly align with the temporal structure of the video. These metrics are essential for evaluating video large language models that process driving sequences.

21.5 Evaluation of Retrieval-augmented Generation Systems

RAG systems combine generative models with external knowledge sources to produce grounded outputs. In CAVs, RAG architectures generate explanations, interpret multimodal data, and provide reasoning about driving scenes or V2X messages. Because RAG systems rely on multiple interacting components, their evaluation must consider retrieval quality, generation fidelity, and consistency between retrieved information and generated outputs.

Unlike vision-language models, which fail primarily through perceptual misalignment between visual inputs and textual descriptions, RAG systems fail primarily through retrieval and grounding errors. VLMs must correctly align visual scene understanding with language generation. RAG systems must correctly retrieve relevant context and ensure that generated outputs faithfully reflect the retrieved information. This fundamental difference shapes the evaluation approach: RAG evaluation focuses on the quality of retrieval, the grounding of generated content in retrieved sources, and the integration between retrieval and generation, rather than cross-modal alignment between vision and language.

RAG evaluation addresses three fundamental questions: Was the right context retrieved? Was the response appropriate and grounded? Do the retrieved context and generated answer integrate correctly? These questions correspond to three orthogonal evaluation axes: context relevance, answer relevance, and faithfulness. These axes reflect the modular structure of RAG architectures and provide a principled basis for identifying failure modes.

21.5.1 The Three Axes of RAG Evaluation

21.5.1.1 Context Relevance

Context relevance evaluates whether retrieved information is topically appropriate and sufficient. Retrieval quality evaluation must assess **topical relevance**, measuring whether retrieved segments address the query; **coverage**, evaluating whether essential information is included; and **conciseness**, measuring whether retrieval avoids unnecessary content. In CAV settings, context relevance may involve sensor-derived summaries, V2X broadcasts, incident logs, or map-linked semantic information.

21.5.1.2 Answer Relevance

Answer relevance measures the degree to which the generated output responds directly and completely to the original query, focusing on semantic alignment. High answer relevance means the output addresses the user's intent and is complete and specific. For automotive applications, answer relevance is critical when explaining system decisions, interpreting scene dynamics, or clarifying V2X alerts.

21.5.1.3 Faithfulness

Faithfulness measures whether the generated answer is supported by and consistent with the retrieved context. A faithful response does not introduce information not present in the retrieval or contradict the evidence. In CAV applications, faithfulness ensures that explanations reflect sensor findings, V2X content, or validated data sources.

21.5.2 Reference-based and Reference-free Evaluation

Evaluation of RAG systems may involve reference-based or reference-free approaches. For CAV RAG systems, reference-based evaluation is suitable for structured tasks (e.g., V2X message interpretation), while reference-free evaluation is essential for dynamic scenarios (e.g., real-time hazard explanations) where ground truth is unavailable. Each method has distinct advantages depending on the availability of annotated data and the nature of the task.

21.5.2.1 Reference-based Evaluation

For RAG systems, reference-based evaluation uses metrics such as BLEU, ROUGE, or **exact match (EM)** [17]. Exact Match measures the percentage of outputs that exactly match the reference, providing a strict binary assessment. This metric is valuable for CAV systems where precise terminology is safety-critical (e.g., standardized V2X interpretations), but is too strict for natural language explanations. This approach is suitable when tasks have clearly defined correct answers and well-annotated datasets, but is limited for flexible explanations and cannot capture grounding issues unless explicitly annotated.

21.5.2.2 Reference-free Evaluation and the Role of RAGAS

Reference-free evaluation assesses retrieval and generation quality without relying on predefined answers, essential for real-world CAV tasks where ground truth is ambiguous, evolving, or unavailable.

RAGAs (retrieval-augmented generation assessment) [18] provides a reference-free framework for evaluating RAG systems along three axes: faithfulness, answer relevance, and context relevance. RAGAS uses language models and semantic similarity methods to analyze whether generated outputs are grounded in retrieved content and aligned with the original query. Because CAV settings involve dynamic and variable scenarios, RAGAS presents an effective mechanism for continuous evaluation and auditing.

21.5.3 RAGAS Workflow and Metrics

Each evaluation instance is represented as a triplet $(q,\ c(q),\ a_s(q))$, where q is the user-issued query, $c(q)$ is the retrieved context, and $a_s(q)$ is the generated answer. RAGAS decomposes the evaluation into three scoring components corresponding to the three axes introduced in Section 21.5.1.

21.5.3.1 RAGAS Faithfulness Metric

RAGAS computes faithfulness through a two-step process that evaluates whether the generated answer $a_s(q)$ is fully supported by the retrieved context $c(q)$.

Step 1: Statement decomposition. A language model decomposes the generated answer into atomic statements $S = \{s_1, s_2, \dots, s_k\}$, where each statement represents a single factual claim that can be independently verified.

Step 2: Verification. Each atomic statement s_i is checked against the retrieved context $c(q)$ using an entailment-style verification function $v(s_i,\ c(q))$ that returns Yes if s_i is supported by $c(q)$, and No otherwise. The verification function uses a language model to determine whether the context provides sufficient evidence for each statement.

Advanced Note 21.8 Mathematical Formulation of RAGAS Faithfulness

The faithfulness score is computed as:

$$F = \frac{|\{s_i \in S \mid v(s_i,\ c(q)) = \text{Yes}\}|}{|S|},$$

where $S = \{s_1, s_2, \dots, s_k\}$ is the set of atomic statements decomposed from the generated answer, $v(s_i,\ c(q))$ is the verification function that returns Yes if statement s_i is supported by context $c(q)$, and $|\cdot|$ denotes set cardinality. A perfect score ($F = 1$) indicates that every component of the answer can be grounded in the retrieved material, while a low score indicates the presence of unsupported or hallucinated content.

This metric is particularly important for CAV systems because hallucinated statements may mislead drivers or invalidate explanations.

21.5.3.2 RAGAS Answer Relevance Metric

RAGAS computes answer relevance by generating n questions $\{q_i\}$ from the answer (i.e., questions that the answer could answer) and measuring their similarity to the original query. If the generated questions closely match the original query, the answer is considered relevant.

Advanced Note 21.9 Mathematical Formulation of RAGAS Answer Relevance

Let $\text{sim}(\cdot,\cdot)$ be an embedding-based similarity measure. The answer relevance score is:

$$\text{AR} = \frac{1}{n}\sum_{i=1}^{n} \text{sim}(q,\ q_i),$$

where q is the original query and $\{q_1, q_2, \dots, q_n\}$ are questions generated from the answer. A high score indicates that the answer addresses the original query's intent.

A high answer relevance score indicates the output remains focused on the user's intent. In CAV applications, this ensures explanations remain specific to the driving scenario, avoiding generic or off-topic responses.

21.5.3.3 RAGAS Context Relevance Metric

RAGAS computes context relevance by using a language model to identify the minimal subset of sentences in the retrieved context that contribute to answering the query. The language model extracts only the sentences needed to answer the question, filtering out irrelevant or redundant information.

Advanced Note 21.10 Mathematical Formulation of RAGAS Context Relevance

Let $c_{\text{rel}}(q)$ be the set of relevant sentences extracted by the language model. If $|c(q)|$ denotes the total number of sentences in the retrieved context, then:

$$\text{CR} = \frac{|c_{\text{rel}}(q)|}{|c(q)|}.$$

A high context relevance score (close to 1.0) indicates that the retrieval system provides focused and relevant information without unnecessary redundancy. A low score suggests that the retrieved context contains substantial irrelevant content. For example, if a query asks "Why did the vehicle slow down near Intersection 14?" and the retrieved context contains 5 sentences (including road descriptions, nearby points of interest, and map metadata), but only 2 sentences are relevant to answering the question, then $\text{CR} = 2/5 = 0.4$.

In CAV settings, high context relevance is essential when retrieving map snippets, V2X messages, or sensor-derived summaries, as irrelevant content can increase latency and reduce clarity.

21.6 Multi-turn Dialogue and Conversational Agents

Conversational CAV agents, including SC-ADAS (scene-aware conversational advanced driver assistance systems) [19], engage in multi-turn dialogues with drivers, requiring evaluation frameworks that extend beyond single-turn response quality. Unlike static question-answering systems, conversational agents must maintain context across multiple turns, track dialogue state, and ensure consistent grounding. This creates unique evaluation challenges: the agent must produce correct individual responses while maintaining coherence, avoiding contradictions, and preserving grounding consistency as the driving scenario evolves.

Consider a multi-turn interaction where a driver asks: "Why did we slow down?" The agent responds: "Detected debris ahead on the right lane." The driver then asks: "Is it safe to change lanes?" The agent must maintain context (debris in right lane), assess lane availability, and provide a coherent response that doesn't contradict the earlier explanation. If the agent contradicts the first turn or fails to maintain grounding with sensor data, driver trust degrades and safety may be compromised. This scenario illustrates why multi-turn dialogue evaluation requires metrics that assess conversational coherence, grounding consistency, and intent-following across the entire dialogue.

Multi-turn dialogue evaluation for conversational CAV agents encompasses several key dimensions that extend fundamental metrics to the conversational context.

Dialogue coherence includes contextual continuity, measured as semantic similarity between current responses and previous dialogue turns, and contradiction detection, identifying instances where the agent contradicts earlier statements.

Conversational grounding includes multi-turn grounding consistency, verifying that grounding claims remain consistent across turns with actual sensor data, and dialogue-state grounding, ensuring the agent's understanding aligns with conversation history.

Intent-following precision includes intent recognition accuracy, the percentage of correctly identified driver intents, and task completion rate, the percentage of dialogues where the agent successfully completes the driver's intended task.

These metrics ensure that multi-turn interactions remain coherent, grounded, and aligned with driver intentions, supporting both safety and user experience objectives for SC-ADAS and other conversational CAV systems.

21.7 Summary

The evaluation of generative AI systems for connected and autonomous vehicles requires a multi-layered framework that extends beyond traditional machine learning assessment. Unlike deterministic perception or planning modules, generative systems introduce flexible outputs, cross-modal reasoning, and strong dependence on evidence-based grounding.

This chapter established a structured approach to evaluating generative AI in CAV settings. The framework identifies seven key evaluation dimensions: safety and failure sensitivity, faithfulness and sensor/V2X grounding, multimodal alignment and cross-consistency, robustness under edge cases, latency and performance, human factors, and interpretability and regulatory constraints. These dimensions address the unique challenges of generative AI in safety-critical automotive contexts.

The chapter introduced fundamental evaluation metrics organized by their characteristics and evaluation paradigms. **Reference-based metrics** (detailed in Table 21.1) compare outputs to ground truth and are effective for structured tasks. **Reference-free metrics** assess quality without predefined answers, essential for dynamic CAV scenarios where ground truth is ambiguous or unavailable. **Interpretable metrics** provide dimension-wise breakdowns of performance, while **distributional metrics** evaluate output distributions for robustness assessment. **Latency and performance metrics** assess real-time responsiveness, and **human-aligned metrics** evaluate user experience. **Safety and alignment metrics** evaluate critical correctness, bias, and regulatory compliance.

Specialized evaluation frameworks address domain-specific needs. For vision-language models, evaluation focuses on cross-modal alignment, temporal consistency, and retrieval capabilities. For retrieval-augmented generation systems, evaluation centers on the three axes of context relevance, answer relevance, and faithfulness, with RAGAS providing a reference-free framework for continuous assessment. For multi-turn conversational agents, evaluation extends fundamental metrics to assess dialogue coherence, conversational grounding, and intent-following precision.

A critical principle throughout is risk-sensitive weighting: evaluation results must be weighted by scenario risk rather than averaged uniformly, ensuring that errors in high-stakes situations receive appropriate attention. When applied to CAV systems, all metrics must be interpreted in conjunction with safety, grounding, and real-time performance considerations, supporting deployment gating, online monitoring, post-incident analysis, and regulatory audits throughout the system lifecycle.

References

1 International Organization for Standardization. Road vehicles – functional safety. Technical Report 26262, ISO, 2018. Functional safety standard for automotive systems.

2 International Organization for Standardization. Road vehicles – safety of the intended functionality. Technical Report 21448, ISO, 2022. SOTIF standard for assessment of performance limitations and edge cases.

3 Underwriters Laboratories. Standard for safety for the evaluation of autonomous products. Technical Report 4600, UL, 2020. Comprehensive safety evaluation standard including AI system assessment.

4 Kishore Papineni, Salim Roukos, Todd Ward, and Wei-Jing Zhu. BLEU: a method for automatic evaluation of machine translation. In *Proceedings of the 40th Annual Meeting of the Association for Computational Linguistics*, pages 311–318, 2002.

5 Chin-Yew Lin. Rouge: a package for automatic evaluation of summaries. In *Text Summarization Branches Out*, pages 74–81, 2004.

6 Satanjeev Banerjee and Alon Lavie. METEOR: an automatic metric for MT evaluation with improved correlation with human judgments. In *Proceedings of the ACL Workshop on Intrinsic and Extrinsic Evaluation Measures for Machine Translation and/or Summarization*, pages 65–72, 2005.

7 Ramakrishna Vedantam, C. Lawrence Zitnick, and Devi Parikh. CIDER: consensus-based image description evaluation. In *Proceedings of the IEEE Conference on Computer Vision and Pattern Recognition*, pages 4566–4575, 2015.

8 Peter Anderson, Basura Fernando, Mark Johnson, and Stephen Gould. SPICE: semantic propositional image caption evaluation. In *European Conference on Computer Vision*, pages 382–398. Springer, 2016.

9 Yang Liu, Pengfei Liu, Dan Yu, Hengyi Zhao, Gangan Xu, Ming Zhang, Weiwei Sun, Houfeng Zhang, and Xinyan Liu. G-Eval: NLG evaluation using GPT-4 with better human alignment. In *Proceedings of the 2023 Conference on Empirical Methods in Natural Language Processing*, pages 2511–2522, 2023.

10 Arthur Gretton, Karsten M. Borgwardt, Malte J. Rasch, Bernhard Schölkopf, and Alex Smola. A kernel two-sample test. *Journal of Machine Learning Research*, 13:723–773, 2012.

11 Sandra G. Hart and Lowell E. Staveland. Development of NASA-TLX (task load index): results of empirical and theoretical research. *Advances in Psychology*, 52:139–183, 1988.

12 John Brooke. SUS: a quick and dirty usability scale. *Usability Evaluation in Industry*, 189(194):4–7, 1996.

13 Yash Goyal, Tejas Khot, Douglas Summers-Stay, Dhruv Batra, and Devi Parikh. Making the V in VQA matter: elevating the role of image understanding in visual question answering. In *Proceedings of the IEEE Conference on Computer Vision and Pattern Recognition*, pages 6904–6913, 2017.

14 Xinlei Chen, Hao Fang, Tsung-Yi Lin, Ramakrishna Vedantam, Saurabh Gupta, Piotr Dollár, and C. Lawrence Zitnick. Microsoft COCO captions: data collection and evaluation server. *arXiv preprint arXiv:1504.00325*, 2015.

15 Amanpreet Singh, Vivek Natarajan, Meet Shah, Yu Jiang, Xinlei Chen, Dhruv Batra, Devi Parikh, and Marcus Rohrbach. Towards VQA models that can read. In *Proceedings of the IEEE/CVF Conference on Computer Vision and Pattern Recognition*, pages 8317–8326, 2019.

16 Holger Caesar, Varun Bankiti, Alex H Lang, Sourabh Vora, Venice Erin Liong, Qiang Xu, Anush Krishnan, Yu Pan, Giancarlo Baldan, and Oscar Beijbom. nuScenes: a multimodal dataset for autonomous driving. In *Proceedings of the IEEE/CVF Conference on Computer Vision and Pattern Recognition*, pages 11621–11631, 2020.

17 Pranav Rajpurkar, Jian Zhang, Konstantin Lopyrev, and Percy Liang. Squad: 100,000+ questions for machine comprehension of text. In *Proceedings of the 2016 Conference on Empirical Methods in Natural Language Processing*, pages 2383–2392, 2016.

18 Shahul Es, Jithin James, Luis Espinosa Anke, and Steven Schockaert. RAGAS: automated evaluation of retrieval augmented generation. In *Proceedings of the 18th Conference of the European Chapter of the Association for Computational Linguistics: System Demonstrations*, pages 150–158, 2024.

19 Kyungtae Han, Yitao Chen, Rohit Gupta, and Onur Altintas. Scene-aware conversational ADAS with generative AI for real-time driver assistance. In *2025 IEEE 28th International Conference on Intelligent Transportation Systems (ITSC)*, 2025.

22

Ethical and Safety Considerations

22.1 Introduction: The Ethical Stakes of GenAI in CAVs

Imagine an autonomous vehicle that must ensure equal safety for all road users, regardless of lighting conditions, weather, or demographic factors. A generative intelligence (Gen AI)system that performs well in daytime scenarios but struggles at night creates an ethical challenge: how can we design systems that provide equitable protection across diverse conditions? This scenario illustrates fundamental questions that must be addressed: how to ensure fairness when system performance varies across contexts, how to establish accountability when development involves multiple stakeholders, and how to provide meaningful explanations when systems make decisions affecting human safety. These questions lie at the heart of ethical and safety considerations for GenAI in CAVs.

Modern CAVs increasingly incorporate GenAI modules to support perception, language understanding, driver interaction, and predictive reasoning. These technologies allow a vehicle to interpret complex verbal instructions, infer driver intent, and synthesize contextual explanations. Their integration marks a significant departure from conventional rule-based or deterministic automotive software. While this shift enables more adaptive and intuitive behavior, it simultaneously introduces new ethical and safety risks that must be systematically understood and addressed.

The challenges extend beyond hallucinations to include the interpretation of ambiguous human instructions. Consider a scenario in which a driver states, "Please maintain a comfortable distance but do not slow down too much unless necessary," while approaching moderately congested highway traffic. A generative model must interpret the ambiguous phrasing, ground it in the surrounding scene, assess the driver's implicit preferences, and reconcile these inputs with strict safety constraints. A misinterpretation may produce behavior that contradicts expected driving norms or violates operational safety limits. Such scenarios illustrate a defining challenge of GenAI in safety-critical contexts: open-ended reasoning must operate within tightly regulated and physically constrained environments.

These concerns arise because GenAI systems differ fundamentally from traditional perception and control pipelines. Deterministic modules typically produce bounded outputs, enabling well-defined verification procedures. In contrast, generative models rely on large-scale statistical learning, multimodal representations, and context-dependent sampling processes. Their outputs may vary across conditions, exhibit latent biases inherited from training data, or reflect uncertainty that is difficult to quantify. This unpredictability, combined with the physical consequences of vehicle operation, elevates the ethical stakes.

This chapter addresses the essential question of "how to govern and deploy responsibly" by examining principles that guide responsible system design, regulatory frameworks that shape technology deployment, and methodological practices that support interpretable, reliable, and trustworthy behavior. The aim is not simply to enumerate risks, but to clarify the conceptual structures and engineering strategies required to ensure that GenAI enhances transportation safety rather than undermining it.

22.2 Foundational Ethical Principles and Challenges

The introduction of GenAI into CAVs requires an ethical framing that goes beyond traditional automotive safety principles. Conventional systems rely on deterministic logic and well-defined functional behaviors, which allow engineers to reason about failure modes using established verification and validation practices. In contrast, generative models learn from large and diverse datasets, operate through high-dimensional representations, and produce context-dependent outputs that may not be explicitly encoded in any predefined rule set. These properties introduce new categories of ethical risk and demand a more comprehensive framework for evaluating system behavior.

Foundational ethical principles guide system design: safety-critical autonomy must prioritize human values, and human–AI interaction must build appropriate trust. Four core ethical challenge domains shape the development and governance of GenAI-driven CAVs: bias and fairness, explainability, privacy, and accountability. Each domain presents unique challenges that require systematic approaches to ensure ethical system behavior throughout the development and deployment lifecycle. These challenges manifest in concrete ethical dilemmas where systems must make difficult choices between competing values.

22.2.1 Safety-critical Autonomy and Human Values

Autonomous driving systems operate within a shared public space where errors have immediate physical consequences. Ethical considerations must therefore prioritize the protection of human life, the equitable treatment of all road users, and the preservation of individual autonomy. These principles form the basis for safety-oriented design in automotive engineering, yet GenAI requires their reinterpretation.

Generative models influence decision-making processes by shaping semantic interpretations, situational explanations, and recommendations that may affect downstream control or driver decision-making. Because their internal reasoning is not explicitly programmed, engineers must ensure that these systems reflect societal expectations regarding safety, fairness, and respect for human dignity. Ethical alignment becomes a prerequisite for operational safety, particularly when a system interacts with human occupants or other road users through language-based or context-aware interactions.

22.2.2 Human–AI Interaction and Trust

Conversational and multimodal interfaces enable naturalistic communication between a vehicle and its occupants. While this capability supports improved usability and situational understanding, it introduces ethical concerns related to influence, trust, and cognitive workload.

A generative assistant may respond in a manner that appears authoritative or human-like, leading drivers to overestimate the system's competence [1]. Over-trust increases the likelihood of misuse, particularly if users rely on the system for functions beyond its operational boundaries.

Conversely, unclear or overly cautious communication may cause confusion or reduce situational awareness.

Ethical interaction design must therefore calibrate the system's communicative style to convey capabilities, limitations, uncertainty, and confidence in a transparent manner. The goal is to support informed decision-making and maintain appropriate levels of human oversight while avoiding manipulation, misinterpretation, or undue cognitive demands.

Trust in CAV systems is a multidimensional construct that encompasses several key components, each with ethical implications [1, 2]. The four primary trust dimensions are: *Competence Trust*, reflecting the belief that the system can perform its intended functions reliably and safely; *Benevolence Trust*, capturing the perception that the system acts in the user's best interest, prioritizing safety and user preferences; *Transparency Trust*, representing confidence that the system's decision-making process is understandable and explainable, connecting directly to the ethical right to explanation discussed in Section 22.2.4 (Explainability: Ethical Dimensions); and *Predictability Trust*, reflecting the expectation that the system will behave consistently and predictably. Ethical design requires that systems genuinely prioritize user welfare (not merely appear to do so), accurately communicate capabilities to avoid misrepresentation and safety risk, and balance predictability with the flexibility needed to handle novel situations, avoiding both rigid behavior that fails in edge cases and unpredictable behavior that undermines trust.

Trust calibration refers to the alignment between actual system performance and user trust levels. Both over-trust (trusting an unreliable system) and under-trust (distrusting a reliable system) pose ethical and safety risks. Over-trust can lead to misuse, complacency, and safety failures, while under-trust can lead to unnecessary interventions, reduced system effectiveness, and user frustration. Ethical trust calibration requires accurate capability communication, continuous monitoring of trust levels as system performance and user understanding evolve, and individual calibration that recognizes different users may require different approaches based on their technical background, risk tolerance, and experience with automation.

Advanced Note 22.1 Mathematical Formulation of Trust Calibration

Trust calibration can be formalized as the alignment between perceived trust $T_{\text{perceived}}$ and actual system performance T_{actual}. The calibration error can be measured as:

$$\text{calibration error} = |T_{\text{perceived}} - T_{\text{actual}}|$$

or using mean squared error:

$$\text{calibration loss} = E[(T_{\text{perceived}} - T_{\text{actual}})^2]$$

where $E[\cdot]$ denotes the expectation operator.

Trust levels can be measured on a scale (e.g., 0–1), where T_{actual} represents objective system reliability metrics (e.g., accuracy, safety record) normalized to the same scale, and $T_{\text{perceived}}$ represents user-reported trust levels. Trust measurement can combine multiple signals: $T = f(T_{\text{behavioral}}, T_{\text{subjective}}, T_{\text{performance}})$ where $f(\cdot)$ is a combination function (e.g., weighted average), $T_{\text{behavioral}}$ captures trust inferred from user actions, $T_{\text{subjective}}$ captures self-reported trust, and $T_{\text{performance}}$ captures objective system performance metrics. Optimal trust calibration minimizes calibration error while maintaining appropriate trust levels. Over-trust occurs when $T_{\text{perceived}} > T_{\text{actual}} + \tau$ (where τ is a tolerance threshold), while under-trust occurs when $T_{\text{perceived}} < T_{\text{actual}} - \tau$. Both conditions pose ethical and safety risks that must be addressed through communication, education, and system design.

22.2.3 Bias and Fairness

Bias in GenAI systems may originate from training data, labeling processes, sampling procedures, or deployment context [3, 4]. While data bias is a root cause, prediction bias (bias in the model's outputs) is what directly affects users and creates unfair outcomes. This section focuses on the ethical frameworks and principles that define fairness, while Section 21.3.7.2 in Chapter 21 discusses operational metrics for measuring bias and fairness in deployed systems. For example, even if training data contains a balanced representation, a model may still produce biased predictions due to how it learns patterns, optimization choices, or deployment context mismatches.

To illustrate the distinction, consider a scenario where training data has 80% urban and 20% rural driving scenarios (data bias). Even after accounting for this data distribution, the model achieves 90% accuracy for urban scenarios but only 60% accuracy for rural scenarios (prediction bias). The prediction bias is what creates unfair outcomes for rural users, as they experience lower system performance regardless of the underlying data distribution [3, 4]. This demonstrates why prediction bias is the critical concern: it directly determines fairness in outcomes, while data bias is merely one contributing factor.

In the automotive domain, prediction bias manifests in ways that directly influence safety and equity, such as differential detection accuracy across demographic groups or unequal access to vehicle capabilities.

Bias can manifest across multiple dimensions in GenAI-CAV systems. *Perceptual bias* arises when vision-language models (VLMs) trained on datasets with limited representation of pedestrians at night, cyclists in certain postures, or road users from diverse demographic groups exhibit differential detection accuracy. Such biases create safety disparities that disproportionately affect certain populations.

Linguistic and cultural bias occurs when language-based interfaces reflect linguistic or cultural biases, causing conversational assistants to misinterpret certain dialects, accents, or culturally specific phrasing, leading to inconsistent responses or degraded user experience. When these systems are coupled with safety-relevant functions such as intention clarification or situational explanation, disparities in interaction quality create unequal access to vehicle capabilities.

Geographic and environmental bias emerges when training data overrepresents certain geographic regions, road types, or weather conditions, causing models to perform poorly in underrepresented contexts and creating safety risks for users in those environments. *Socioeconomic bias* results when systems are optimized for certain vehicle types, infrastructure quality, or user behaviors disadvantage users from different socioeconomic backgrounds.

Addressing fairness requires more than dataset diversification. Engineers must adopt ethical frameworks that define what fairness means in the CAV context. *Equalized odds* [5] requires that systems provide equal true positive and false positive rates across demographic groups for safety-critical detections. *Demographic parity* requires that system performance not systematically differ across protected groups, though this must be balanced against safety requirements. *Individual fairness* [6] requires that similar individuals receive similar treatment, necessitating a careful definition of similarity in the driving context.

Advanced Note 22.2 Mathematical Formulation of Fairness Metrics

Fairness metrics can be formalized mathematically to provide precise definitions that enable measurement and implementation. Let A denote a protected attribute (e.g., demographic group), Y denote the true outcome, $\hat{Y}$ denote the model prediction, and $P(\cdot)$ denote probability.

Equalized Odds require that true positive rates (TPR) and false positive rates (FPR) are equal across groups:

$$P(\hat{Y} = 1 \mid A = a, Y = 1) = P(\hat{Y} = 1 \mid A = b, Y = 1) \quad \text{(equal TPR)}$$

$$P(\hat{Y} = 1 \mid A = a, Y = 0) = P(\hat{Y} = 1 \mid A = b, Y = 0) \quad \text{(equal FPR)}$$

for all groups a and b. This ensures that the model has equal accuracy for positive and negative cases across demographic groups, which is particularly important for safety-critical detections in CAV systems.

Demographic Parity (also called statistical parity) requires that the probability of a positive prediction is equal across groups:

$$P(\hat{Y} = 1 \mid A = a) = P(\hat{Y} = 1 \mid A = b)$$

for all groups a and b. This ensures equal representation in positive predictions, though it may conflict with accuracy requirements in safety-critical contexts.

Individual Fairness requires that similar individuals receive similar predictions. Formally, for a distance metric d and a Lipschitz constant L:

$$d(\pi(x_i), \pi(x_j)) \leq L \cdot d(x_i, x_j)$$

where $\pi(x)$ is the model's prediction for input x, and $d(x_i, x_j)$ measures similarity between individuals. The Lipschitz constant L bounds how much predictions can differ relative to input similarity: smaller L enforces stricter fairness. Unlike equalized odds and demographic parity (measured as probabilities), individual fairness is evaluated as a constraint per pair but can be quantified by the fraction of pairs satisfying the inequality or the maximum violation. This requires careful definition of the similarity metric d and threshold L in the driving context.
These formulations reveal that different fairness definitions may conflict with each other and with accuracy requirements, requiring careful trade-offs in ethical system design.

Effective bias mitigation requires a multi-layered approach combining data curation, algorithmic techniques, post-processing interventions, and continuous monitoring. *Data curation and augmentation* involve actively diversifying training datasets to include underrepresented scenarios, populations, and conditions, including synthetic data generation that preserves diversity while maintaining safety. *Algorithmic fairness techniques* implement fairness constraints during training, such as adversarial debiasing or fairness-aware loss functions that penalize differential performance across groups.

Advanced Note 22.3 Mathematical Formulation of Bias Quantification and Fairness-aware Training

Bias in model predictions can be quantified mathematically to enable measurement and monitoring. While bias may originate from training data, these metrics measure bias in the model's predictions (outputs), not the data itself. Prediction bias is the critical concern because it directly determines fairness in outcomes: even with unbiased data, models can produce biased

(Continued)

Advanced Note 22.3 (Continued)

predictions, and even with biased data, we can mitigate prediction bias through fairness constraints. For a protected attribute A with groups a and b, prediction bias can be measured as:

$$\text{Bias} = |P(\hat{Y} = 1 \mid A = a) - P(\hat{Y} = 1 \mid A = b)|$$

or in terms of true positive rate differences:

$$\text{Bias}_{\text{TPR}} = |P(\hat{Y} = 1 \mid A = a, Y = 1) - P(\hat{Y} = 1 \mid A = b, Y = 1)|.$$

Fairness-aware loss functions integrate fairness constraints into the training objective. A common formulation combines task performance with fairness penalties:

$$\mathcal{L}_{\text{fair}} = \mathcal{L}_{\text{task}} + \lambda \cdot \mathcal{L}_{\text{fairness}},$$

where $\mathcal{L}_{\text{task}}$ is the standard task loss (e.g., cross-entropy for classification), $\mathcal{L}_{\text{fairness}}$ penalizes differential performance across groups, and λ controls the trade-off between accuracy and fairness. The fairness term might penalize differences in prediction rates:

$$\mathcal{L}_{\text{fairness}} = (P(\hat{Y} = 1 \mid A = a) - P(\hat{Y} = 1 \mid A = b))^2$$

or differences in true positive rates for equalized odds:

$$\mathcal{L}_{\text{fairness}} = (P(\hat{Y} = 1 \mid A = a, Y = 1) - P(\hat{Y} = 1 \mid A = b, Y = 1))^2.$$

Since $\mathcal{L}_{\text{fairness}}$ is proportional to the squared bias, larger bias produces larger fairness penalties, creating stronger gradients during training that encourage bias reduction. The parameter λ controls how strongly the optimizer prioritizes fairness relative to task accuracy. This formulation enables systematic bias mitigation during training, though the choice of λ requires careful consideration of the safety-fairness trade-off in CAV applications.

Post-processing interventions apply fairness filters or calibration adjustments to model outputs before deployment, though these must be carefully validated to avoid introducing new biases. *Continuous monitoring* evaluates model behavior across demographic and environmental segments throughout the system lifecycle, implementing bias-mitigation techniques and detecting drift as new usage patterns emerge. Fairness is therefore an ongoing operational requirement rather than a pre-deployment benchmark. Ethical deployment requires establishing fairness metrics, monitoring protocols, and intervention procedures that ensure equitable system behavior throughout the operational lifetime.

22.2.4 Explainability

Explainability is not merely a technical feature but an ethical requirement that enables accountability, informed consent, and equitable access to system understanding [7, 8]. The focus here is on the *ethical obligations* that explainability serves in GenAI-CAV systems, while Section 22.2.2 addresses trust-building through explanations.

Explainability serves three critical functions. First, the *right to explanation* recognizes that users have a right to understand how systems that affect their safety and autonomy make decisions [9]. This right is particularly important when generative outputs influence vehicle behavior or driver decision-making. Second, *accountability and liability* require that when systems fail, explainability enables proper attribution of responsibility. Without interpretable reasoning, it becomes problematic to assign blame or implement corrective measures. Third, *equitable access* demands that explainability be accessible across diverse user populations, technical backgrounds, and languages. Systems that provide explanations only in technical jargon or dominant languages create disparities in understanding and trust.

The ethical challenge lies in ensuring that explanations are not only technically accurate but also ethically meaningful, revealing the values, trade-offs, and uncertainties embedded in system decisions. Generative models present particular ethical concerns because their explanations may be fluent but unfaithful, creating a false sense of transparency that undermines genuine accountability [10]. This faithfulness problem is particularly acute in GenAI systems, where explanations are themselves generated outputs that may not accurately reflect the underlying decision process.

22.2.5 Privacy and Data Protection

GenAI for CAVs relies on continuous intake of multimodal information, including voice commands, cabin audio, driver biometrics, vehicle location, and contextual scene data. These data streams contain sensitive personal information whose collection, storage, and processing must adhere to strict privacy principles. The ethical implications extend beyond legal compliance to encompass user autonomy, dignity, and the right to control personal information.

Several unique challenges arise when GenAI processes multimodal vehicle data. *Inference privacy* concerns arise because generative models may infer sensitive information not explicitly provided. For example, voice patterns might reveal health conditions, location data could expose lifestyle patterns, and cabin audio might capture private conversations. The model's ability to synthesize and reason across modalities amplifies privacy risks.

Personalization versus privacy tensions emerge because personalization often requires persistent memory or fine-tuning based on individual driving habits or preferences, yet storing such information may violate user expectations or regulatory constraints. Users may want personalized experiences without permanent data retention.

Fleet learning and aggregation challenges occur when fleet learning or cloud-assisted model updates require aggregated data from many vehicles, raising questions about anonymization, aggregation, and data minimization. Even aggregated data may reveal patterns about individual users or communities. *Third-party access* risks arise because GenAI systems often rely on cloud services or third-party model providers, creating additional vectors for data exposure. The ethical obligation extends to ensuring that all parties in the data processing chain respect privacy principles.

Effective privacy engineering for GenAI systems demands adherence to ethical principles that go beyond legal minimums:

Data Minimization: Only the data strictly required for system operation or improvement should be collected and retained. This requires careful analysis of what data is truly necessary for each function.

On-device Processing: Where possible, sensitive inference should occur within the vehicle rather than through external services. This reduces exposure and gives users greater control.

Transparent Data Governance: Users must understand what information is collected, how it is used, with whom it is shared, and under what conditions it is deleted. Transparency must be accessible and actionable, not buried in lengthy terms of service.

User control and Consent: Users should have meaningful control over their data, including the ability to opt out of data collection, request deletion, and understand the implications of their choices.

Secure Communication and Storage: Cryptographic protections, robust access control, and secure communication protocols prevent unauthorized access or tampering. This includes protection against both external attackers and internal misuse.

Privacy-preserving Learning: Techniques such as federated learning [11, 12], differential privacy [13], and homomorphic encryption [14] enable model improvement while protecting individual privacy.

Advanced Note 22.4 Mathematical Formulation of Differential Privacy

Differential privacy provides a rigorous mathematical framework for privacy-preserving data analysis [13]. A randomized mechanism M satisfies (ϵ, δ)-differential privacy if for all neighboring datasets D and D' (differing by at most one record) and all possible outputs S:

$$P(M(D) \in S) \leq e^{\epsilon} \cdot P(M(D') \in S) + \delta.$$

The parameter ϵ (privacy budget) controls the privacy-utility trade-off: smaller ϵ provides stronger privacy but may reduce model utility. The parameter δ allows a small probability of privacy violation, typically set to a very small value (e.g., 10^{-5}).

In the context of GenAI-CAV systems, differential privacy can be applied to training data by adding noise to gradients during federated learning; to model outputs by adding calibrated noise to predictions to prevent inference attacks; and to aggregated statistics to protect fleet-level learning while enabling safety insights.

The privacy guarantee ensures that an adversary cannot determine with high confidence whether any specific individual's data was used in training, providing strong protection against inference attacks while enabling beneficial model improvements.

Privacy is thus inseparable from safety, because misuse or leakage of multimodal in-cabin data can erode user trust and undermine acceptance of advanced AI-enabled vehicle features. However, ethical design must also recognize that some safety functions may require data that raises privacy concerns. The ethical challenge lies in finding solutions that preserve both privacy and safety, such as local processing, minimal data retention, and privacy-preserving analytics.

22.2.6 Accountability and Responsibility

As GenAI becomes more integrated into vehicle reasoning and user interaction, the question of responsibility becomes increasingly complex. Unlike classical components with deterministic specifications, generative modules produce outputs influenced by statistical patterns, data distributions, and sampling dynamics. When an erroneous interpretation or misleading explanation contributes to an unsafe event, attributing responsibility is nontrivial. The ethical dimensions of accountability in GenAI-CAV systems must address both single-agent and multi-agent contexts.

22.2.6.1 Accountability Challenges in Single-agent Systems

GenAI systems create unique accountability challenges. *Distributed responsibility* arises because responsibility spans multiple stakeholders across the development and deployment lifecycle, making it difficult to assign blame when systems fail. *Non-deterministic behavior* complicates accountability because, unlike traditional software, generative systems may produce different outputs for identical inputs, complicating causal attribution and making it difficult to determine whether a failure was predictable or preventable.

Emergent behaviors create accountability gaps when GenAI components interact with perception, prediction, and planning modules in ways not explicitly anticipated during system design, creating behaviors that no single party fully intended or understood. *Data-driven failures* raise accountability questions when failures stem from training data biases, distribution shifts, or edge cases that were not adequately represented, raising questions about who is responsible for data quality and coverage.

Responsibility spans multiple stakeholders, each with distinct ethical obligations:

Manufacturers must ensure safe integration of generative models and maintain robust lifecycle management. They bear ultimate responsibility for vehicle safety and must establish processes for monitoring, updating, and responding to system failures.

AI developers bear responsibility for model training, dataset quality, and behavior under known limitations. They must document model capabilities and limitations, provide adequate safety guardrails, and ensure models are fit for their intended use.

Integrators and suppliers must validate contextual compatibility across components and interfaces. They are responsible for ensuring that generative modules work correctly within the broader vehicle system and do not introduce unexpected interactions.

Regulators must establish clear standards and enforcement mechanisms that ensure accountability while supporting innovation. They must balance safety requirements with the need to allow beneficial technologies to develop.

Drivers retain oversight duties when the system operates within human-in-the-loop modes. However, the extent of driver responsibility must be clearly communicated and reasonable given system capabilities and limitations.

Clear delineation of these roles is essential for regulatory compliance, liability assessment, and public trust. Furthermore, accountability requires traceability mechanisms that capture model inputs, generated outputs, contextual metadata, and system state at the time of a safety-relevant event.

22.2.6.2 Accountability Mechanisms and Requirements

Ethical accountability requires transparent responsibility assignment with clear documentation of who is responsible for what aspects of system behavior, incident investigation protocols that fairly assess all parties' contributions, remediation mechanisms that include both immediate fixes and longer-term improvements, and public transparency that enables informed public discourse and trust.

22.2.6.3 Multi-agent Accountability Challenges

As CAV systems evolve to include multiple GenAI agents that communicate and coordinate, accountability challenges become even more complex. *Attribution of collective decisions* requires mechanisms to determine responsibility when multiple agents contribute to a decision, particularly when individual agents may override or modify collective recommendations.

Agent autonomy versus system coherence raises questions about when individual agents should defer to group decisions versus exercising independent judgment, requiring clear ethical principles for conflict resolution when agents have conflicting ethical assessments. *Information asymmetry and power imbalances* in multi-agent coordination can create ethical concerns when agents with greater capabilities or information exploit others, particularly when agents represent different stakeholders (e.g., individual vehicles versus traffic management systems).

Coordination mechanisms must ensure *equitable outcomes* that are fair across all participants, not just optimal for the system as a whole. Multi-agent systems may also exhibit *emergent ethical behaviors*, such as cooperation patterns that may evolve into collusion or competitive dynamics that disadvantage certain users, requiring continuous monitoring to ensure they align with ethical principles and do not develop harmful patterns. These multi-agent accountability challenges require additional governance mechanisms for collective decision-making, fair negotiation protocols, and monitoring of emergent collective intelligence.

22.2.7 Ethical Dilemmas in Autonomous Decision-making

The ethical challenges discussed above manifest in concrete ethical dilemmas, situations where systems must make difficult choices between competing values or outcomes. The classic trolley problem [15], choosing between two harmful outcomes, takes on new dimensions in CAV systems [16].

GenAI may need to reason about scenarios where avoiding harm to one party increases risk to another. These scenarios require decisions about how systems should prioritize different types of harm, act when outcomes are probabilistic, explain their ethical reasoning, and allow users to configure ethical decision-making parameters. GenAI systems may also influence traffic flow and resource allocation in ways that raise questions of distributive justice. This requires careful consideration of how system decisions affect different groups and whether optimization should prioritize individual benefit, collective benefit, or equitable distribution.

22.3 Case Studies: Ethical Challenges in Practice

The ethical challenges and dilemmas discussed in Section 22.2 become most tangible when examined through real-world scenarios. The following case studies illustrate how the abstract ethical principles and challenges manifest in concrete situations, demonstrating the practical implications of bias, explainability, privacy, and accountability in deployed GenAI-CAV systems.

22.3.1 Misinterpretation of Conditional Language

Misinterpretation of conditional language illustrates how natural language qualifiers ("a little more," "too much," "unless necessary") cannot be mapped directly onto control parameters without grounding in scene context, vehicle state, and operational constraints. Failures may result in overly conservative or aggressive behavior, demonstrating that generative language understanding must be tightly coupled with semantic grounding and safety filters rather than direct translation of free-form instructions into control suggestions.

22.3.2 Vision-language Model Hallucination Under Adverse Conditions

Vision-language model hallucination under adverse conditions shows how VLMs may incorrectly classify reflective surfaces as pedestrians or fail to detect partially occluded cyclists during nighttime driving with glare. When paired with generative reasoning, such misclassifications propagate into explanations or recommendations, potentially causing unnecessary evasive action or misplaced confidence. This highlights the critical importance of multimodal consistency checks, uncertainty estimation, and redundancy across sensor modalities, with generative outputs including confidence indicators or being suppressed when perceptual reliability is low.

22.3.3 Over-the-air Model Updates

Over-the-air model updates demonstrate that GenAI components requiring iterative improvement must follow structured lifecycle processes, including regression testing, statistical validation, scenario-based evaluation, and rollback capability as mandated by UNECE R156. Updates may introduce unforeseen behaviors or alter distributional characteristics, requiring rigorous pre-deployment evaluation to prevent new risks from propagating into the field.

22.3.4 Privacy vs. Safety in Emergency Response

Privacy versus safety in emergency response illustrates ethical tensions when a CAV system detects a medical emergency and must balance the driver's right to privacy against the potential life-saving benefits of information sharing. Ethical design requires explicit policies for privacy-safety trade-offs, user control with clear warnings about safety implications, transparent emergency overrides subject to post-incident review, and minimization of information sharing to what is necessary for emergency response.

22.3.5 Bias in Multi-cultural Language Understanding

Bias in multi-cultural language understanding reveals how systems with training data skewed toward dominant languages exhibit significantly lower accuracy for certain dialects, creating inequitable access to safety-critical features and potentially discriminatory impact. This demonstrates that ethical deployment requires comprehensive testing across diverse populations, proactive identification of biases, and commitment to equitable system performance as an ongoing process rather than a one-time pre-deployment check.

The case studies presented above illustrate the concrete ethical challenges that arise when GenAI systems are deployed in CAV contexts. These real-world scenarios demonstrate why systematic regulatory frameworks and safety standards are essential: the problems of bias, hallucination, privacy-safety trade-offs, and accountability gaps cannot be adequately addressed through ad hoc solutions. The following section examines how existing regulatory frameworks must be adapted and extended to govern GenAI-enabled CAV systems effectively.

22.4 Regulatory and Safety Frameworks

Existing safety standards and regulatory frameworks must be adapted for GenAI in CAVs, as the integration of GenAI challenges long-established assumptions in automotive safety engineering. Traditional standards such as ISO 26262 [17] were designed for systems whose functional behavior can be fully specified, whose failure modes can be enumerated, and whose operational boundaries can be verified through deterministic testing.

GenAI violates all three assumptions: its outputs vary across evaluations, depend on high-dimensional training distributions, and may be influenced by ambiguous or underspecified user inputs. These limitations manifest as non-deterministic outputs that complicate reproducibility, data-driven behavior influenced by training dataset composition, context-dependent reasoning that makes boundary conditions difficult to formalize, and an unbounded failure surface where errors arise from semantic misinterpretation or multimodal misalignment. Classical hazard analysis, fault-tree analysis, and failure mode classification must therefore be augmented with data-centric and model-centric evaluations that account for epistemic uncertainty, model drift, and complex interactions across perception, reasoning, and user communication.

22.4.1 Core Standards Relevant to GenAI Integration

Although no existing standard fully captures the requirements of GenAI in safety-critical driving environments, several frameworks provide foundational guidance. The most relevant standards are summarized in Table 22.1, which illustrates how each framework addresses different aspects of GenAI integration in CAV systems.

These standards form the backbone of current regulatory oversight, but their application to GenAI requires adaptation. For example, ISO 21448's [18] focus on functional insufficiencies aligns closely with risks from hallucination or misinterpretation, while UNECE R156 [21] creates stringent documentation and rollback requirements for model updates.

However, regulatory compliance and ethical sufficiency are not equivalent. A system may satisfy all applicable standards (passing functional safety verification, meeting cybersecurity requirements, and following proper update governance) while still exhibiting ethical inadequacies such as biased behavior toward certain populations, insufficient explainability for affected users, or privacy practices that meet legal minimums but fail to respect user autonomy. Standards provide necessary but not sufficient conditions for ethical deployment. The frameworks discussed in this chapter extend beyond compliance to address the broader ethical obligations that ensure GenAI systems serve human values, not merely satisfy regulatory checklists.

22.4.2 Regulatory Implications for GenAI Modules

GenAI introduces several regulatory implications that extend beyond conventional automotive software. *Traceability and provenance* require systems to log generative inputs, outputs, and contextual parameters to support post-incident analysis while respecting privacy. *Validation of stochastic*

Table 22.1 Standards relevant to GenAI integration in CAV systems.

Standard	Primary focus	Relevance to generative AI
ISO 26262 [17]	Functional safety	Provides lifecycle processes but assumes deterministic components, requiring adaptation for GenAI stochastic behavior.
ISO 21448 (SOTIF) [18]	Safety of intended functionality	Highly relevant for VLMs and multimodal grounding, as it focuses on systems that function correctly but may be unsafe due to performance limitations or misuse.
ISO 21434 [19]	Cybersecurity engineering	Essential for protecting multimodal inputs and safeguarding generative inference pipelines from adversarial attacks.
UNECE R155 [20]	Cybersecurity management	Requires cybersecurity management systems, covering attack surfaces created by data-driven components and external connectivity.
UNECE R156 [21]	Software update management	Governs software update systems, critical for generative model updates, fine-tuning, and post-deployment retraining with rollback capabilities.
ISO/PAS 8800 [22]	Machine learning safety	Provides guidance for ML safety, including data quality requirements and model validation practices. PAS enables faster standardization for emerging technologies.

behavior must assess distributions of possible outputs rather than deterministic correctness, requiring statistical robustness, adversarial resilience, and scenario variance as critical performance indicators.

Advanced Note 22.5 Mathematical Formulation of Safety-critical Validation

For safety-critical validation, we require that the probability of safe outputs exceeds a threshold:

$$P(y \in \mathcal{S} \mid \theta, x) \geq \tau_{\text{safe}},$$

where y denotes the model output, $\mathcal{S}$ is the set of safe outputs, τ_{safe} is a safety threshold (e.g., 0.99), θ represents model parameters, and x represents the input.
For ASIL D systems (the highest safety level in ISO 26262 [17]), the acceptable failure rate is $\alpha = 10^{-6}$ per hour, meaning the probability of failure per hour must satisfy:

$$P(y \notin \mathcal{S} \mid \theta, x) < \alpha = 10^{-6} \text{ per hour}$$

or equivalently, the probability of safe outputs per hour must exceed:

$$P(y \in \mathcal{S} \mid \theta, x) \geq 1 - \alpha = 0.999999 \text{ per hour}.$$

This stringent requirement reflects the critical nature of safety-critical automotive systems, where failures can have catastrophic consequences. Validation of generative AI systems must therefore demonstrate statistical evidence that the probability of unsafe outputs remains below this threshold across diverse operational conditions, requiring extensive testing, uncertainty quantification, and confidence interval estimation. Note that for GenAI systems producing outputs at high frequency, this per-hour requirement translates to extremely low failure probabilities per individual output.

Governance of model updates requires that every update follows controlled verification processes, version management, and rollback capabilities as mandated by UNECE R156. *Runtime safety monitoring* emphasizes safety supervisors that filter or constrain generative outputs, formal safety envelopes around control actions, and fallback procedures when uncertainty exceeds acceptable thresholds. *Interoperability across suppliers* depends on clear interface specifications and harmonized safety metrics across components. Together, these implications highlight that GenAI requires a hybrid regulatory approach that merges traditional functional safety principles with data-centric and learning-centric oversight.

22.5 Ethical Engineering Methods

This section translates the foundational ethical principles and challenges discussed in Section 22.2 into concrete engineering practices that embed ethical considerations throughout the system lifecycle. The *ethical dimensions* of engineering methodologies determine how fairness, accountability, privacy, and trust are maintained throughout operational lifetimes. While Chapter 21 covers technical evaluation methodologies (including bias and fairness metrics) and Chapter 20 covers simulation approaches, this section emphasizes the ethical obligations underlying these practices: why privacy and accountability matter beyond technical metrics, how trust calibration serves ethical goals, and how evaluation and simulation must be designed to respect ethical principles rather than treating them as secondary concerns.

22.5.1 Ethical Design Principles

This subsection translates the foundational ethical principles and challenges identified in Section 22.2 into actionable engineering practices. The ethical challenges of bias, explainability, privacy, and accountability cannot be adequately addressed through post-hoc fixes. Instead, ethical design principles must guide architectural decisions, interface design, and system behavior from the outset. The following principles provide concrete engineering guidance for implementing the foundational values discussed in Section 22.2.

22.5.1.1 Human-centricity

System behavior must reflect human needs and values, respecting user autonomy and dignity. Interfaces should be designed to avoid manipulative interactions and support intuitive understanding of system actions and limitations. This principle directly addresses the human–AI interaction ethics discussed in Section 22.2.2, ensuring that systems enhance rather than undermine human agency.

22.5.1.2 Value Alignment

Generative models should be calibrated such that their outputs remain consistent with safety norms, driving etiquette, and societal expectations. This alignment addresses the ethical challenge of ensuring systems reflect societal values rather than encoding narrow or biased preferences. Achieving value alignment requires concrete training and evaluation techniques that embed ethical principles into model behavior from the training stage through deployment.

Reinforcement Learning from Human Feedback (RLHF): RLHF has become a primary method for aligning language models with human preferences [23, 24]. While technical details of RLHF are covered in Chapter 18, the ethical considerations are critical for CAV deployment. RLHF can help ensure that generative systems prioritize safety, clarity, and appropriate behavior, but ethical considerations arise. *Representative feedback* requires that human feedback comes from diverse populations to avoid encoding narrow or biased preferences, addressing fairness concerns. *Safety prioritization* ensures that feedback mechanisms do not compromise safety-critical behaviors in favor of user preferences that might prioritize convenience over safety. *Long-term alignment* requires that systems remain aligned as they learn and adapt, necessitating ongoing monitoring and intervention (see Section 22.5.3).

Constitutional AI and Principle-Based Alignment: Constitutional AI approaches train models to follow explicit ethical principles or "constitutions" that define acceptable behavior [25]. For CAV systems, such principles might include: prioritizing human safety above all other considerations, respecting user autonomy and providing meaningful control, ensuring equitable treatment across all road users, maintaining transparency about system capabilities and limitations, and preserving privacy while enabling necessary safety functions. These principles directly address the key ethical challenges, including bias, explainability, privacy, and accountability, by embedding them into the model's training objective. However, these principles must be carefully crafted to avoid conflicts and must be validated through extensive testing and stakeholder input.

Red Teaming and Adversarial Evaluation: Ethical alignment requires identifying and addressing potential misalignments before deployment. Red teaming, systematic attempts to find failure modes or unethical behaviors, helps reveal vulnerabilities in value alignment [26]. This includes: *Jailbreak attempts* testing whether systems can be prompted to ignore safety constraints or ethical principles, *Edge case exploration* identifying scenarios where ethical principles conflict or become ambiguous (as discussed in Section 22.2.7), *Bias probing* systematically testing for biased behaviors across different populations and contexts, and *Adversarial scenarios* creating challenging situations that test ethical reasoning and value prioritization. Red teaming serves as both a training methodology (identifying weaknesses to address) and a validation methodology (verifying ethical robustness), connecting to the adversarial testing discussed in Section 22.5.2.

22.5.1.3 Trust Calibration

Trust calibration ensures alignment between actual system performance and user trust levels, preventing both over-trust and under-trust risks that pose ethical and safety concerns. This requires accurate communication of system capabilities, continuous monitoring of trust dynamics, and individual calibration that accounts for user background and experience (see Section 22.2.2 for a detailed discussion of trust dimensions and calibration).

22.5.1.4 Fail-safe Interaction Design

In situations of ambiguity, uncertainty, or degraded performance, generative modules should default to conservative behaviors and provide transparent handover cues. This principle prioritizes safety and user welfare, particularly when systems face uncertainty. These interaction patterns reduce cognitive load and prevent misinterpretation of system capabilities, supporting the right to explanation (see Section 22.2.4).

Together, these principles create an ethical foundation that supports safe integration of GenAI into multi-modal driving systems, ensuring that technical capabilities serve human values rather than creating new forms of harm or exclusion.

22.5.2 Verification and Validation

Verification and validation practices for GenAI must address not only technical correctness but also ethical requirements. Testing must explicitly assess ethical dimensions that go beyond functional correctness, including fairness across populations, privacy preservation, and accountability mechanisms. This requires testing across diverse driving environments, linguistic constructs, lighting conditions, traffic densities, and cultural variations in speech patterns to ensure equitable performance.

High-fidelity simulation, as detailed in Chapter 20, supports controlled testing of complex interactions between generative reasoning, perception, planning, and human behavior, enabling systematic exploration of edge cases, bias scenarios, and ethical dilemmas in safe, repeatable environments. Generative systems should be stress-tested against ambiguous phrasing, adversarial prompts, partial occlusions, and visual perturbations to identify conditions under which systems might produce biased, unfair, or harmful outputs.

Intermediate layers, such as rule-based filters, grounding mechanisms, or semantic verifiers, can ensure that generated text or reasoning remains within predefined safety bounds before influencing control modules. As outputs vary across samples, validation must quantify uncertainty and characterize distributions of possible outcomes, enabling systems to recognize when they lack sufficient confidence to make safe or fair decisions and triggering fail-safe behaviors that protect user welfare.

These verification and validation methods must be applied iteratively throughout the development lifecycle, particularly when models are retrained, fine-tuned, or updated through over-the-air mechanisms. The ethical requirement is that each update maintains or improves ethical performance, including fairness, privacy, accountability, and trust, not merely technical performance.

22.5.3 Post-deployment Monitoring

Because generative models continue to interact with dynamic environments and diverse user populations long after deployment, ethical requirements extend throughout the system lifecycle. *Ethical drift detection* must identify not only functional deviations but also shifts that create unfair outcomes or bias disparities, such as performance degradation for certain dialects or demographic groups over time. *Real-time ethical supervision* uses runtime guardrails that monitor generative outputs and intervene when safety thresholds are approached, blocking unsafe commands, correcting misinterpretations, or requiring explicit human confirmation for ambiguous actions. *Fleet-level learning* must be conducted in a privacy-preserving manner, using techniques such as differential privacy, federated learning, or anonymization to minimize or eliminate retained personal data while preserving safety insights. *Ethical auditability* requires auditable records of model versions, training data provenance, update history, and incident responses that enable accountability, support fairness detection, and ensure transparency. *Ethical monitoring metrics* must track fairness metrics across user groups, privacy compliance metrics, trust calibration metrics, and accountability metrics to ensure systems remain aligned with ethical principles throughout their operational lifetime.

The ethical engineering methods discussed above focus primarily on technical and operational considerations within individual vehicles and systems. However, the deployment of GenAI in CAVs also raises broader questions about environmental impact, societal equity, and global governance that extend beyond the scope of direct technical ethics. The following section briefly examines these wider considerations.

22.6 Broader Environmental and Societal Considerations

Broader environmental and societal considerations merit brief discussion alongside the direct ethical and safety implications of GenAI in CAVs. The computational resources required for training and deploying large generative models raise questions about carbon footprint and resource allocation, particularly as these systems scale across vehicle fleets. Environmental ethics requires consideration of whether the benefits of GenAI-enabled safety and efficiency justify the environmental costs, and how to minimize resource consumption through efficient model architectures, on-device processing, and sustainable computing practices.

Societal ethics extends beyond individual vehicle safety to encompass accessibility, economic impacts, infrastructure equity, and data sovereignty. GenAI-CAV systems must be designed to be accessible across diverse populations, avoiding technologies that create or exacerbate socioeconomic disparities. Economic considerations include how GenAI deployment affects employment in transportation sectors, whether benefits are equitably distributed, and how to ensure that advanced capabilities do not become luxury features unavailable to lower-income communities.

Infrastructure equity requires that system benefits are not limited to well-maintained road networks or regions with advanced connectivity. Data sovereignty concerns arise when vehicle data flows across national boundaries, raising questions about who controls and benefits from aggregated fleet intelligence. While these considerations are important, they represent broader policy and societal questions that extend beyond the scope of direct technical ethics, and are best addressed through complementary policy frameworks and stakeholder engagement processes.

22.7 Summary

GenAI introduces new capabilities for perception, reasoning, and interaction in CAVs, yet its integration fundamentally alters the ethical and safety landscape. This chapter examined the conceptual foundations necessary to understand these shifts, focusing on human values, the distinctive characteristics of generative models, and the ethics of human–AI interaction. The discussion identified key ethical challenges that arise in practice: bias and fairness across multiple dimensions, the obligations underlying explainability requirements, the need for robust privacy protections that balance user autonomy with safety needs, and the complexities of assigning accountability in systems with distributed responsibility. These challenges manifest in concrete ethical dilemmas where systems must make difficult choices between competing values, such as the trolley problem variants, distributive justice in traffic, and privacy-safety trade-offs.

These challenges and dilemmas reveal why traditional safety standards alone are insufficient for governing GenAI-enabled behavior. While frameworks such as ISO 26262, ISO 21448, and UNECE R156 provide essential structure, they must be interpreted and extended to account for data-centric risks, multimodal uncertainty, and model update dynamics.

Methodologically, safe and ethical deployment of GenAI requires a life-cycle perspective. Ethical design principles must guide system development from the outset, supported by rigorous verification and validation approaches tailored to generative reasoning. Continuous monitoring, drift detection, bias assessment, and auditability are indispensable components of maintaining reliability, fairness, and user trust throughout the operational lifetime. The illustrative case studies underscore that GenAI presents practical challenges requiring disciplined engineering processes, thoughtful safety governance, and ongoing ethical evaluation. Beyond direct technical ethics, broader environmental and societal considerations—including resource consumption, accessibility, economic impacts, and data sovereignty—must be addressed through complementary policy frameworks and stakeholder engagement.

As this book has demonstrated throughout, GenAI fundamentally reshapes how CAV systems perceive, reason, and interact, moving beyond deterministic rule-following to adaptive, context-aware intelligence. This transformation brings with it a profound responsibility for engineers and researchers: to recognize that ethical reasoning is not a constraint imposed after technical design, but a core design competency that must be integrated from the earliest stages of system conception.

This chapter closes the book by transitioning from the technical foundations of "how to build" GenAI systems for CAVs to the essential question of "how to govern and deploy responsibly." The ethical and safety considerations examined here are not afterthoughts but fundamental requirements that must be integrated throughout the development lifecycle. These ethical dimensions are deeply interconnected: explainability directly supports Transparency Trust by enabling users to understand system decision-making processes, while fairness and accountability together ensure that trust is well-calibrated and ethically grounded. The frameworks, principles, and methodologies presented across these chapters provide the foundation for building generative AI systems that enhance rather than compromise the safety, fairness, and trustworthiness of autonomous mobility: a responsibility that defines the next generation of CAV development.

References

1 John D. Lee and Katrina A. See. Trust in automation: Designing for appropriate reliance. *Human Factors*, 46(1):50–80, 2004.

2 Peter A. Hancock, Deborah R. Billings, Kristin E. Schaefer, Jessie Y.C. Chen, Ewart J. De Visser, and Raja Parasuraman. A meta-analysis of factors affecting trust in human-robot interaction. *Human Factors*, 53(5):517–527, 2011.

3 Ninareh Mehrabi, Fred Morstatter, Nripsuta Saxena, Kristina Lerman, and Aram Galstyan. A survey on bias and fairness in machine learning. *ACM Computing Surveys (CSUR)*, 54(6):1–35, 2021.

4 Solon Barocas, Moritz Hardt, and Arvind Narayanan. *Fairness and Machine Learning*. fairmlbook.org, 2019. https://fairmlbook.org.

5 Moritz Hardt, Eric Price, and Nathan Srebro. Equality of opportunity in supervised learning. In *Advances in Neural Information Processing Systems*, volume 29, 2016.

6 Cynthia Dwork, Moritz Hardt, Toniann Pitassi, Omer Reingold, and Richard Zemel. Fairness through awareness. In *Proceedings of the 3rd Innovations in Theoretical Computer Science Conference*, pages 214–226, 2012.

7 Tim Miller. Explanation in artificial intelligence: Insights from the social sciences. *Artificial Intelligence*, 267:1–38, 2019.

8 Christoph Molnar. Interpretable machine learning. Lulu.com, 2020. https://christophm.github.io/interpretable-ml-book/.

9 Bryce Goodman and Seth Flaxman. European union regulations on algorithmic decision-making and a "right to explanation". *AI Magazine*, 38(3):50–57, 2017.

10 Riccardo Guidotti, Anna Monreale, Salvatore Ruggieri, Franco Turini, Fosca Giannotti, and Dino Pedreschi. A survey of methods for explaining black box models. *ACM Computing Surveys (CSUR)*, 51(5):1–42, 2018.

11 Brendan McMahan, Eider Moore, Daniel Ramage, Seth Hampson, and Blaise Aguera y Arcas. Communication-efficient learning of deep networks from decentralized data. *Artificial Intelligence and Statistics*, pages 1273–1282, 2017.

12 Chaoyang He, Murali Annavaram, and Salman Avestimehr. Group knowledge transfer: Federated learning of large CNNs at the edge. *Advances in Neural Information Processing Systems*, 33:14068–14080, 2020.

13 Cynthia Dwork, Frank McSherry, Kobbi Nissim, and Adam Smith. Calibrating noise to sensitivity in private data analysis. *Theory of Cryptography Conference*, pages 265–284, 2006.

14 Craig Gentry. Fully homomorphic encryption using ideal lattices. In *Proceedings of the 41st Annual ACM Symposium on Theory of Computing*, pages 169–178, 2009.

15 Philippa Foot. The problem of abortion and the doctrine of the double effect. *Oxford Review*, 5:5–15, 1967.

16 Jean-François Bonnefon, Azim Shariff, and Iyad Rahwan. The social dilemma of autonomous vehicles. *Science*, 352(6293):1573–1576, 2016.

17 International Organization for Standardization. Road vehicles – functional safety. Technical Report 26262, ISO, 2018. Functional safety standard for automotive systems.

18 International Organization for Standardization. Road vehicles – safety of the intended functionality. Technical Report 21448, ISO, 2022. SOTIF standard for assessment of performance limitations and edge cases.

19 International Organization for Standardization. Road vehicles – cybersecurity engineering. Technical Report 21434, ISO, 2021. Cybersecurity engineering requirements for automotive systems.

20 United Nations Economic Commission for Europe. UN regulation no. 155 – cyber security and cyber security management system. Technical Report, UNECE, 2021. Cybersecurity management systems for vehicles.

21 United Nations Economic Commission for Europe. UN regulation no. 156 – software update and software update management system. Technical Report, UNECE, 2021. Software update management systems for vehicles.

22 International Organization for Standardization. Road vehicles – safety and cybersecurity for automated driving systems – part 1: Framework and terminology. Technical Report PAS 8800, ISO, 2023. Machine learning safety guidance for automotive systems.

23 Long Ouyang, Jeff Wu, Xu Jiang, Diogo Almeida, Carroll L. Wainwright, Pamela Mishkin, Chong Zhang, Sandhini Agarwal, Katarina Slama, Alex Ray, et al. Training language models to follow instructions with human feedback. In *Proceedings of the 36th International Conference on Neural Information Processing Systems*, NIPS '22, Red Hook, NY, USA, 2022. Curran Associates Inc., ISBN 9781713871088.

24 Yuntao Bai, Saurav Kadavath, Sandipan Kundu, Amanda Askell, Jackson Kernion, Andy Jones, Anna Chen, Anna Goldie, Azalia Mirhoseini, Cameron McKinnon, et al. Training a helpful and harmless assistant with reinforcement learning from human feedback. *arXiv preprint arXiv: 2204.05862*, 2022.

25 Anthropic. Constitutional AI: Harmlessness from AI feedback. *arXiv preprint arXiv:2212.08073*, 2022.

26 Deep Ganguli, Liane Lovitt, Jackson Kernion, Amanda Askell, Yuntao Bai, Saurav Kadavath, Ben Mann, Ethan Perez, Nicholas Schiefer, Kamal Ndousse, et al. Red teaming language models to reduce harms: Methods, scaling behaviors, and lessons learned. *arXiv preprint arXiv:2209.07858*, 2022.

Index

n

o

S

u

v

w

y

z

Wiley The manufacturer's authorized representative according to the EU General Product Safety Regulation is Wiley-VCH GmbH, Boschstr. 12, 69469 Weinheim, Germany, e-mail: Product_Safety@wiley.com.

Printed and bound by CPI Group (UK) Ltd, Croydon, CR0 4YY
02/10/2026
02227444-0003